Study and Interpretation of the Chemical Characteristics of Natural Water

By JOHN D. HEM

GEOLOGICAL SURVEY WATER-SUPPLY PAPER 1473

A review of chemical, geologic, and hydrologic principles and processes that control the composition of natural water, with methods for studying and interpreting chemical analyses

University Press of the Pacific
Honolulu, Hawaii

Study and Interpretation of the Chemical Characteristics of Natural Water

by
John D. Hem

for
U.S. Geological Survey

ISBN: 1-4102-2308-6

Copyright © 2005 by University Press of the Pacific

Reprinted from the 1970 edition

University Press of the Pacific
Honolulu, Hawaii
http://www.universitypressofthepacific.com

All rights reserved, including the right to reproduce this book, or portions thereof, in any form.

PREFACE

Most of the text of the first edition of this publication, U.S. Geological Survey Water-Supply Paper 1473, was written in 1955 and 1956. When the plans were being made for the report a few years earlier, the expectation was that it would be used within the U.S. Geological Survey for the most part as a guide for field investigators to help in determining and interpreting the chemical quality of water resources. At the same time, however, a substantial increase of interest and activity in the fields of natural-water chemistry and water-pollution control was beginning to be noticeable, and, as a result, when the report was published in 1959, it soon attained a considerably wider circulation than had originally been anticipated. A series of reprintings was necessary.

This revision of the earlier book has been prepared because of the continued and still-growing interest in the subject matter. The first edition is now seriously out of date. It no longer gives a satisfactory indication of present-day techniques of water-analysis interpretation or of the state of development of the field of water chemistry.

This revision has two principal objectives: (1) to outline more completely the principles and processes, chemical and environmental, which shape and control the chemical composition of natural water and (2) to bring the discussions of chemistry of individual constituents, methods of data interpretation, and associated bibliographic references up to date.

It is to be expected that most potential users of this book will have some chemical training. In any event, it seems inappropriate for anyone without such training to specialize in the interpretation of chemical data. Accordingly, a somewhat more extensive use of theoretical chemistry has been introduced here as compared to the earlier edition. The book should still be useful to the nonchemist, however, and it is nowhere very demanding in either a chemical or a mathematical sense.

The interdisciplinary nature of the field requires the introduction of much material which can be discussed only briefly. Many of the concepts which are introduced will be familiar to the reader, and supplementary material referred to in the bibliography should enable him to obtain more information on various topics.

The state of knowledge of the field covered here is rapidly improving, but broad areas remain where little serious research has been done.

The reader will note occasional comments to this effect, and hopefully he will be able to draw some conclusions of his own both from what is contained here and from what has been left out, as to what subjects need more careful study.

Discussions and correspondence with many of my friends and colleagues within the U.S. Geological Survey and elsewhere have probably had even more influence on my thoughts and their expression here than I realize, but certainly I owe much to the aid of others which cannot be fully acknowledged. Colleagues who helped review the manuscript include D. C. Culbertson, Eugene Brown, J. H. Feth, B. B. Hanshaw, R. O. Fournier, V. C. Kennedy, M. W. Skougstad, and T. D. Steele, all of the U.S. Geological Survey. The tireless labor of Mrs. Mary E. Grisak in preparing the manuscript, reviewing, and repairing it where necessary also is gratefully acknowledged.

CONTENTS

	Page
Preface	III
Abstract	1
Scope and purpose	2
Arrangement of topics	3
Properties and structure of water	4
Composition of the earth's crust	5
The hydrosphere	10
The atmosphere	10
Principles and processes controlling composition of natural water	12
Atomic and nuclear alterations	14
Induced radioactivity	14
Chemical reactions	15
Types of reactions in water chemistry	16
Chemical equilibrium	18
Temperature and pressure effects	22
Solution of calcite	22
Application of chemical thermodynamics	24
Electrochemical equilibrium	28
Chemical reaction rates	30
Reaction order	30
Effect of temperature on reaction rate	33
Solubility concepts	34
Solubility product	34
Reactions at interfaces	35
Gas-liquid interfaces	35
Liquid-solid interfaces	36
Adsorption and ion-exchange calculations	38
Membrane effects	39
Clays as semipermeable membranes	40
Environmental influences	40
Climate	41
Geologic effects	41
Biochemical factors	42
Ecology applied to natural water	43
Influence of soil and soil-forming processes	44
Aquatic biota	45
The hydrologic cycle	47
Sources of solutes in the atmosphere	47
Composition of atmospheric precipitation	48
Influence of man	50
A systems approach to study of water chemistry	51
Open versus closed systems	51
The phase rule	52
Influence of water movement	54
Some characteristics of ground-water systems	54

CONTENTS

	Page
Principles and processes controlling composition of natural water—Continued	
Surface water systems—rivers	55
Lakes and reservoirs	56
Geochemical cycles	56
Buffer capacity	59
Evaluation of water composition	60
Collection of water samples	60
Sampling of river water	61
Lake and reservoir samples	68
Ground-water samples	68
Completeness of sample coverage	71
Analysis of water samples	74
Field testing of water	74
Electric logs as indicators of ground-water quality	75
Laboratory procedures	76
Expression of water analyses	77
Hypothetical combinations	77
Analyses reported in terms of ions	77
Determinations included in analyses	78
Units used in reporting analyses	79
Weight-per-weight units	79
Weight-per-volume units	79
Equivalent-weight units	81
Composition of anydrous residue	82
Concentration in terms of calcium carbonate	84
Comparison of units of expression	85
Significance of properties and constituents reported in water analyses	86
Nature of the dissolved state	86
Hydrogen-ion activity (pH)	88
Buffered solutions	92
Range of pH of natural water	93
Measurement and interpretation of pH	95
Specific electrical conductance	96
Units for reporting conductance	96
Physical basis of conductance	96
Range of conductance values	102
Accuracy and reproducibility	103
Silica	103
Forms of dissolved silica	105
Solubility controls	107
Reaction mechanisms	108
Occurence in natural water	108
Aluminum	109
Sources of aluminum in water	109
Species in solution	110
Solubility controls	111
Reaction mechanisms and their implications	112
Occurrence of aluminum in water	113

CONTENTS

	Page
Significance of properties and constituents reported in water analyses—Continued	
Iron	114
Sources of iron	114
Species of iron in natural water	116
Solubility controls	117
Occurrence of iron in water	121
Manganese	126
Sources of manganese	126
Manganese-solubility controls	128
Manganese-reaction rates and mechanisms	130
Occurrence of manganese in water	130
Calcium	131
Sources of calcium	131
Solute species	132
Chemical controls of calcium concentration	132
Occurrence of calcium in water	138
Magnesium	140
Sources of magnesium	141
Form of dissolved magnesium	141
Solubility controls	141
Occurrence of magnesium in water	143
Sodium	145
Sources of sodium	146
Solubility and form of dissolved species	147
Occurrence of sodium in water	149
Potassium	150
Sources of potassium	150
Occurrence of potassium in water	151
Alkalinity	152
Sources of alkalinity	153
Occurrence and determination of alkalinity	154
Occurrence of bicarbonate and carbonate	157
Acidity	159
Sources of acidity	159
Origin of acid waters	160
Sulfate and other sulfur species	161
Sources of sulfate in natural water	162
Circulation of sulfate in natural water	164
Forms of dissolved sulfate	166
Reduced sulfur species	168
Chloride	170
Sources of chloride	171
Occurrence and chemistry of chloride in water	172
Accuracy of determination	176
Fluoride	176
Sources of fluoride in water	177
Chemistry of fluoride in water	177
Range of concentration	178

CONTENTS

	Page
Significance of properties and constituents reported in water analyses—Continued	
Nitrogen	180
Sources of nitrate in water	181
Reduced nitrogen species	183
Phosphorus	183
Sources of phosphate	184
Chemistry of phosphate in water	185
Nitrogen and phosphorus as nutrients	185
Boron	187
Sources of boron	187
Chemistry of boron in natural water	188
Minor and trace constituents	188
Availability of analytical data	189
Analysis techniques	190
Sample collection and treatment	190
Sources of general information	191
Minor elements in sea water	192
Occurrence of minor constituents	193
Alkali metals	193
Alkaline earths	194
Beryllium	194
Strontium	195
Barium	197
Other metallic elements	197
Titanium	198
Vanadium	198
Chromium	199
Molybdenum	199
Cobalt	200
Nickel	201
Copper	202
Silver	202
Gold	203
Zinc	203
Cadmium	204
Mercury	205
Germanium	205
Lead	205
Nonmetals	206
Arsenic	206
Selenium	207
Bromine	208
Iodine	209
Radioactive elements	209
Uranium	211
Radium	212
Radon	213
Thorium	213
Other radioactive nuclides	213

CONTENTS

	Page
Significance of properties and constituents reported in water analyses—Continued	
Minor and trace constituents—Continued	
Organic constituents	215
Detergents	216
Color	216
Nature of colored material in water	216
Chemically related properties	218
Residue on evaporation	218
Significance of determination	218
Total dissolved solids—computed	219
Chemical factors in dissolved determination	220
Dissolved oxygen	221
Oxygen demand and other evaluations of organic pollution load	222
Biochemical oxygen demand	223
Chemical oxygen demand	223
Total organic carbon	223
Hardness	224
Range of hardness concentration	225
Redox potential	226
Range of redox potential	228
Other units of expression	228
Sodium-adsorption-ratio (SAR)	228
Density	229
Stable nuclides	229
Organization and study of water-analysis data	230
Evaluation of water analysis	231
Accuracy and reproducibility	232
Accuracy checks	233
Significant figures	236
General evaluations of areal water quality	236
Inspection and comparison	237
Ion ratios and water types	237
Statistical treatment of water-quality data	239
Use of averages	239
Frequency distributions	247
Correlations	250
Ion and solute inventories	255
Graphical methods for representation of analyses	256
Ion-concentration diagrams	257
Trilinear plotting systems	264
Methods for extrapolation of records	270
Water-quality hydrographs	270
Water quality in relation to stream discharge	271
Water-quality maps	280
Map symbols	281
Isogram maps	284
Water-quality profiles	286

	Page
Relationship of water quality to lithology	287
Igneous rocks	290
Sedimentary rocks	294
Resistates	295
Hydrolyzates	300
Precipitates	303
Evaporites	306
Metamorphic rocks	306
Influence of activities of man	309
Salt-water intrusion	311
Application of water-quality measurements to quantitative hydrology	315
Relationship of quality of water to use	320
Domestic uses and public supplies	321
Agricultural use	324
Industrial use	333
Recreational and esthetic uses	336
Water-management concepts and problems	336
Selected references	338
Index	359

ILLUSTRATIONS

		Page
PLATE	1. Nomograph for computing ionic strength of natural water	In pocket
	2. Map of part of lower Safford Valley, Ariz., showing dissolved-mineral content of ground water in alluvial fill of the inner valley, 1944	In pocket
FIGURES	1–4. Graphs showing:	
	1. Relation of activity coefficients for dissolved ions to ionic strength of solution	21
	2. Bicarbonate, sulfate, hardness, and pH of samples collected in cross section of Susquehanna River at Harrisburg, Pa., July 8, 1947	62
	3. Conductance of daily samples and mean daily discharge of Rio Grande at San Acacia, N. Mex., 1945	66
	4. Conductance and discharge at times of sampling of Rio Grande at San Acacia, N. Mex., under typical summer flow conditions	67
	5. Geologic section showing changes in conductance of ground water, Pinal County, Ariz	70
	6–11. Graphs showing:	
	6. Specific conductance and altitude of water table for three typical observation wells, Safford Valley, Ariz	72
	7. Dissolved solids and depth to water in two irrigation wells, Wellton-Mohawk area, Yuma County, Ariz	73

CONTENTS

		Page
FIGURES 6–11.	Graphs showing—Continued	
	8. Specific conductance of potassium chloride solutions	97
	9. Specific electrical conductance of a 0.01 molar solution of potassium chloride at various temperatures	98
	10. Dissolved solids and specific conductance of composites of daily samples, Gila River at Bylas, Ariz., October 1, 1943, to September 30, 1944	100
	11. Relation of conductance to chloride, hardness, and sulfate concentrations, Gila River at Bylas, Ariz., October 1, 1943, to September 30, 1944	101
12.	Schematic diagram of hydrated aluminum ion, $Al(H_2O)_6^{+3}$	110
13–30.	Graphs showing:	
	13. Equilibrium activities of free aluminum for three forms of aluminum hydroxide and calculated activity of $Al^{+3}+AlOH^{+2}$	112
	14. Fields of stability for solid and dissolved forms of iron as function of Eh and pH at 25°C and 1 atmosphere pressure. Activity of sulfur species 96 mg/l as SO_4^{-2}, carbon dioxide species 1,000 mg/l as HCO_3^-, and dissolved iron 0.0056 mg/l	118
	15. Solubility of iron in moles per liter as function of Eh and pH at 25°C and 1 atmosphere pressure. Activity of sulfur species 96 mg/l as SO_4^{-2}; carbon dioxide species 61 mg/l as HCO_3^-	119
	16. Fields of stability of solids and solubility of manganese as a function of Eh and pH at 25°C and 1 atmosphere pressure. Activity of dissolved carbon dioxide species 100 mg/l as HCO_3^-. Sulphur species absent	128
	17. Equlibrium pH in relation to calcium and bicarbonate activities in solutions in contact with calcite, 25°C and pressure 1 atmosphere	134
	18. Solubility of calcium carbonate (calcite) in water at 25°C in presence of carbon dioxide	135
	19. Percentages of total dissolved carbon dioxide species in solution as a function of pH, 25°C and pressure 1 atmosphere	155
	20. Fields of dominance of sulfur species at equilibrium at 25°C and 1 atmosphere pressure. Total dissolved sulfur activity 96 mg/l as SO_4^{-2}	163
	21. Solubility of gypsum in sodium chloride solutions at 25°C and 1 atmosphere pressure	169
	22. Percentages of total dissolved phosphate species in solution as a function of pH at 25°C and 1 atmosphere pressure	186
	23. Cumulative frequency curve of dissolved solids for Colorado River above and below Hoover Dam, Ariz. and Nev	248

CONTENTS

FIGURES 13–30. Graphs showing—Continued

		Page
24.	Cumulative frequency curve of specific conductance, Allegheny, Monongahela, and Ohio River waters, Pittsburgh area, Pennsylvania, 1944–50	249
25.	Number of samples having percent-sodium values within ranges indicated, San Simon artesian basin, Arizona	250
26.	Sodium chloride relationship, Gila River at Bylas, Ariz., October 1 1943, to September 30, 1944	253
27.	Calcite and gypsum equilibrium solubility limits, 25° C and 1 atmosphere pressure	254
28.	Analyses represented by bar lengths in milliequivalents per liter	258
29.	Analyses represented by bar lengths in milliequivalents per liter; also shows hardness values in milligrams per liter	259
30.	Analyses represented by bar lengths in milliequivalents per liter; also shows dissolved silica in millimoles per liter	260

31–35. Diagrams showing:

31.	Analyses represented by vectors based on milliequivalents per liter	261
32.	Analyses represented by patterns based on milliequivalents per liter	262
33.	Analyses represented by circles subdivided on the basis of percentage of total milliequivalents per liter	263
34.	Analyses represented by patterns based on combined anion and cation concentrations	264
35.	Analyses represented by logarithmic plotting of concentration in milligrams per liter	265
36.	Graph showing analyses represented by linear plotting of cumulative percentage composition based on milligrams per liter	266
37.	Trilinear diagram showing analyses represented by this three-point plotting method	269
38.	Graph showing specific conductance of daily samples and daily mean discharge, San Francisco River at Clifton, Ariz., October 1, 1943, to September 30, 1944	276
39.	Graph showing weighted-average dissolved solids and annual mean discharge, Rio Grande at San Acacia, N. Mex., and Pecos River near Artesia, N. Mex	279
40.	Map showing iron content of ground water from the principal artesian aquifer and topographic regions, southwestern Georgia	282
41.	Map showing chemical quality of water and dissolved solids in the Minot aquifer, North Dakota	283
42.	Map showing distribution of iron and pH in water from the "500 foot" sand, western Tennessee	285

		Page
FIGURES 43–50	Graphs showing	
	43. Temperature and dissolved solids of water in Lake Mead in Virgin and Boulder Canyons, 1948	287
	44. Composition of ground water obtained from igneous rocks	295
	45. Composition of ground water obtained from resistate sedimentary rock types	299
	46. Composition of water obtained from hydrolyzate sedimentary rock types	302
	47. Average composition of water from Rio Grande during two periods in 1945 and 1946	303
	48. Composition of ground water obtained from precipitate sedimentary rock types	305
	49. Composition of ground water obtained from evaporite sedimentary rock types	307
	50. Composition of ground water obtained for metamorphic rocks	309
51.	Diagram for use in interpreting the analysis of irrigation water	332

TABLES

		Page
TABLE 1.	Average composition of igneous and some types of sedimentary rocks	7
2.	The composition of sea water	11
3.	Composition of river water	12
4.	Mean composition of the atmosphere	13
5.	Values for parameter a_i in Debye-Huckel equation	20
6.	Composition of rain and snow	50
7.	Average residence time of elements in the ocean	58
8.	Conversion factors for quality-of-water data	81
9.	Conversion factors: milligrams per liter to milliequivalents per liter or millimoles per liter	83
10.	Chemical analysis of a water sample expressed in six ways	85
11.	Analyses of waters whose pH is unstable	94
12.	Analyses of waters containing high proportions of silica	106
13.	Analyses of waters high in dissolved aluminum or manganese	115
14.	Analyses of waters containing iron	123
15.	Analyses of waters in which calcium is a major constituent	138
16.	Analyses of waters in which magnesium is a major constituent	144
17.	Analyses of waters in which sodium is a major constituent	148
18.	Analyses of waters having various alkalinity-acidity-pH relationships	154
19.	Analyses of waters containing fluoride, nitrogen, phosphorus or boron in unusual amounts	179
20.	Analyses of waters containing unusual concentrations of metals and other constituents	196

		Page
TABLE	21. Hypothetical chemical analyses compared by means of ratios	238
	22. Average of chemical analyses computed by different methods for Rio Grande at San Acacia, N. Mex., for 1941–42 and 1945–46 water years	243
	23. Discharge-weighted average of chemical analyses for Rio Grande at San Acacia, N. Mex., representing periods in the 1941–42 and 1945–46 water years in which different sources of runoff predominated	245
	24. Analyses showing effects of sea-water contamination in Gaspur water-bearing zone, Dominguez Gap, Los Angeles County, Calif	315
	25. Analyses of water from Clifton Hot Springs and from San Francisco River above and below Clifton, Ariz	317
	26. Recommended limits of fluoride concentration in drinking water	322
	27. Relative tolerance of crop plants to boron	329
	28. Rating of irrigation water for various crops on the basis of boron concentration in the water	329
	29. Water-quality requirements for selected industries and processes	334

STUDY AND INTERPRETATION OF THE CHEMICAL CHARACTERISTICS OF NATURAL WATER

By John D. Hem

ABSTRACT

The chemical composition of natural water is derived from many different sources of solutes, including gases and aerosols from the atmosphere, weathering and erosion of rocks and soil, solution or precipitation reactions occurring below the land surface, and cultural effects resulting from activities of man. Some of the processes of solution or precipitation of minerals can be closely evaluated by means of principles of chemical equilibrium including the law of mass action and the Nernst equation. Other processes are irreversible and require consideration of reaction mechanisms and rates. The chemical composition of the crustal rocks of the earth and the composition of the ocean and the atmosphere are significant in evaluating sources of solutes in natural fresh water.

The ways in which solutes are taken up or precipitated and the amounts present in solution are influenced by many environmental factors, especially climate, structure and position of rock strata, and biochemical effects associated with life cycles of plants and animals, both microscopic and macroscopic. Taken all together and in application with the further influence of the general circulation of all water in the hydrologic cycle, the chemical principles and environmental factors form a basis for the developing science of natural-water chemistry.

Fundamental data used in the determination of water quality are obtained by the chemical analysis of water samples in the laboratory or onsite sensing of chemical properties in the field. Sampling is complicated by changes in composition of moving water and the effects of particulate suspended material. Most of the constituents determined are reported in gravimetric units, usually milligrams per liter or milliequivalents per liter.

More than 60 constituents and properties are included in water analyses frequently enough to provide a basis for consideration of the sources from which each is generally derived, most probable forms of elements and ions in solution, solubility controls, expected concentration ranges and other chemical factors. Concentrations of elements that are commonly present in amounts less than a few tens of micrograms per liter cannot always be easily explained, but present information suggests many are controlled by solubility of hydroxide or carbonate or by sorption on solid particles.

Chemical analyses may be grouped and statistically evaluated by averages, frequency distributions, or ion correlations to summarize large volumes of data. Graphing of analyses or of groups of analyses aids in showing chemical relationships among waters, probable sources of solutes, areal water-quality regimen, and water-resources evaluation. Graphs may show water type based on chemical composition, relationships among ions, or groups of ions in individual waters or many waters considered simultaneously. The relationships of water quality to hydrologic parameters, such as stream discharge rate or ground-water flow patterns, can be shown by mathematical equations, graphs, and maps.

About 75 water analyses selected from the literature are tabulated to illustrate the relationships described, and some of these, along with many others that are not tabulated, are also utilized in demonstrating graphing and mapping techniques

Relationships of water composition to source rock type are illustrated by graphs of some of the tabulated analyses. Activities of man may modify water composition extensively through direct effects of pollution and indirect results of water development, such as intrusion of sea water in ground-water aquifers.

Water-quality standards for domestic, agricultural, and industrial use have been published by various agencies. Irrigation project requirements for water quality are particularly intricate.

Fundamental knowledge of processes that control natural water composition is required for rational management of water quality.

SCOPE AND PURPOSE

The chemistry of natural water is concerned for the most part with impurities much more than with the great bulk of relatively peaceful and inoffensive molecules of H_2O. At first glance, this preoccupation with what usually is a very minor fraction of the water studied by hydrologists might seem unsound or trivial. But natural waters are never pure H_2O—they are in actual fact dilute aqueous solutions of a mixed and often complex character. And even dilute solutions may behave differently in many ways from pure water.

Although a natural water whose quality is suitable for drinking is a more dilute solution than most specialists in solution chemistry are accustomed to working with, the general principles of solution chemistry are certainly applicable to natural water. These principles are the framework on which the material presented here is based. Ways in which theoretical considerations can be used in practical study of natural water chemistry are demonstrated by citing and discussing actual examples wherever possible.

Water chemistry can be applied in various ways in planning water utilization and treatment. The study of natural-water chemistry also involves other disciplines including geology, hydrology, and biological sciences. This book is intended to provide an introduction to the field of natural-water chemistry, with particular emphasis on inorganic geochemistry, which will be intelligible to scientists and engineers whose interests may lie within some part of this area and also to technically trained individuals whose interests may lie in related fields. Some knowledge of chemistry is assumed, and applications of chemical principles that are discussed require general familiarity with the subject. For the most part, however, the treatment is not as chemically demanding as would be appropriate if the book were intended for the use of chemists alone.

As the chemical composition of natural water is controlled by many interrelated processes, it follows that some understanding of these processes is needed before one can speak or act intelligently toward the aim of water-quality control and improvement. It is hoped this book will help in providing impetus toward the needed understanding or suggest ways by which the present understanding of this subject may be improved.

ARRANGEMENT OF TOPICS

The arrangement of topics in this book is similar to that used in the earlier edition. In the introduction there are some basic data from the literature on composition of the lithosphere, hydrosphere, and atmosphere. Those chemical principles and processes which control natural-water composition are then enumerated and described.

Natural-water composition is evaluated by chemical analyses. This discussion is not concerned with analytical procedures, which are amply described elsewhere. A major part of the book, however, is concerned with the completed analysis. To this end, the subjects of sampling, units, and terminology are considered. The constituents and properties reported in the analyses are discussed individually to show what is known or can reasonably be assumed about the forms of various dissolved substances that are most likely to occur, the factors influencing solubility of various ions, and the probable sources of certain solutes. In these discussions, actual chemical analyses of natural waters are extensively used as illustrative material.

The treatment of chemical topics here is more extensive than in the first edition of this book, owing in part to the large amount of research activity in natural-water chemistry which has occurred in recent years. Although the discussion of each constituent is necessarily short, references to current research papers are given wherever possible. This section of the book will soon become out of date as additional research is done in this very active field.

The concluding sections of the book are intended to provide aids for water-analysis intepretation by hydrologists, geologists, or others who are less interested in strictly chemical aspects of the subject. Included here are techniques for analysis classification, graphing, simple statistical correlations, and data extrapolation. Special attention also is given to the correlation of water composition with geology. The use of water analyses in measuring and estimating quantities and rate of movement of water and the relationship of water composition to its use by man are the last topics covered.

Many examples of water-analysis interpretations in published literature have been cited in the bibliography. These were some of the examples which came to the writer's attention; for lack of space, many similar citations had to be eliminated. The large number of references to U.S. Geological Survey publications is due at least in part to the fact that these were readily available to the writer. In the United States, however, there is no other body of material on this subject that is comparable in scope, and every practitioner in the field of natural-water chemistry should be familiar with this report series.

Some references have been made to papers published in other countries, but these by no means constitute a complete survey of literature published elsewhere. A very large amount of material has been published in the U.S.S.R., for example, but for the most part it has not been translated into English; copies are thus not always easy to obtain. References not readily available or not available in English probably are not of major interest to most readers of this book. The predominance of references to papers published in the United States is not meant to imply any monopoly of the field. Natural-water chemistry has been studied throughout the world, and contributions to the field have been made in many countries.

PROPERTIES AND STRUCTURE OF WATER

Water is a chemical compound of hydrogen and oxygen and in the gaseous state, at least, has the molecular formula H_2O. Although the same formula also represents the compositions of liquid water and ice, the molecules in these two forms are associated structurally, and it is a good idea to think of the condensed phases in terms of these associations rather than as simple aggregates of molecules. Because three isotopes of hydrogen and three of oxygen exist in nature, 18 varieties of water molecules are possible. Consideration will be given to some hydrologic applications of the isotopic composition later in this paper.

The physical properties of water are unique in a number of respects, and these departures from what might be considered normal for such a compound are of great importance, both with respect to the development and continued existence of life forms and to the shape and composition of the earth's surface. The boiling and freezing points of water are both far higher than would be expected for a compound with such a low molecular weight, and the surface tension and dielectric constant of liquid water are also very much greater than might be expected.

The principal reason for the unusual properties of liquid water can be discerned from the structure of the H_2O molecule. The two bonds between the oxygen and the hydrogens form an angle of 105°. Therefore, both hydrogens are on the same side of the molecule and that side of the molecule has a net positive charge relative to the other side. This configuration of the water molecule is sometimes represented as. a sphere with two rather prominent bulges or blisters attached to it to indicate the positions of the hydrogen ions. The nature of atoms is not completely represented by models which use spheres of finite diameters but for the present purpose the spherical model will suffice. The distances between the center of the oxygen and the center of each hydrogen are 0.9 angstrom unit (1 angstrom$=10^{-8}$ centimeter), and the hydrogens actually are partly drawn into the area usually considered to be occupied by the oxygen atom.

When the molecules approach each other closely, as they do in the liquid state, the attractive forces between positive and negative sides of adjacent molecules bring about a considerable degree of coherence. The energy required to separate the molecules is indicated by the high heat of vaporization of water and in another way by the high surface tension. The intermolecular forces in water are called hydrogen bonds.

The presence of dissolved ions in water changes some of its physical properties, notably its ability to conduct electricity. The dipolar nature of the water molecule, however, is an important factor in the behavior of the solute ions as well as the solvent. The details of the structure of liquid water are still far from being fully understood. The present state of knowledge has been summarized in recent papers by Drost-Hansen (1967) for pure water and by Kay (1968) for water containing dissolved ions. An extensive review of literature and discussion of concepts has been prepared by Kavanaugh (1964).

The dipolar water molecules are strongly attracted to most mineral surfaces, form sheaths arranged in an orderly pattern around many forms of dissolved ions, and insulate the electrical charges on the ions from other charged species. The effectiveness of water as a solvent is related to such activities. The effectiveness of water in weathering of rocks is also increased by the ability of this cohesive liquid to wet mineral surfaces and penetrate into small openings.

COMPOSITION OF THE EARTH'S CRUST

The average composition of the crustal material of the earth has been a subject of much interest to geochemists for many years. Although the subject of natural-water chemistry is only indirectly concerned with these averages, a knowledge of rock composition is essential to the understanding of the chemical composition of natural water, and it is therefore desirable to discuss the subject briefly.

The earth is generally considered to be made up of an iron-rich core surrounded by a thick mantle made up of magnesium- and iron-rich silicates and a thin outer crust made up of the most common silicates and other minerals. This outer crust, where it is exposed above the level of the oceans, is the only part exerting a direct influence on the composition of natural terrestrial water. It is also the part for which the best data exist for computing an average composition. The bottom of the crust is considered to be at the Mohorovicic discontinuity, which occurs at a depth of 30–50 kilometers beneath most of the continental areas. The influence of material more than a few kilometers below the surface on composition of water circulating in the hydrologic cycle is slight.

Estimates of average composition of the earth's crust by Clarke (1924a) and by Clarke and Washington (1924) are still extensively quoted, although more recent estimates have the advantage of many more analyses and better values for minor constituents. Among the most recent estimates and compilations are those of Fleischer (1953; 1954), Turekian and Wedepohl (1961), Taylor (1964), and Parker (1967). Data on concentrations of some of the rarer elements still are incomplete, however, and some further revisions of the abundance estimates can be expected in the future as better analytical values become available. The amount of extrapolation and inference required to extend the analyses to large volumes of rock that cannot be sampled is obvious.

Table 1 is taken principally from a compilation by Horn and Adams (1966), which is a synthesis by electronic computer of estimates published by others from Clarke's time to the present. Values are given here for the 65 elements covered by Horn and Adams and for two others omitted by them which are of particular interest in natural-water chemistry. These two are nitrogen and carbon. Data on which these two values are based were taken from tabulations by Parker (1967). Oxygen is the most abundant of all the elements in the crustal rocks, and according to Goldschmidt (1954) it constitutes 466,000 parts per million, or 46.6 percent of the weight of the lithosphere. It is not included in table 1. The other omissions include the elements produced in the radioactive decay of uranium and thorium, the elements produced artificially in nuclear reactions, the noble gases, hydrogen, and a few elements for which data are inadequate to make any abundance estimates. The values of Horn and Adams were arbitrarily carried to three significant figures by the computer program from which they were produced. In preparing table 1, all concentrations reported below 100 parts per million were rounded to two significant figures. Because of the uncertainties in the estimates, they can hardly be expected to be accurate enough even to justify two significant figures for many elements, and the reader should not attribute the accuracy that the values seem to imply; for many of the less common elements, especially in sediments, the estimates may well be inaccurate by more than an order of magnitude.

TABLE 1.—*Average composition, in parts per million, of igneous and some types of sedimentary rocks*

[After Horn and Adams (1966)]

Element	Igneous rocks	Sedimentary rocks		
		Resistates (Sandstone)	Hydrolyzates (Shale)	Precipitates (Carbonates)
Si	285,000	359,000	260,000	34
Al	79,500	32,100	80,100	8,970
Fe	42,200	18,600	38,800	8,190
Ca	36,200	22,400	22,500	272,000
Na	28,100	3,870	4,850	393
K	25,700	13,200	24,900	2,390
Mg	17,600	8,100	16,400	45,300
Ti	4,830	1,950	4,440	377
P	1,100	539	733	281
Mn	937	392	575	842
F	715	220	560	112
Ba	595	193	250	30
S	410	945	1,850	4,550
Sr	368	28	290	617
C	320	13,800	15,300	113,500
Cl	305	15	170	305
Cr	198	120	423	7.1
Rb	166	197	243	46
Zr	160	204	142	18
V	149	20	101	13
Ce	130	55	45	11
Cu	97	15	45	4.4
Ni	94	2.6	29	13
Zn	80	16	130	16
Nd	56	24	18	8.0
La	48	19	28	9.4
N	46	----------	600	----------
Y	41	16	20	15
Li	32	15	46	5.2
Co	23	.33	8.1	.12
Nb	20	.096	20	.44
Ga	18	5.9	23	2.7
Pr	17	7.0	5.5	1.3
Pb	16	14	80	16
Sm	16	6.6	5.0	1.1
Sc	15	.73	10	.68
Th	11	3.9	13	.20
Gd	9.9	4.4	4.1	.77
Dy	9.8	3.1	4.2	.53
B	7.5	90	194	16
Yb	4.8	1.6	1.6	.20
Cs	4.3	2.2	6.2	.77
Hf	3.9	3.0	3.1	.23
Be	3.6	.26	2.1	.18
Er	3.6	.88	1.8	.45
U	2.8	1.0	4.5	2.2
Sn	2.5	.12	4.1	.17
Ho	2.4	1.1	82	.18
Br	2.4	1.0	4.3	6.6
Eu	2.3	.94	1.1	.19
Ta	2.0	.10	3.5	.10
Tb	1.8	.74	.54	.14
As	1.8	1.0	9.0	1.8
W	1.4	1.6	1.9	.56
Ge	1.4	.88	1.3	.036
Mo	1.2	.50	4.2	.75
Lu	1.1	.30	.28	.11
Tl	1.1	1.5	1.6	.065
Tm	.94	.30	.29	.075
Sb	.51	.014	.81	.20
I	.45	4.4	3.8	1.6
Hg	.33	.057	.27	.046
Cd	.19	.020	.18	.048
In	.19	.13	.22	.068
Ag	.15	.12	.27	.19
Se	.050	.52	.60	.32
Au	.0036	.0046	.0034	.0018

According to Clarke and Washington (1924), 95 percent of the crust of the earth to a depth of 16 kilometers (10 miles) is igneous rock. Therefore, the average composition of the 16-kilometer crust closely approaches the average for igneous rocks. In the consideration of natural water and its relation to rock composition, however, this predominance of igneous rock is not of overriding importance. Most recoverable ground water occurs at depths of less than 2 kilometers below the land surface, and in the portion of the crust near the surface, sedimentary rocks are more prevalent than igneous ones. As a rule, igneous rocks are poor aquifers, so they transmit little water; also they do not present large areas of active mineral surface to be contacted by relatively small volumes of water as do more porous rock types. In the headwater areas of many streams, igneous rocks are at the surface, and they may contribute solutes to surface runoff both directly and through leaching of partly decomposed minerals in overlying soil. The areas where igneous rocks are exposed to attack by surface streams are not a predominant part of the earth's surface. Therefore, the sedimentary rocks and the soil assume major importance as the immediate sources of soluble matter to be taken up by circulating underground and surface water. Reactions between water and the minerals of igneous rocks, however, are of fundamental importance in studies of geochemical processes, and they will be considered in some detail later in this paper.

The three classes into which sedimentary rocks are divided in table 1 are adapted from Goldschmidt (1933). This classification is based on the chemical composition and the degree of alteration of the minerals making up the rocks. It is probably better suited to studies where chemical composition is involved than are the usual geologic classifications of sediments by means of mineral character, texture, and stratigraphic sequence.

For the purpose of this paper, the following definitions are applicable:

   Resistate: A rock composed principally of residual minerals not chemically altered by the weathering of the parent rock.
   Hydrolyzate: A rock composed principally of relatively insoluble minerals produced during the weathering of the parent rock.
   Precipitate: A rock produced by chemical precipitation of mineral matter from aqueous solution.

A fourth rock type, the evaporites, consists of relatively soluble minerals deposited as a result of evaporation of the water in which they were dissolved. Quantitative data on composition of evaporite rocks has been given by Stewart (1963). The evaporites influence the composition of some natural water, but their average content of most of the minor elements is still not accurately known; data for this class of rocks are not included in table 1.

The severity of chemical attack in weathering has a wide range. Under severe attack, the residue from a given igneous rock might consist almost wholly of quartz sand. Under less severe attack, an arkose containing unaltered feldspar along with the quartz might be produced from the same original rock. Some types of weathering could leave less stable minerals of the original rock in the residue.

Another important factor in determining the compostion of sediments is the process of comminution and the mechanical sorting accompanying weathering and transport of weathering products. Resistates, as the term is usually interpreted, are relatively coarse grained. Some of the resistant mineral particles, however, may be converted to a very fine powder and depostied with the naturally fine grained hydrolyzates.

Chemical precipitation commonly occurs in a saline environment, and the differentation between precipitate and evaporite rock is somewhat arbitrary; so precipitate and evaporite components may be interbedded.

Geochemists often add other classifications such as oxidates, typified by iron ore, and reduzates, for material of biological origin such as coal. The dividing lines, however, between these and the classes already considered are quite inexact.

Because so many of the sedimentary rocks contain mixtures of weathering products, any division into classes must be somewhat arbitrary. Thus, although one might think of a pure quartz sand as the ideal representative of the resistates, for the purpose of this paper the class also includes sandstone, conglomerate, arkose, graywacke, and even unconsolidated alluvium. Likewise, although clay is the ideal representative of the hydrolyzates, the class includes shale which often contains high percentages of constituents other than clay minerals. Both classes of rock usually contain chemically precipitated minerals as coatings or discrete particles. The precipitate rocks, limestone and dolomite, generally are aggregates of calcitic or dolomitic particles, with many impurities, rather than massive crystalline precipitates.

Although classification is a useful scientific activity, it is by no means the most useful result of the scientific study, and the subject of sedimentary rock classification will not be pursued further here. More extensive discussions of classification and identification are contained in texts on sedimentary petrography, notably that of Carozzi (1960).

THE HYDROSPHERE

The hydrosphere is generally defined by geochemists as the liquid and solid water present at and near the land surface, and its dissolved constituents. Water vapor and condensed water of the atmosphere are usually included, but water which is immoblized by incorporation into mineral structures in rocks is usually not thought of as a part of the hydrosphere. The term has no great utility for this discussion in any event.

The oceans constitute about 98 percent of the hydrosphere, and its average composition then is for all practical purposes that of sea water. The water of the ocean basins is generally fairly well mixed with regard to major constituents, although contents of most minor elements are not very uniform with depth or areally. The average chemical composition of sea water is indicated in table 2. The data are mostly from the compilation of Goldberg (1963) and cover most of the elements given in table 1. A knowledge of the principal form in which each element occurs in sea water, as indicated by Goldberg, is helpful in understanding the chemistry of sea water.

The average composition of river waters of the world has been of interest to geochemists, and one of the more recent computations by Livingstone (1963) is given in table 3. The composition of water from a river, however, changes with time, and the composition of water from different rivers may cover a wide range. Therefore, a single average analysis does not convey much useful information. Also included in table 3 are analyses for two large rivers. The sample from the Amazon, the largest of the world's rivers, was obtained at a high stage of flow. Other analyses for this stream show a higher concentration at lower discharge, but neither the average flow rate nor the composition of the water of this stream is adequately known. The other analysis in table 3 represents an average for 1 year for the Mississippi River near New Orleans, La. The average discharge for the Mississippi into the Gulf of Mexico was stated by Livingstone (1963, p. 38) to be about 650,000 cfs (cubic feet per second).

THE ATMOSPHERE

The composition of the atmosphere in terms of volume percent and partial pressures of the gaseous components is given in table 4. Local variations in atmospheric composition are produced by the activities of man, plant and animal metabolism and decay, and exuded gases from volcanoes and other geothermal areas. Particulate matter carried into the air by wind, discharged from smokestacks,

TABLE 2.—*The composition of sea water*
[After Goldberg, 1963]

Constituent	Concentration (mg/l)	Principal forms in which constituent occurs
Cl	19,000	Cl^-
Na	10,500	Na^+
SO_4	2,700	SO_4^{-2}
Mg	1,350	Mg^{+2}, $MgSO_4$ aq
Ca	400	Ca^{+2}, $CaSO_4$ aq
K	380	K^+
HCO_3	142	[a]HCO_3^-, H_2CO_3 aq, CO_3^{-2}
Br	65	Br^-
Sr	8	Sr^{+2}, $SrSO_4$ aq
SiO_2	6.4	H_4SiO_4 aq, $H_3SiO_4^-$
B	4.6	H_3BO_3 aq, $H_2BO_3^-$
F	1.3	F^-
N	.5	[b]NO_3^-, NO_2^-, NH_4^+
Li	.17	Li^+
Rb	.12	Rb^+
P	.07	HPO_4^{-2}, $H_2PO_4^-$, PO_4^{-3}, H_3PO_4 aq
I	.06	IO_3^-, I^-
Ba	.03	Ba^{+2}, $BaSO_4$ aq
Al	.01	(c)
Fe	.01	$Fe(OH)_3$(c)
Mo	.01	MoO_4^{-2}
Zn	.01	Zn^{+2}, $ZnSO_4$ aq
Se	.004	SeO_4^{-2}
As	.003	$HAsO_4^{-2}$, $H_2AsO_4^-$, H_3AsO_4 aq, H_3AsO_3 aq.
Cu	.003	Cu^{+2}, $CuSO_4$ aq
Sn	.003	
U	.003	$UO_2(CO_3)_3^{-4}$
Mn	.002	Mn^{+2}, $MnSO_4$ aq
Ni	.002	Ni^{+2}, $NiSO_4$ aq
V	.002	$VO_2(OH)_3^{-2}$
Ti	.001	
Co	.0005	Co^{+2}, $CoSO_4$ aq
Cs	.0005	Cs^+
Sb	.0005	
Ce	.0004	
Ag	.0003	$AgCl_2^-$, $AgCl_3^{-2}$
La	.0003	
Y	.0003	
Cd	.00011	Cd^{+2}, $CdSO_4$ aq
W	.0001	WO_4^{-2}
Ge	.00007	$Ge(OH)_4$ aq, $H_3GeO_4^-$
Cr	.00005	
Th	.00005	
Sc	.00004	
Ga	.00003	
Hg	.00003	$HgCl_3^-$, $HgCl_4^{-2}$
Pb	.00003	Pb^{+2}, $PbSO_4$ aq
Bi	.00002	
Nb	.00001	
Au	.000004	$AuCl_4^-$
Be	.0000006	
Pa	2×10^{-9}	
Ra	1×10^{-10}	Ra^{+2}, $RaSO_4$ aq

[a] Reported HCO_3 also includes some carbon present in organic compounds.
[b] Total N also includes some dissolved nitrogen gas.
[c] Probably present as $Al(OH)_4^-$, AlF^{+2}, and AlF_2^+ (Hem, 1968).

TABLE 3.—*Composition of river water*

[Date under sample number is date of collection. Source of data: 1, Oltman (1968, p. 13); 2, U.S. Geol. Survey Water-Supply Paper 1950 (p. 316); 3, Livingstone (1963, p. 41)]

Constituent	1 July 16, 1963		2 Oct. 1, 1962– Sept. 30, 1963		3	
	mg/l	meq/l	mg/l	meq/l	mg/l	meq/l
Silica (SiO_2)	7.0		6.7		13	
Aluminum (Al)	.07					
Iron (Fe)	.06		.04		.67	
Calcium (Ca)	4.3	0.214	42	2.096	15	0.750
Magnesium (Mg)	1.1	.090	12	.987	4.1	.342
Sodium (Na)	1.8	.078	25	1.088	6.3	.274
Potassium (K)			2.9	.074	2.3	.059
Bicarbonate (HCO_3)	19	.311	132	2.163	58	.958
Sulfate (SO_4)	3.0	.062	56	1.166	11	.233
Chloride (Cl)	1.9	.054	30	.846	7.8	.220
Fluoride (F)	.2	.011	.2	.011		
Nitrate (NO_3)	.1	.002	2.4	.039	1	.017
Dissolved solids	28		256		90	
Hardness as $CaCO_3$	15		155		55	
Noncarbonate	0		47		7	
Specific conductance (micromhos at 25° C)	40		421			
pH	6.5		7.5			
Color			10			
Dissolved oxygen	5.8					
Temperature °C	28.4					

1. Amazon at Obidos, Brazil. Discharge, 7,640,000 cfs (high stage).
2. Mississippi at Luling Ferry, La., (17 mi west of New Orleans). Time weighted mean of daily samples.
3. Mean composition of river water of the world.

or entering the atmosphere from outer space provides a number of atmospheric components that may influence the composition of water but that cannot be readily evaluated in terms of average contents.

Among the minor constituents of air are certain nuclides produced in the outer reaches of the atmosphere by cosmic-ray bombardment and by other processes. Some of these nuclides are radioactive, notably tritium and carbon-14. Naturally produced radioactive materials are present in the atmosphere in very small quantity, however, and can be detected only by very sensitive techniques.

PRINCIPLES AND PROCESSES CONTROLLING COMPOSITION OF NATURAL WATER

Some of the well-known principles of the chemistry of aqueous solutions have been used extensively in recent studies of natural-water composition. The book "Mineral Equilibria at Low Temperature and Pressure," by R. M. Garrels (1960), and a revised and enlarged edition by Garrels and Christ entitled "Solutions, Minerals and Equilibria" (1964) are concerned with applications of chemical principles in the behavior of minerals and metallic elements in aqueous environments and thus apply directly to natural-water chemistry.

TABLE 4.—*Mean composition of the atmosphere*
[After Mirtov (1961)]

Gas	Percentage by volume	Partial pressure (atm)
N_2	78.1	0.781
O_2	20.9	.209
A	.93	.0093
H_2O	0.1–2.8	.001–.028
CO_2	.03	.0003
Ne	1.8×10^{-3}	1.8×10^{-5}
He	5.2×10^{-4}	5.2×10^{-6}
CH_4	1.5×10^{-4}	1.5×10^{-6}
Kr	1.1×10^{-4}	1.1×10^{-6}
CO	$(.06–1) \times 10^{-4}$	$(.06–1) \times 10^{-6}$
SO_2	1×10^{-4}	1×10^{-6}
N_2O	5×10^{-5}	5×10^{-7}
H_2	$\sim 5 \times 10^{-5}$	$\sim 5 \times 10^{-7}$
O_3	$(.1–1.0) \times 10^{-5}$	$(.1–1.0) \times 10^{-7}$
Xe	8.7×10^{-6}	8.7×10^{-8}
NO_2	$(.05–2) \times 10^{-6}$	$(.05–2) \times 10^{-8}$
Rn	6×10^{-18}	6×10^{-20}

Papers in which equilibrium solubility calculations, complex ion effects, and electrochemical equilibria and chemical kinetics are explored as means to explain the chemical composition of natural water have been published by a growing number of researchers trained in geochemistry and interested in those aspects concerned with water. This field of research owes much to the pioneering work of Prof. Garrels and his associates.

The trend toward increased application of chemical theory and principles in the whole field of geochemistry is evident in recent textbooks such as that of Krauskopf (1967).

Although much still needs to be done before the chemical principles controlling natural-water composition can be fully defined, a general summary of what now seems most important is a necessary part of any comprehensive discussion of natural-water chemistry. The material presented here is only a brief outline, but references cited provide opportunity to pursue the subject much further, should it be desired to do so.

The principles and processes to be described include (1) the fundamental alterations of atoms and nuclei, (2) the chemical reactions of various kinds involving water, solutes, and solids or gases, and (3) the means used to study these reactions systematically, especially as they relate to environmental factors that influence the rate and path of each reaction as water moves through the hydrologic cycle. The theoretical basis of the study of natural-water chemistry consists of these fundamental factors.

ATOMIC AND NUCLEAR ALTERATIONS

Certain of the combinations of fundamental particles which go to make up the nuclei of atoms are inherently unstable. The nuclides which exhibit this instability are naturally radioactive; that is, they emit energy when unstable nuclei change toward more stable configurations. The degree of instability can be represented in terms of half life, which is the length of time required for half of the unstable nuclide present at a given time to disappear.

The energy emitted by radioactive nuclides is of three types. Alpha radiation consists of the emission of alpha particles, which are helium nuclei. Beta radiation consists of streams of electrons, which are unit negative charges or their positive equivalents, positrons. Gamma radiation is short wavelength electromagnetic radiation equivalent to X-rays.

Certain elements predicted by the periodic table have only unstable isotopes with short halflives. Thus, element 43, technetium, is unknown in nature. Other unstable nuclides, such as potassium-40, have long halflives and still occur naturally. Four naturally occurring radioactive nuclides, uranium-238 and -235, thorium-232, and neptunium-237, each pass through a disintegration series of relatively shortlived species, finally ending in a stable isotope of lead or bismuth. As a result of these decay sequences, certain relatively shortlived and highly radioactive species occur naturally. The most common such forms are isotopes of the noble gas radon and the element radium. The most abundant isotopes of these elements are the ones produced in the uranium-238 decay scheme, including principally radium-226 and radon-222.

INDUCED RADIOACTIVITY

Bombardment of nuclei of atoms by energetic particles can produce an unstable structure which later undergoes radioactive decay. Several species are produced by cosmic rays acting on atoms in the upper atmosphere. Among the most important ones formed this way are hydrogen-3 (tritium) and carbon-14, although the naturally occurring quantities of these are small in normal air or water and can be detected only by very sensitive techniques. The half life of carbon-14 (5,730 yr) is such that it is a convenient dating tool for archeologic and geologic material containing carbon that was derived from the atmosphere; research is currently underway to use carbon-14 to date ground water. Tritium, with a half life of 12.3 years, is useful for hydrologic studies.

With the advent of man-induced nuclear fission in the 1940's, large quantities of radioactive nuclides covering the entire periodic table have been produced. The byproducts of nuclear power plants will continue to be a source of these nuclides.

CHEMICAL REACTIONS

The chemical reactions in which elements participate do not involve changes in the nuclei of atoms, but represent changes in the arrangement and interactions among electrons, which surround the nucleus. The field of natural-water chemistry is principally concerned with reactions that occur in relatively dilute aqueous solution, although some natural waters have rather high concentrations. Over the past century, an extensive body of theory regarding the behavior of dilute solutions has been developed, and methods of evaluating aqueous-solution chemistry have attained a high degree of precision. It is still true that some aspects are poorly understood, but powerful methods and techniques are available from physical and inorganic chemistry that can be directly applied in studies of natural water. The most significant development in the field of natural-water chemistry during the past 10-15 years has been the increased application of these techniques to help explain and understand reactions between water and rock minerals. As a result of this work, theoretical evaluations of the behavior of some elements now are available which are in reasonable agreement with conditions observed in the field.

Among studies which can be cited as examples are the application of calcite solubility equilibria in studies of the hydrology of limestone (Back, 1963; Holland and others, 1964), silica solubility and kinetics (Krauskopf, 1956a; Van Lier and others, 1960), electrochemical equilibria in studies of behavior of iron and manganese in water (Hem and Cropper, 1959; Hem, 1963a), studies of silicate mineral dissolution (Feth, Roberson, and Polzer, 1964), and various studies of adsorption-desorption equilibria (Robinson, 1962). The chemistry of dilute solutions of geochemical interest has been effectively summarized by Garrels and Christ (1964).

Some of the progress in understanding the chemistry of natural water has stemmed from related work in prospecting for mineral deposits. Much of the pioneering work on use of natural-water composition in mineral prospecting was done in the U.S.S.R. (Fersman and others, 1939). This subject has been summarized in English by Hawkes and Webb (1962).

Studies abroad more directly related to natural-water chemistry have also been extensive. A particularly large volume of literature is from the U.S.S.R. Studies there, for example, have included electrochemical equilibria involving iron (Germanov and others, 1959), and there are many variations on the general theme of the alteration of water that occurs during movement underground and the related chemical properties of ground water. Examples are papers by Valyashko (1958, 1967). A paper in English by Chebotarev (1955) shows some of the approaches which have been used. Many papers on classification of water into types on the basis of the analysis appear in the Soviet literature. The emphasis on classification may be related to the pioneering work of V. I. Vernadsky, who was a very prominent Russian geochemist a generation ago and who was very much interested in this subject.

Papers on aqueous geochemistry from other countries have been cited as appropriate. No attempt, however, has been made to provide a complete review of the literature.

TYPES OF REACTIONS IN WATER CHEMISTRY

Two different types of chemical reactions are important in establishing and maintaining the composition of natural water. The first of these categories includes reactions whose energy relationships and rates are favorable for establishment of chemical equilibrium in conditions likely to be encountered at or near the earth's surface. These reversible reactions may be further subdivided into three groups:

1. Reversible solution and deposition reactions in which water is not chemically altered. Reactions of this type include reversible solution of solid crystalline material, such as

$$NaCl(c) \rightleftarrows Na^+ + Cl^-$$

and

$$CaSO_4(c) \rightleftarrows Ca^{+2} + SO_4^{-2},$$

solution of gases and organic or other solids which form aqueous molecular dispersions, for example,

$$N_2(g) \rightleftarrows N_2(aq)$$

and

$$\text{sugar (c)} \rightleftarrows \text{sucrose (aq)},$$

and reactions among aqueous species or complexes, such as

$$Ca^{+2} + SO_4^{-2} \rightleftarrows CaSO_4(aq).$$

Sorption-desorption and ion-exchange reactions, such as

$$Na_2X + Ca^{+2} \rightleftarrows CaX + 2Na^+,$$

could also be included in this class. Water participates in none of these reactions except as individual molecules incorporated into hydration sheaths around charged ions.

2. Reversible solution and deposition reactions in which water molecules are broken down into H^+ and OH^- Reactions of this type include simple hydrolysis reactions such as solution of carbonates or

dissociation of acid-forming solutes:

$$CaCO_3(c) + H_2O \rightleftharpoons Ca^{+2} + HCO_3^- + OH^-$$

$$CO_2(aq) + H_2O \rightleftharpoons HCO_3^- + H^+.$$

Another type of reaction involves hydrolysis among dissolved weakly basic ions (water shown here is in the hydration sheath of the ion):

$$Al(H_2O)_6^{+3} \rightleftharpoons AlOH(H_2O)_5^{+2} + H^+.$$

3. Reversible solution and deposition reactions or ion reactions involving changes in oxidation state, such as the reduction of ferric iron:

$$Fe^{+3} + e^- \rightleftharpoons Fe^{+2}.$$

Such reactions also may involve hydrolysis:

$$Fe(OH)_3(c) + e^- + 3H^+ \rightleftharpoons Fe^{+2} + 3H_2O.$$

These three classes of reversible reactions generally trend toward chemical equilibrium and can be studied by means of chemical techniques that are appropriate for equilibrium conditions.

Reactions less readily reversible than the ones cited above represent the second general type. These reactions may include relatively complicated processes such as the weathering of albite (sodium feldspar), for example, which does not seem to be reversible under ordinary weathering conditions:

$$2NaAlSi_3O_8(c) + 2H^+ + 9H_2O = 2Na^+ + H_4Al_2Si_2O_9 + 4Si(OH)_4(aq).$$
$$\text{(albite)} \qquad\qquad\qquad\qquad \text{(kaolinite)}$$

Barriers to reversing the reaction may result from slow reaction rates or the need for an energy input. Life processes of organisms may reverse some reactions by expending energy derived from other sources, for example, the oxidation of cellulose, the reversal of which is accomplished in photosynthesis:

$$C_6H_{10}O_5 + 6O_2(aq) = 6CO_2 + 5H_2O.$$

Many nitrogen and sulfur oxidations or reductions are also involved in biologic processes, for example,

$$NO_3^- + 8e^- + 10H^+ = NH_4^+ + 3H_2O.$$

Even though some of these reactions require no energy input, they can be very slow to take place if organisms are absent.

Reactions that are not readily reversible can be evaluated by study of rates and mechanisms of reaction, but obviously do not lend themselves to interpretations requiring the existence of equilibrium.

CHEMICAL EQUILIBRIUM

The substances entering into a chemical reaction are not normally completely converted to products, although a reaction may approach completion if products leave the reaction site or if the original reacting substances are highly unstable with respect to the products under the conditions of the reaction. The hypothetical reaction in which substance A reacts with substance B to form the new species C and D may be written

$$aA + bB = cC + dD,$$

where the lowercase letters represent the multiples of the reactants needed to balance the equation. If the reaction is allowed to proceed in a closed system where none of C or D escapes and a reverse reaction sets in after a time in which C combines with D to form A and B; the rates of the forward and reverse reactions ultimately become equal, and the system may then be said to have attained a state of chemical equilibrium.

The law of mass action, proposed about 100 years ago by Guldberg and Waage, states that at equilibrium the following relation will hold:

$$\frac{[C]^c[D]^d}{[A]^a[B]^b} = K.$$

The bracketed quantities represent effective molar concentrations of the reactants and products, and K is the equilibrium constant, a characteristic value for any given set of reactants and products. Although it was not realized at the time the law was first proposed, solution chemists soon found that K was not constant for all concentrations when analytically measured values were used.

The law of mass action, therefore, applies strictly to actual solution concentrations only if they are equal to, or can be accurately corrected to, the "effective concentrations" specified by the mass law. The effective concentration also is commonly referred to as the "activity," or "thermodynamic concentration," and is expressed in moles per liter for solutes and in atmospheres of pressure for gases. For reactions in dilute solution, the activities of the solvent and any solids that are present are assigned a value of unity. The behavior of the system is influenced both by temperature and pressure. Standard conditions, to which equilibrium constants generally are referred, are 25.0° Celsius and 1 atmosphere of pressure.

The departure of dissolved participants in chemical reactions from "ideal" behavior (concentration=activity) is largely attributable in dilute solution to electrostatic effects. Ions carrying like charges repel each other, and those with unlike charges attract each other. The strength of the fields around the ions is a function of the dielectric, or insulating, properties of the solvent. The mobility of ions also is influenced by their physical dimensions, as well as by the concentration of charged particles.

Some measurement techniques are available for determining the activity of ions directly. Ion-sensitive electrodes, for example, are available which attain electrical potentials in proportion to the activity of a particular ion in solution. The first to be developed and best known of these is the glass electrode, which has been used for many years to determine the hydrogen-ion activity, or pH, of solutions. Electrodes are now available which can determine the activities of a considerable number of the cations and anions of interest in natural water.

If a direct measurement of the activity of an ion is impossible, there are methods for calculating a correction factor, the activity coefficient, which when multiplied by the analytically determined concentration of the ionic species in question will give the activity. For dilute solutions, activity coefficients of single ions can be computed by means of the Debye-Hückel equation. Various forms of this equation exist. It is based on an assumption that ions behave as charged particles of finite sizes in an electrostatic field of uniform intensity. Several of the parameters in the equation have been empirically determined, but is seems generally agreed the equation works satisfactorily for solutions whose total concentration is not much over 0.10 mole per liter of uni-univalent salts. This would be equivalent to a concentration of about 5,800 mg/l (milligrams per liter) of total dissolved ions in a sodium chloride solution. Ions with charges greater than 1 give a more intense effect, and the maximum concentration permissible is somewhat lower.

The form of the Debye-Hückel equation used here is

$$-\log \gamma_i = \frac{A z_i^2 \sqrt{I}}{1 + B a_i \sqrt{I}},$$

where

γ_i is the activity coefficient of the ion,

A is a constant relating to the solvent (for water at 25° C it is 0.5085),

z_i is the ionic charge,

B is a constant relating to the solvent which for water at 25° C is 0.3281.

a_i is a constant related to the effective diameter of the ion in solution, and

I is the ionic strength of the solution.

The values of a_i for various ions of interest are given in table 5 and were adapted from Kieland (1937).

The ionic strength of a solution is a measure of the strength of the electrostatic field caused by the ions and is computed from the expression

$$I = \Sigma \frac{m_i z_i^2}{2},$$

where m is the concentration of a given ion in moles per liter and z is the charge on that ion. The terms in the summation include one for each ionic species present. A nomograph which simplifies calculation of ionic strength from analytical data in milligrams per liter published earlier by the writer (Hem, 1961a) is reproduced in modified form here as plate 1. Figure 1 is a graph for determining γ when I is known for the various major ions of natural water. Butler (1964, p. 473) gave a nomograph for computing γ which can be used for solutions somewhat more concentrated than those considered in figure 1, and there are values for mean activity coefficients for inorganic salts in solutions where ionic strength exceeds 0.10 in Latimer (1952, p. 349–356). Mean activity coefficients, however, cannot be directly substituted for single ion activity coefficients. Garrels and Christ (1964, p. 20–73) have described methods for calculating activity coefficients for a wide range of ion concentrations.

In solutions containing less than 50 mg/l of dissolved ions, the ionic strength normally is less than 10^{-4}, and activity coefficients for most ions are 0.95 or more. In solutions this dilute, activity values are equal to measured concentrations within ordinary analytical error. If concentrations are near 500 mg/l of dissolved solids, the

TABLE 5.—*Values for parameter a_i in Debye-Hückel equation*

[After Kielland (1937)]

a_i	Ion
11	Th^{+4}, Sn^{+4}
9	Al^{+3}, Fe^{+3}, Cr^{+3}, H^+
8	Mg^{+2}, Be^{+2}
6	Ca^{+2}, Cu^{+2}, Zn^{+2}, Sn^{+2}, Mn^{+2}, Fe^{+2}, Ni^{+2}, Co^{+2}, Li^+
5	$Fe(CN)_6^{-4}$, Sr^{+2}, Ba^{+2}, Cd^{+2}, Hg^{+2}, S^{-2}, Pb^{+2}, CO_3^{-2}, SO_3^{-2}, MoO_4^{-2}
4	PO_4^{-3}, $Fe(CN)_6^{-3}$, Hg_2^{+2}, SO_4^{-2}, SeO_4^{-2}, CrO_4^{-2}, HPO_4^{-2}, Na^+, HCO_3^-, $H_2PO_4^-$
3	OH^-, F^-, CNS^-, CNO^-, HS^-, ClO_4^-, K^+, Cl^-, Br^-, I^-, CN^-, NO_2^-, NO_3^-, Rb^+, Cs^+, NH_4^+, Ag^+

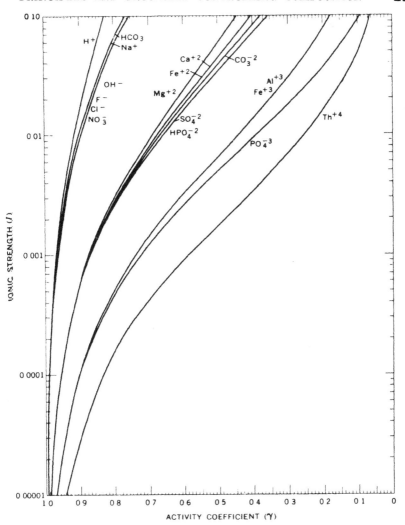

FIGURE 1 — Relation of activity coefficients for dissolved ions to ionic strength of solution

value of γ may be as low as 0.70 for divalent ions. At the maximum ionic strength where the Debye-Hückel equation can be accurately used, the activity coefficients of some divalent ions may be less than 0.40.

The equilibrium constants for many chemical reactions have been determined and are published in standard references. Most values are specified as applying to solutions of zero ionic strength. Before laboratory analytical values can be used to compute a comparable constant, they must be converted to activities. Some chemists have preferred to conduct solubility experiments in the presence of constant and rather high concentrations of inert ions. The equilibrium constants resulting are strictly applicable only at the ionic strength of the system which was investigated. For the concentration ranges of principal interest in natural-water chemistry, the Debye-Hückel equation and equilibrium constants applicable at zero ionic strength provide the most versatile approach. Chemists working with more highly concentrated solutions such as sea water may prefer to use constants directly applicable to these solutions.

TEMPERATURE AND PRESSURE EFFECTS

As noted above, the behavior of dissolved ions is influenced by temperature and pressure. The treatment in this paper has been simplified by assuming standard conditions of 25° C and 1 atmosphere. The effects of moderate departures from standard conditions (a few atmospheres pressure and $\pm 10°$–$15°C$) are not large enough to preclude the direct application of standard conditions to most natural-water environments. Unavoidable errors of sampling and analysis generally can be expected to affect equilibrium calculations at least as much as these small departures from standard temperature and pressure. Effects of temperature and pressure on equilibrium calculations and methods of compensating for them are given by Garrels and Christ (1964, p. 306–354) and will not be considered here. They may need to be taken into account in certain kinds of investigations, however.

SOLUTION OF CALCITE

The use of mass-law calculations in natural-water chemistry can be conveniently illustrated for a system containing no gas phase, where solid calcite and water are present. The solution of calcite here follows the chemical equation

$$CaCO_3(c) + H^+ = Ca^{+2} + HCO_3^-.$$

An equilibrium will be attained with the H^+ derived from the water or other source attacking the solid to give calcium and bicarbonate ions. In mass-law form, the equation for the equilibrium constant will be

$$\frac{[Ca^{+2}][HCO_3^-]}{[CaCO_3][H^+]} = K.$$

The quantities in square brackets represent activities in moles per liter. By convention, the activity of the solid is taken as unity. It is essential that some solid calcite be present to have equilibrium.

The equilibrium constants which have been published for a great many reactions have been compiled by Sillén and Martell (1964). Most of these are applicable at 25°C, but some are for other temperatures. The value for a particular reaction, however, may not always be available. Sometimes a simple series of additions or subtractions of equilibria for which constants are given will yield the desired value, for example,

$$CaCO_3(c) = Ca^{+2} + CO_3^{-2} \qquad K_1 = 4.70 \times 10^{-9}$$

$$HCO_3^- = CO_3^{-2} + H^{+1} \qquad K_2 = 4.84 \times 10^{-11}.$$

If the second equilibrium is subtracted from the first, we obtain

$$CaCO_3(c) + H^+ = Ca^{+2} + HCO_3^-.$$

The equilibrium constant for the combined reaction is obtained by dividing the equation for K_1 by the equation for K_2:

$$\frac{[Ca^{+2}][CO_3^{-2}]}{[CaCO_3]} \div \frac{[CO_3^{-2}][H^+]}{[HCO_3^-]} = \frac{4.70 \times 10^{-9}}{4.84 \times 10^{-11}} = 0.97 \times 10^2.$$

It is also possible to calculate equilibrium constants by another procedure, to be given in the next section.

At equilibrium in the calcite + water system the activities of solutes are related by the equation

$$\frac{[Ca^{+2}][HCO_3^-]}{0.97 \times 10^2} = [H].$$

If values are assigned to any two of the ion activities, the other can be computed. If concentrations of Ca^{+2} and HCO_3^- are known from the analysis, corresponding activities can be calculated by using the Debye-Hückel equation, and the ionic strength and the hydrogen-ion activity representing saturation, or equilibrium with respect to calcite, can be calculated. The chemical analysis thus can be used to calculate whether or not a given ground water is at saturation with respect to calcite. The analytical data and especially the pH measurement must be obtained in such a way that they represent the conditions to which the results are being applied, however. The pH attained by samples stored in the laboratory for a time, for example, may differ considerably from that attained by the water in its natural habitat, because in the sample bottle, solid calcite may not be present, and there is generally a gas phase.

Garrels and Christ (1964, p. 74–92) gave a number of sample calculations for different equilibrium conditions where calcium carbonate is present in a natural aqueous system. A comparison of the calculated hydrogen-ion activity with the measured pH of the solution will show whether or not saturation has been reached.

The Langelier index, used in evaluating the stability of water supplies after treatment, is a representation of the degree of supersaturation or undersaturation with respect to calcium carbonate (Langelier, 1936). The Langelier index number is the arithmetical difference between the observed pH and the calculated saturation pH. Thus, a value of $+1.0$ represents supersaturation by a factor of 10.

APPLICATION OF CHEMICAL THERMODYNAMICS

Thermodynamics is the study of energy in its various forms. Energy is defined as capacity to perform work. Three broad generalizations known as the first, second, and third laws of thermodynamics form the basis for an extensive development of thermodynamic principles. The first and second laws are of principal concern here.

The first law of thermodynamics is the principle of conservation of energy. Although energy can take various forms, such as heat, electricity, motion, or energy stored in chemical compounds, these forms are interchangeable without any net loss or gain in total energy.

The second law deals with energy transfers and states that energy will only be transferred spontaneously when it can move from a high energy level to a lower one. For example, heat moves spontaneously from a hot body to a cooler one, never the reverse. A result of this law is that energy generally tends to become evenly distributed.

In applying thermodynamic principles to chemistry, the interest is principally in the forms of energy stored in chemical compounds and the mechanisms of alteration or transfer of such energy. The total energy content plus the product of pressure and volume of a particular substance is referred to as its enthalpy and is commonly represented by the symbol H. In a chemical change, however, only a part of this energy is normally available to perform work if the temperature is held constant. The unavailable part of the energy is termed the entropy, designed by the symbol S, and the available part is called free energy, denoted by symbol G. (The symbol F is also sometimes used to designate this quantity, but because some thermodynamicists have used F to refer to the free energy defined somewhat differently, the International Union of Pure and Applied Chemistry has recommended the use of G for the quantity defined as it is here. This definition was first proposed by J. Willard Gibbs.)

The entropy of a chemical substance is sometimes described in terms of the amount of randomness or disorder displayed in the material. Entropy also is defined as the degree of probability of occurrence of a particular form of a material related to the readily observable fact that, left to themselves, systems tend to become disordered, whether they are geochemical systems or are almost any of the works of man. A corollary of the second law of thermodynamics thus points out that in spontaneous processes the entropy of the products is greater than that of the reactants. In a closed system, therefore, free energy tends to decrease and entropy to increase.

It is well to be cautious in applying broad thermodynamic generalizations to geochemical systems, because these systems are not generally closed. The earth receives a considerable energy input from the sun, and the degree to which such processes as weathering cause changes in the total stored energy of natural materials remains imperfectly understood. However, the value of application of the second law in the study of many natural processes is obvious.

The quantities given above are related by the equation

$$G = H - TS$$

where T is the temperature in degrees Kelvin. Absolute measurements of total energy are not possible; hence, to show that measurements are referred to an arbitrary zero datum, or reference state, the equation is written

$$\Delta G = \Delta H - \Delta(TS).$$

If a constant temperature is maintained,

$$\Delta G = \Delta H - T\Delta S.$$

The third law of thermodynamics states that at 0°K (absolute zero) the entropy of any substance is zero. Calculations using the foregoing equations and the third law may be useful for determining ΔG, when some of the other quantities are known.

The present discussion is principally concerned with free energy, which is the energy available from a chemical reaction occurring isothermally. From the second law it can be deduced that a chemical reaction can be expected to occur spontaneously if free energy contained in the products is less than the amount contained in the reactants. It also follows from the second law that at chemical equilibrium the energy contents will be uniformly distributed and that that there should be a rigorous relationship between free energy change and the equilibrium constant postulated by the law of mass action.

For simplest application of these principles, standard conditions of 1 atmosphere of pressure and a temperature of 25°Celsius are specified. The standard Gibbs free energies of formation, $\Delta G°$, of a great many chemical compounds, elements, and ions have been determined and are reported in the literature (Robie and Waldbaum, 1968). The text of Latimer (1952), although somewhat out of date, remains a very useful compendium of such data, as does the compilation by the National Bureau of Standards (Rossini and others, 1952). Standard free energies are expressed in kilocalories per mole and can be determined both by thermal and by chemical methods. The plus or minus sign indicates the amount by which $\Delta G°$ is more or less than the reference state. The reference state for most substances is the pure element in the form to be expected at 25°C.

The net change in free energy occurring in a chemical reaction under standard conditions can be determined from $\Delta G°$ values for participating substances. For example, for the solution of calcite by aqueous H^+,

$$CaCO_3(c) + H^+ = Ca^{+2} + HCO_3^-,$$

the $\Delta G°$ values are

$CaCO_3(c)$ (calcite)	-269.78 kilocalories per mole,
H^+	0 (reference state).
Ca^{+2}	-132.18, and
HCO_3^-	-140.31.

The net change occurring when one mole of reaction products is produced is $-140.31 + (-132.18) - (-269.78) = -2.71$. The negative sign of this $\Delta G_R°$ (representing the standard free energy change in the reaction) indicates that energy is released, and the reaction can be expected to be spontaneous. The energy relationships do not, however, predict the rate at which the reaction will take place.

As shown in the earlier discussion where the mass law was applied to the solution of calcite, if the activities of the products Ca^{+2} and HCO_3^- are allowed to rise, the reverse reaction will eventually attain the same rate as the forward reaction. When this state of chemical equilibrium is attained, the free energies of components of the system also will be in balance.

The equilibrium constant and the standard free energy change of a reaction are, therefore, different ways of expressing the same thing, and they are related by the equation

$$-\Delta G_R° = RT \ln K,$$

where

R is the gas constant of physical chemistry,
T is the temperature in degrees Kelvin, and
$\ln K$ is the natural logarithm of the equilibrium constant.

When a temperature of 25°C is specified (298.15°K) and base-10 logs are used, the expression becomes

$$-\Delta G_R° = 1.364 \log K.$$

This is an important and useful relationship. Standard free energies of formation are available for all the simpler ions and many other components of interest in water chemistry and for many rock minerals, although better values for many of these are probably needed. At any rate, the chair-borne investigator can write any chemical reaction he might choose, representing such processes as the solution and deposition of single minerals, conversion of one mineral to another by exchange of solutes, or other kinds of chemical alteration, and can compute its equilibrium constant from the standard free energies of the participating materials. He can thus obtain an indication as to the stability of minerals in various environments, the probable importance of different reactions in bringing about the composition of natural water, the limits of solubility of various ions, and other features of the behavior of complicated natural systems that can be evaluated by means of the mass law. An interesting example of application to relationships between rock weathering and the composition of sea water has been developed by Sillén (1967a).

The logic and mathematics of thermodynamics are exact and precise, and with the aid of electronic computers it is possible to make detailed computations that at an earlier time would not have been practicable. Thus, rather comprehensive mathematical modeling of natural-water systems may be within reach. Any thermodynamic calculation of the type described here, however, is only valid for systems that are at chemical equilibrium. Although a strict interpretation might hold that complete equilibrium is never attained in natural systems, the possibility certainly exists that many systems are close enough to equilibrium that one can assume it exists and use thermodynamic approaches without serious error. At this point, human judgment enters and arguments may begin, for there is no way of knowing how far away from equilibrium many natural systems may be. The fit of one or two measured properties of the system into an assumed equilibrium model may not be complete proof that equilibrium does indeed exist.

Some investigators tend to become preoccupied with calculations and make only a minimal number of actual observations of the system they are studying, whether in the laboratory or in nature. Satisfactory progress in the application of equilibrium thermodynamics in natural-water chemistry requires that investigators measure all the significant properties of these systems with as great an accuracy as is practicable. These measurements will serve to verify or refute the theoretical models. Although some systems in geochemistry provide few opportunities for making actual measurements, the aqueous solutions and associated solids and gases of concern in natural-water chemistry are by no means inaccessible to the investigator.

ELECTROCHEMICAL EQUILIBRIUM

Chemical reactions where a participating element changes valence number, losing or gaining orbital electrons, are referred to as oxidations or reductions. In oxidation an element loses electrons, and in reduction it gains them. The reduction process can be represented by an expression such as

$$Fe^{+3}+e^{-}=Fe^{+2},$$

where ferric iron is reduced to the ferrous state. A further reduction,

$$Fe^{+2}+2e^{-}=Fe^{0},$$

would carry the ferrous ion to metallic iron. The symbol "e^{-}" represents the electron, or unit negative charge.

These representations are "half reactions," or redox couples. To take place, a reduction requires a source of electrons. This source may be another element which is simultaneously oxidized, or it could be an actual source of electric current.

Under standard conditions, 25°C and 1 atmosphere, and with unit activity of reactants, it follows that at equilibrium a certain electric potential would be present in a couple such as

$$Fe^{+2}+2e^{-}=Fe^{0},$$

and this standard potential is conventionally represented by the symbol E^{0}. The potential is given in volts, with the potential of the hydrogen electrode,

$$2H^{+}+2e^{-}=H_{2},$$

taken as zero.

The sign of the potential associated with a reaction written as a reduction is negative if the system is reducing and positive if the system is oxidizing. The magnitude of the positive or negative value is a measure of the oxidizing or reducing tendency of the system. Tables of standard potential for couples which have been evaluated are given in standard reference books such as Latimer (1952) and Sillén and Martell (1964). The sign convention used by Latimer results in his data having signs opposite the ones given in other standard references.

When the activities of participating species in a system differ from unity, the potential observed at equilibrium is termed the "redox potential." The redox potential, represented by the symbol Eh, is related to the standard potential and the activities of participating substances by the Nernst equation:

$$Eh = E^{0} + \frac{RT}{nF} \ln \frac{[oxidized]}{[reduced]},$$

where
> R is the gas constant,
> T is the temperature in degrees Kelvin,
> n is the number of electrons appearing in the balanced redox couple or half-reaction, and
> F is the Faraday constant.

The bracketed quantities are the activities of participating solutes. At the standard temperature of 25°C, with base-10 logarithms instead of natural logarithms, this expression becomes

$$\text{Eh} = E^0 + \frac{0.0592}{n} \log \frac{[\text{oxidized}]}{[\text{reduced}]}.$$

The Nernst equation is essentially an extension of the law of mass action and is applicable only to solutions and associated species when chemical equilibrium has been established. The equation is applicable to half-reactions of the type cited, when they are written as reductions. Such half-reactions contain a term indicating the number of electrons required to accomplish the reduction of the species which changes.

As noted, for a reduction actually to occur, a source of electrons must be available. For example, ferric iron might be reduced to the ferrous state in a reaction where organic carbon is being oxidized. The complete reaction can be evaluated by means of the mass law.

Standard potentials are not always available for half-reactions of interest, but usually they can be computed. The free-energy change in a chemical reaction in which electrons are gained or lost (reduction and oxidation) can also be equated to the standard potential for an oxidation or reduction half-reaction. Basically, this represents only a conversion of units from kilocalories per mole to volts. The equation involved is

$$-\Delta G^\circ = nFE^\circ,$$

where n is the number of electrons shown in the balanced half-reaction, F is the Faraday constant, and E^0 is the standard potential. If E^0 is to be in volts,

$$E^0 = \frac{-\Delta G^\circ}{23.06}.$$

For this to give the correct sign, the reaction must be written as a reduction.

With these equations and free-energy data from the literature, one can estimate equilibrium ion activities and redox potentials for a great many systems of geochemical interest. As in similar calculations based on the law of mass action, the application of such estimates to natural systems may furnish useful guidelines as to what to expect. Calculations based on the Nernst equation have been particularly useful in studying the chemistry of iron in ground water.

As in mass-law calculations, a condition of equilibrium must exist to apply the Nernst equation. Practical measurement of redox potential involves some important problems. Measurement of Eh in the ground water environment (actually in the aquifer) is generally not feasible with presently available instrumentation. Measurement of Eh of pumped ground water that has not contacted the air, which would cause it to become oxidizing, requires special equipment and exercise of great care. Such measurements are discussed later under "Redox potential."

CHEMICAL REACTION RATES

In applications of thermodynamics, the conditions of significance are those finally attained after a system has reached equilibrium. One need not be directly concerned with the processes by which equilibrium is attained or with the rate at which these processes occur. The shortcomings of such an approach for the more complicated reactions of dissolution of rock minerals are evident—the rate may be too slow to reach equilibrium, and there may be no possible pathway by which the reaction can be reversed.

The study and evaluation of aqueous geochemical processes by investigating reaction rates are in an early state of development, but many processes can be evaluated in no other way; thus, an increased emphasis on geochemical kinetics is to be expected. The study of kinetics requires a stimultaneous investigation of reaction mechanisms as well as rates and, therefore, insures proper attention to details of the processes.

This discussion will introduce a few concepts of reaction order and mechanism and means of expressing reaction rates. The topic is considered by all standard texts in physical chemistry and more completely in specialized treatises such as that of Benson (1960). However, only a small number of specific uses of kinetic concepts in aqueous geochemistry exist in published literature.

REACTION ORDER

In the simple reaction
$$aA = bB,$$
there is only one reactant and one product. The rate of the forward reaction, R, can be represented as the change in concentration of either A or B per unit time, or in differential form as

$$R = -\frac{dA}{dt} = -\frac{dC_A}{dt};$$

the negative sign indicates that the rates of decrease of A is being observed. C_A represents concentration of reactant A.

By proper manipulation of experimental conditions in the laboratory, it is often possible to observe the rate at which a single reactant changes in concentration without significantly affecting concentrations of any of the other reactants or products. In the example shown, if the effect of increasing concentration of B during the reaction can be minimized or eliminated, the effect of various starting concentrations of A can be evaluated. The relationship of this initial concentration to the rate of reaction can provide insights into the way the reaction occurs, which are not easily deduced from observations of natural systems where all components are varying.

In the hypothetical reaction $aA \rightarrow bB$, if the effect of increasing concentrations of B (the reverse of the reaction shown) can be ignored or overcome, the experimental observations of concentration of A when plotted against time since the beginning of the reaction will typically yield a "die away" curve that can be evaluated to determine reaction order.

The order of a reaction is an important concept and is defined as the actual number of atoms, ions, or molecules whose concentrations determine the rate of the reaction being considered. This is a property which must be determined experimentally. Although in the reaction shown it might be expected that the order would be equal to a, it is not invariably true, especially in the more complex reactions, that the multiples of reactants shown in the summarizing equation are a reliable indication of the reaction order.

A more useful statement of the rate equation for the disappearance of reactant A is

$$-\frac{dC_A}{dt} = kC_A^n,$$

where the order of the reaction with respect to A is the exponent n and k is a quantity termed the "rate constant." A reaction of the first order would be one in which the rate was proportional to the first power of the concentration of A. If the rate were found to be proportional to the square of the concentration of A, the reaction would be of the second order. Third-order reactions also are possible, but are rather unusual.

Integration of the first-order rate equation leads to the expression of the rate constant as

$$k = \frac{1}{t} \ln \frac{C_{A_0}}{C_A},$$

where

C_{A0} = concentration of A at time 0

and

C_A = concentration of A at time t.

This relationship may be expressed in various ways. It is evident that for any given period of time a certain specific fraction of the reactant A will be used up. A convenient measure of rate for a first-order reaction is the half time $t_{1/2}$. This is defined as the length of time required for half the reactant present at the beginning of the specified time period to disappear.

Although strictly not a chemical reaction, the radioactive decay of an unstable nuclide follows first-order kinetics. The half life of such a nuclide is the half time as defined above.

The second-order rate equation, where only one component is involved, integrates to

$$k = \frac{1}{t}\left(\frac{1}{C_A} - \frac{1}{C_{A_0}}\right).$$

Reactions in which a particular component follows second-order kinetics are also common. The oxidation of ferrous iron in oxygenated water has second-order dependence on hydrogen-ion activity, as shown by Stumm and Lee (1961).

Reactions of zero order are also known. In such a reaction

$$\frac{dC_A}{dt} = K;$$

that is, the rate is constant and independent of the concentration. Such reaction kinetics generally indicate some factor, such as the rate of diffusion of reactant to a surface or the area of reactant surface exposed limits the rate.

In a more complex hypothetical reaction such as

$$aA + bB + cC = \text{products,}$$

it is common practice to refer to an "overall order" of the reaction. Here the rate might be specified

$$R = k C_A{}^{n_1} C_B{}^{n_2} C_c{}^{n_3},$$

and the overall order n as

$$n = n_1 + n_2 + n_3.$$

Thus, for example, a reaction which is first order with respect to A and B and zero order with respect to C would have an overall second order. The integrated rate equation for a second-order reaction of this type would be

$$k=\frac{1}{t(C_{A_0}-C_{B_0})}\ln\frac{C_{B_0}C_{A_t}}{C_{A_0}C_{B_t}},$$

where the C terms represent concentrations of reactants at time zero and at time t. The "rate law" for a chemical process is a statement of the rate in terms of all the reactants. Some writers use this term in preference to "overall reaction order."

The most complex reactions commonly occur in a series of steps, one of which often occurs at a slow rate compared to the others, constituting what is usually termed the "rate-determining step." For some reactions there may be several possible reaction schemes, having different rate-determining steps. The route followed, however, by most of the reactants will be the one having the most rapid rate. The existence of stepwise reactions having differing rates may sometimes account for the persistence of nonequilibrium species and may reveal details of chemical reactions that can be observed readily but that are difficult to explain on the basis of chemical equilibrium alone.

EFFECT OF TEMPERATURE ON REACTION RATE

The generalization is commonly made that a 10°C change in temperature changes the rate of reaction by a factor of about 2. The term "reaction" here includes biochemical processes as well as inorganic reactions. The generalization is only a very rough approximation. By measuring experimentally the effect of temperature changes on reaction rate, however, it is possible to obtain some thermodynamic insight into reaction steps.

The Arrhenius equation, which relates the rate constant to temperature, is commonly written

$$k = Ae^{-E/RT},$$

where A is a term representing probability of reaction occurrence, often called the frequency factor, e is the base of natural logs, E is the energy of activation, and R and T represent the gas constant and the temperature in degrees Kelvin. The activation energy represents an amount of energy that must be supplied to bring about a reaction. In a reaction which has an overall favorable free-energy relationship but which proceeds slowly, the explanation may be that one step in the reaction path has a high energy of activation. This is equivalent to a reaction path which goes from high to low energy levels but has to surmount a divide between these levels, and so part of the path leads "uphill" before a "downhill" spontaneous process may begin.

In this very brief summary, only a few of the principles and concepts of kinetics have been introduced. The general importance of the subject in studies of natural-water chemistry, however, is obvious, and progress in understanding natural systems will require the extensive use of kinetic models.

SOLUBILITY CONCEPTS

The chemist generally defines solubility in terms of equilibrium. It represents the total amount of solute species which can be retained permanently in solution under a given set of conditions (fixed temperature, pressure) and in the presence of an excess of undissolved material of definitely known composition and crystal structure from which the solute is derived. A full elaboration of the concept and its application to mixed solutions, such as natural water, is beyond the scope of this book. However, the factors that control the amounts of the major and minor inorganic constituents of natural water are a subject of fundamental concern in this discussion, and the term "solubility" will be used frequently in discussing the behavior of individual constituents.

Water analyses are expressed in terms of concentration of elements or of ions, sometimes implying a single species or form of constitutent and sometimes indicating only the total of an element present without regard to the species. Actually, however, the common analytical procedures are designed to determine total concentrations and do not necessarily provide indications of species. For example, the determination of bicarbonate also includes any bicarbonate that is present in complexes or ion pairs with other ions. Where the term is used in this discussion, solubility will include all forms of a particular element or ion that can be considered to be present as solutes, but will not include amounts present as suspended solid particles.

SOLUBILITY PRODUCT

The solubility product is a form of equilibrium constant for the solution of a specific compound. For example, the chemical equation representing the dissolution of gypsum is

$$CaSO_4 \cdot 2H_2O = Ca^{+2} + SO_4^{-2} + 2H_2O,$$

and the corresponding solubility product expression is

$$[Ca^{+2}][SO_4^{-2}] = K_{s0}.$$

The determined concentration of calcium in a solution cannot be used directly as an equivalent of [Ca] because of activity effects and because some of the calcium may be in other forms. For example, there is an ion pair $CaSO_4^0$ which may contribute to the total solubility, and other ion pairs containing sulfate might also occur if other cations are present.

If the only forms of calcium present are the free ion and the sulfate ion pair, the analytical concentration of calcium C_{Ca} will represent the sum of these two species:

$$C_{Ca} = C_{Ca^{+2}} + C_{CaSO_4^0}.$$

To test a solution to see whether it approaches saturation with respect to gypsum, the activity coefficients for the two dissolved species will be needed. The concentrations of the species can then be expressed in terms of activities:

$$C_{Ca} = \frac{[Ca^{+2}]}{\gamma_{Ca^{+2}}} + \frac{[CaSO_4]}{\gamma_{CaSO_4^0}}.$$

Also, the equilibrium constant for the formation of the ion pair is needed:

$$\frac{[Ca^{+2}][SO_4^{-2}]}{[CaSO_4]} = K.$$

If the two equilibrium constants and the total concentration of calcium are known, the activities of all three solute species can be calculated and the product of $[Ca^{+2}]$ and $[SO_4^{-2}]$ compared with K_{sp}.

Ion pairs and complexes influence the solubility of most dissolved constituents in waters that have high dissolved-solids concentrations. In waters in the concentration range below 1,000 mg/l their influence is of less importance.

Solutions can be simultaneously in equilibrium with two or more solids having a common ion, and combined mass-law equations relating solute species may be written. For example, at saturation with both calcite and gypsum

$$CaCO_3(c) + H^+ + SO_4^{-2} = HCO_3^- + CaSO_4(c)$$

$$\frac{[HCO_3^-]}{[H^+][SO_4^{-2}]} = K_{eq}.$$

REACTIONS AT INTERFACES

The surfaces at which water is in contact with solid phases or gases represent sites of importance in both physical and chemical processes.

GAS-LIQUID INTERFACES

The surface of a body of water in contact with the atmosphere at standard pressure and temperature is rather rigidly maintained by intermolecular forces, as shown by the surface tension of water, which is 72.2 dynes per centimeter at 20°C, a value higher than that of most liquids. Water molecules, however, are able to pass through this surface into the air, and gas molecules from the air can diffuse into the water. Both processes tend to produce mutual saturation near the interface. Rates of absorption of gases by water or rates of evaporation of water are functions of the characteristics of the system. Important factors in both kinds of rates are the total area of interface, the degree of departure from saturation just beyond the interface, and the rate at which the molecules of the dissolved or vaporized phase are transported away from the interface. The transport rate could be slow if it depended solely on molecular diffusion. In most natural systems, however, motion of the gas or liquid phase helps to move the evaporated or dissolved material away from the interface.

The processes by which gases from the atmosphere dissolve in water are of direct concern in water quality. The solution of oxygen in stream or lake water is dependent on such physical and chemical parameters as area of interface, mechanisms of transport away from interface, temperature, and pressure. In a secondary sense, because dissolved oxygen is essential to the clean-water organisms that thrive only under oxidizing conditions, the occurrence, solution, and transport of oxygen are important to the study of biochemical processes relating to water pollution. The process of photosynthesis is a major source of oxygen in some water bodies and is not directly dependent on assimilation of atmospheric gases at a liquid-gas interface. However, nonbiological exchange across the interface is of fundamental importance. Langbein and Durum (1967) reviewed some parameters of stream-channel geometry and stream-flow rates as they apply to the effectiveness of rivers in absorbing oxygen from the air. Understanding systems of this kind entails consideration of rates and the way in which the rate of one process may limit the rate of another one.

LIQUID-SOLID INTERFACES

Solution-deposition reactions obviously are processes occurring at solid liquid interfaces. Although the details of processes of this kind are not well understood, they doubtlessly control the rates at which many alterations of rock minerals proceed. Research on rates of reactions will require increased attention to these processes.

The surface of a solid which is in contact with a gas or a liquid is covered with a thin layer of tightly held molecules or ions taken up from the nonsolid phase. The layer is held to the solid surface by physical or chemical forces and in either instance is tenaciously retained. The phenomenon is called adsorption.

Finely divided solids have a large surface area per unit volume and hence a considerable capacity for physical adsorption. Finely divided charcoal, for example, may be used to adsorb gases or organic materials that produce colors or odors in water or to adsorb and concentrate organic compounds from polluted water. The intermolecular attractions that produce physical adsorption and that do not involve the strong electrical forces of chemical bonding are termed "van der Waals forces." The adsorbed material is retained in a layer on the order of one molecule in thickness. Such layers are often referred to as monolayers. The quantity of material so adsorbed is a function of the surface area of the solid and the effective area occupied by one molecule of the adsorbed material. Natural waters are exposed to various types of finely divided solids that have an adsorptive capacity. Van der Waals effects, however, are most likely to be significant for organic constituents. The charges on inorganic ions and even on water molecules are intense enough that the adsorption effects are more nearly chemical than physical and are not approximated by the monolayer model. A brief discussion of physical adsorption processes of interest in water chemistry has been published by Osipow (1967).

The type of adsorption most significant in natural-water chemistry involves the forces of chemical bonding in some degree and is commonly termed "chemisorption." The replacement of adsorbed ions by ions in solution is termed "ion exchange." The capacity of a solid for chemisorption is generally the result of local unsaturation of the chemical bonding within a mineral structure. In most of the important natural exchange media there is an excess of negative charge in certain areas of the crystal lattices, and positively charged ions are attracted to those areas. In general, divalent cations are more strongly adsorbed than monovalent ones, and in a solution in contact with the exchange medium the concentrations of adsorbable solutes will approximately obey the law of mass action.

In some natural minerals positively charged sites exist, and anion exchange reactions may occur. Relatively little seems to be known about anion-exchange processes, however, although they often are credited with helping retain plant nutrients in soils. The anions most abundant in natural water, bicarbonate and sulfate, are large units which would have low charge densities and are weakly adsorbed. Chloride ions are smaller, but still are considerably larger than any of the common cations. Anion-exchange processes were mentioned by Wayman (1967), who also discussed the general process of sorption by clays, by Halevy (1964), and by Yamabe (1958).

Cation-exchange capacities are commonly determined in studies of soil chemistry and are reported in terms of milliequivalents per 100 grams of solid. Some types of clay have exchange capacities of as much as 100 meq per 100 g. Nearly all soil and rock minerals have some ion-exchange capacity. Kennedy (1965) reported exchange capacities of stream sediments collected from a wide variety of environments in the United States. His values ranged from 14 to 65 meq per 100 g for clay-size particles (less than 4 microns in diameter), from about 4 to 30 meq per 100 g for particles between 4 and 61 microns in diameter, and from 0.3 to 13 meq per 100 g for particles from 61 to 1,000 microns in diameter.

The adsorption capacity of certain minerals is influenced by other characteristics of the adsorbed ion besides its charge. Important effects may be related to ion radius and to degree of hydration. The spacing of exchange positions in the clay mineral illite, for example, favors the entry of potassium or the heavier alkali metals into positions between structural layers. Once adsorbed, potassium is very difficult to remove from such positions.

ADSORPTION AND ION-EXCHANGE CALCULATIONS

Several different mathematical approaches are available for evaluating adsorption effects. The Freundlich adsorption isotherm was developed for adsorption of a gas by a solid but also can be used for solute adsorption. It is written

$$\frac{x}{m} = kc^n,$$

where

$\frac{x}{m}$ = weight of adsorbed ions per weight of adsorbent,

c = concentration of adsorbate remaining in solution, and

k and n are constants.

The Langmuir isotherm is less empirical. It assumes adsorption proportional to surface area not already occupied by adsorbed material and can be written

$$\frac{x}{m} = \frac{ksc}{1+kc},$$

where s is the total exchange capacity of weight m of adsorbent and other symbols have the same significance as in the Freundlich isotherm.

These expressions may be useful for evaluating adsorptive capacities for single ions, but have little value in ion-exchange studies. Where two or more types of ions are competing for exchange positions, the mass-law expression can sometimes be used. For example, in a system where

$$Ca^{+2} + 2NaX \sim CaX_2 + 2Na^+,$$

one might write
$$\frac{[CaX_2][Na^+]^2}{[Ca^{+2}][NaX]^2}=K.$$

The bracketed quantities are activities. The terms containing X represent adsorbed fractions, and the assignment of an activity coefficient to adsorbed material can be a problem. Wahlberg, Baker, Vernon, and Dewar (1965), however, reported some success in using this type of equation in laboratory experiments with pure clays and Sr^{+2}, Ca^{+2}, Mg^{+2}, Na^+, and K^+.

The partition coefficient, which is the ratio of free to adsorbed ion concentrations, was found by Wahlberg and Fishman (1962) to be a straight-line function of the logarithm of the concentration of a competing ion where the latter was in excess.

A general survey of the literature on ion-exchange phenomena of particular interest in water chemistry was prepared by Robinson (1962). The ion-exchange process and mathematical means of evaluating it were described in greater detail by Helfferich (1962).

MEMBRANE EFFECTS

If two aqueous solutions of different solute concentrations are separated by a semipermeable membrane, water molecules tend to pass through the membrane into the more concentrated solution by osmosis and thus give rise to a pressure differential across the membrane. In terms of the vapor pressures of pure water and the less concentrated solution, the osmotic pressure is

$$P=\frac{RT}{V}\ln\frac{p_0}{p},$$

where

P is osmotic pressure,
R is the gas constant,
T is temperature in degrees Kelvin,
V is molar volume of water,
p_0 is vapor pressure of water, and
p is vapor pressure of solution.

The vapor-pressure lowering effect of a given concentration of solutes can be calculated from Raoult's law

$$N=\frac{p_0-p}{p_0},$$

where

N is $\dfrac{\text{moles solute per liter}}{\text{moles solute}+\text{moles solvent per liter}}.$

Osmotic effects are involved in the movement and assimilation of water by plant and animal cells, and the ion content of a water can be evaluated in terms of its effect on osmotic pressure. According to U.S. Department of Agriculture Handbook 60 (U.S. Salinity Laboratory Staff, 1954), a solution having a specific conductance of 3,000 micromhos per cm will give rise to a pressure differential of 1.0 atmosphere, and if the conductance is 30,000 the pressure becomes 12.5 atmospheres. Thus, as salinity of the water in the root zone of the soil increases it becomes more difficult for plants to obtain moisture from the soil, because the plants must overcome this pressure.

Besides the osmotic effect, some membranes also show different permeabilities for different ions. An electrical charge imposed on a membrane increases this effect. The reverse is also to be expected—that is, there may be a difference in electrical potential across a semipermeable membrane which separates two solutions.

CLAYS AS SEMIPERMEABLE MEMBRANES

In many ground-water circulation systems, horizontal strata of high permeability may be separated by clay layers of much lower permeability. These clay layers can act as semipermeable membranes and thus give rise to anomalous effects on the pressure head of the water and the concentrations of dissolved ions. There also may be electrical-potential differences from layer to layer. Several other kinds of electrochemical influences in the subsurface may also bring about measurable potentials. The spontaneous potential measured by well-logging equipment is an indicator of such effects.

Although considerable study has been given some aspects of the behavior of clays as a factor influencing ground-water movement and composition, a great deal more remains to be learned before this subject can be considered well understood. Some aspects of the membrane effects of clay have been discussed by Bredehoeft, Blyth, White, and Maxey (1963), Hanshaw (1964), and Hanshaw and Zen (1965).

ENVIRONMENTAL INFLUENCES

The principles and processes discussed so far are the direct and fundamental ones that determine the composition of natural water. The manner in which the processes operate, however, and their relative importance in any one environment are controlled by numerous nonchemical forces. The hydrologist is usually directly concerned with most of these, and their general significance in natural-water chemistry is considered here.

CLIMATE

The processes of rock weathering are strongly influenced by temperature and by amount and distribution of precipitation The influence of climate on water quality goes beyond these direct effects. Climatic patterns tend to produce characteristic plant communities and soil types, and the composition of water draining from such areas could be thought of as a product of the ecologic balance. A somewhat similar concept seems to have been used in the study and classification of water composition in some areas of the U.S.S.R., but has not been successfully applied to water by investigators in the United States.

Certain of the major ionic constituents of natural water are influenced more strongly than others by climatic effects. Bicarbonate, for example, tends to predominate in water in areas where vegetation grows profusely. Some metals are accumulated by vegetation and may reach peak concentrations when plant-decay cycles cause extra amounts to enter the circulating water.

Humid temperate climates and warm wet climates generally are the most favorable for growth of vegetation. Runoff from tropical rain-forest areas commonly is low in dissolved-solids concentration. An example is the water of the Amazon for which an analysis is given in table 3. An arid climate is unfavorable for rapid rates of solvent erosion, but concentration of dissolved weathering products by evaporation can give rise to water that is high in dissolved-solids content. On the other hand, the occasional flood runoff in such regions can be very low in dissolved material because soluble weathering products are not available in major quantity.

Climates characterized by alternating wet and dry seasons may favor weathering reactions that produce considerably larger amounts of soluble inorganic matter at some seasons of the year than at other seasons. Streams in regions having this kind of climate may fluctuate greatly in volume of flow, and the water may have a wide range of chemical composition. The influence of climate on water quality may thus be displayed not only in amounts and kinds of solute ions, but also in the annual regime of water-quality fluctuation.

GEOLOGIC EFFECTS

The ultimate source of most dissolved ions is the mineral assemblage in rocks near the land surface. This topic will be discussed in more detail later in this paper, with the aim of developing some indications of the rock type that might have been associated with a given water. The importance of rock composition, however, is only part of the story. The purity and crystal size of minerals, the rock texture and porosity, the regional structure, the degree of fissuring, the length of previous exposure time, and a good number of other factors might influence the composition of water passing over and through the rock.

Rock temperatures increase with depth below the land surface. Where water circulates to a considerable depth it normally attains a substantially higher temperature than water near the land surface. Increased temperature raises both the solubility of most inorganic solutes and the rate of dissolution of rock minerals.

Most thermal ground waters (hot springs) are found in areas where the temperature gradient with depth is relatively steep. The solute content of such water is commonly higher than that of cooler surficial water. Some thermal waters may be very high in dissolved-solids concentration and contain unusual amounts of metal ions. The brines from deep wells in Imperial Valley, Calif. (White, 1965, p. 346) and from deep basins at the bottom of the Red Sea (Miller and others, 1966) are interesting examples.

A factor about which there has been much speculation but about which little is known is the contribution to the circulating water of the earth made by water released from rock minerals as the result of changes in crystal structure or chemical alteration. Water from such sources is termed "juvenile" and is defined as water which has not previously been involved in the circulating system of the hydrologic cycle. "Meteoric" is a term commonly used to refer to circulating water. The difficulty of ascertaining whether a particular water would be wholly or partly of juvenile origin is obvious. The entrapped water of igneous rock that might potentially be released on fusion of the rock minerals is a relatively substantial quantity; it was estimated by Clarke and Washington (1924) to be a little over 1 percent of the weight of average igneous rock. There is great doubt as to the validity of this figure (Goldschmidt, 1954, p. 126), but it would appear that a very large amount of water is held in igneous rock in the 16-kilometer crust of the earth.

The data reported by White and Waring (1963) on the composition of volcanic gases show that water is usually predominant. However, the degree to which the water associated with volcanic activity is of magmatic origin rather than meteoric origin is very difficult to determine. White (1957a, b) studied the composition of water from many thermal-spring areas and concluded that in such areas there are few, if any, conclusive indications that any of the water is juvenile.

BIOCHEMICAL FACTORS

Life forms and the chemical processes associated with them are intimately related to water and to the solutes contained in water. Extensive discussions of this relationship can be found in the output of various branches of the life sciences. Although the principal concern of this book is with inorganic aspects of water chemistry, the biological ones cannot be avoided. In fact, the water chemist will find that biological factors are important in almost all aspects of natural-water composition. Much of the support for his work derives from the importance of use of water by man and the standards required for safe drinking water. Water-pollution control programs commonly aim to benefit desirable forms of aquatic life. Beyond these factors, the composition of all water is influenced to greater or lesser degree by the life processes of plants and animals. The following brief discussion points out some of these processes and shows how they fit into and complement other factors that control natural-water composition.

ECOLOGY APPLIED TO NATURAL WATER

Ecology is the study of relationships between organisms and their environment. It thus implies what might be considered a study of biological systems and the way in which the various parts of the system influence each other. For example, the development of a particular set of plant and animal species, soil type, and general land form can be thought of as the end result of a particular set of climatic and geologic factors, given time to attain a state of equilibrium. An equilibrium of this kind, is not directly comparable to thermodynamic equilibria in chemical systems. For some purposes it may be useful to think of a steady state as being attained or attainable. But biological systems in any regional sense are subject to such an enormous number of feedbacks and variations, some cyclic and some random or one-time-only, that conditions are in a continual state of dynamic flux. Studies in ecology, therefore, probably should be more concerned with mechanisms and rates than with final equilibrium.

In a sense, the study of natural-water composition involves concepts of ecology, because a large number of factors and processes are interrelated in bringing about the composition of the water. As in ecologic systems, changes in one factor may bring about a considerable number of other changes that can influence the particular variable being observed. Also as in ecologic systems, the separation of cause from effect can become difficult.

Whether the biochemical processes described here represent as fundamental an influence as the inorganic processes mentioned earlier is partly a matter of opinion, but their importance is unquestionable The life processes of principal interest in water chemistry can be classified in a general way on the basis of energy relationships to include the following.

1 Processes that use energy captured from the sun or some other source for promotion of chemical reactions which require a net energy input.

2. Processes that redistribute chemically stored energy.
3. Processes that convert chemically stored energy to other forms of energy.
4. Processes without significant energy transfer.

The first type of process is represented by photosynthesis, in which carbon dioxide, water, and radiant energy are used to manufacture carbohydrate and gaseous oxygen is liberated. This process in turn provides the fuel for most of the other processes which would be classified under 2 and 3.

Processes of metabolism and decay are included in types 2 and 3. These chemical reactions may involve a net release of energy from the original materials, but not all the reactions can be made to proceed at observable rates in the absence of life forms. For such reactions, the biological processes seem to offer pathways for reactions that do not have the high stepwise energy barriers which prevent the reaction from occurring spontaneously or which cause it to be slow in the absence of biota.

Reactions that promote chemical reduction or oxidation might be thought of as belonging to type 3, although no actual electrical energy may be produced. It is possible, however, to make a cell that is capable of producing a current as a result of biochemical processes. The production of heat, chemiluminescence, and motion are biological manifestations of the third type of reaction.

Effects which come within class 4 can be represented by indirect influences such as stabilization of inorganic colloids in water by soluble organic matter or the release of various waste products to water.

The processes that sustain life are particularly strongly developed in water bodies exposed to air and sunlight. In environments where neither is present, as in ground-water aquifers, biological activity normally is much less important. At some stage in its movement through the hydrologic cycle, however, all water is influenced by biochemical processes, and their residual effects are widely discernible, even in ground water.

INFLUENCE OF SOIL AND SOIL-FORMING PROCESSES

The systems of classification of soils in common use emphasize strongly the effect of climate and vegetation and do not place as much emphasis on the nature of the original rock from which the soil was obtained. Soils of high productivity generally contain a considerable amount of organic debris, and in most, the mineral-species distribution inherited from the parent rock has been extensively altered. The minerals themselves also have often been changed. Discussions of rock-weathering processes leading to soil formation are plentiful. As examples might be cited papers by Keller (1957) and Reiche (1950), who have discussed the processes with some attention to soluble products and from a geologically oriented viewpoint.

Many of the soil-forming processes are the same ones that control natural-water composition, and a considerable area of common interest exists between water chemistry and soil chemistry. This fact, however, has not been realized very extensively by workers in the two fields. A large part of the atmospheric precipitation that reaches the land surface falls on soil surfaces; generally the fraction that ultimately appears as runoff or ground water has had some contact with the soil, and much of it has spent a considerable period of time as soil moisture. The chemical composition of soil moisture and the processes that go on in soil to dissolve or precipitate minerals or otherwise to alter the composition of soil moisture have not received adequate attention from workers in the field of natural-water chemistry.

Among the factors influencing chemical composition of soil moisture are the solution or alteration of silicate and other minerals, precipitation of sparingly soluble salts (notably calcium carbonate), selective removal and circulation of nutrient elements by plants, biochemical reactions producing carbon dioxide, sorption and desorption of ions by mineral and organic surfaces, concentration of solutes by evapotranspiration, and conversion of gaseous nitrogen to forms available for plant nutrition. Of these, one of the most important is the production of carbon dioxide. The air in soil interstices is commonly 10–100 times richer in CO_2 than ordinary air (Buckman and Brady, 1960, p. 246). Water moving through soil dissolves some of this CO_2, and the H^+, HCO_3^-, and CO_3^{-2} ions are potent forces in controlling the pH of the water and in attacking rock minerals.

AQUATIC BIOTA

Those life forms that occur in water bodies or in close association with them form an ecologic system that has been widely studied. The science of limnology is concerned to a high degree with fresh-water ecology. Hutchinson's (1957) well-known text covers this subject in considerable detail.

Many of the chemical processes occurring in soil also occur in freshwater bodies. Photosynthesis of plant species which are rooted in the pond or stream bottom, as well as of floating species, produces oxygen and consumes carbon dioxide, and respiration and decay consume oxygen and produce carbon dioxide. A well-defined diuranl cyclic fluctuation of pH often can be observed in near-surface water of lakes and streams (Livingstone, 1963, p. 9). The photosynthesizing species provide some of the food and oxygen used by other life forms which require oxygen. The various species, among themselves, can attain an ecologic equilibrium, which, however, is generally subject to seasonal and other types of fluctuation. Algal blooms for example are caused by rapid increases in algae population and result in large concentrations of algae. The bloom may be followed by an equally rapid die away. The cause of such fluctuations is not always obvious, although frequently it is associated with a change in supply of dissolved nutrient elements, especially nitrogen or phosphorus.

The usual approach in studies of algal growth has been to try to determine critical concentration levels of the nutrients, below which growth is impaired. For this kind of system the principles of kinetics seem particularly appropriate as a means of defining the controls over rates of growth. The rate at which a particular nutrient is supplied may be the rate-controlling step of the whole process. The very rapid rates attained when a bloom appears represent a system temporarily out of control, which like other exponential processes cannot continue long without colliding with a new set of limits.

Maintaining a proper balance among life forms in lakes and in streams is a critical aspect of pollution control. Water bodies have a considerable capacity to absorb organic material. Extensive populations of biota can be generated in such processes. Lakes are sometimes characterized in terms of their productivity—that is, the amount of organic material synthesized per unit of surface area in a given time. Water bodies with high productivity are sometimes termed "eutrophic." This word was coined from Greek terms equivalent to "nutrient rich." The similarly coined word for water with low productivity "oligotrophic" means "nutrient poor." Both words, however, have been rather widely used in other contexts in recent literature.

Lakes in environments where growing conditions are favorable and nonaquatic vegetation is abundant generally are highly productive and may have short careers in the geologic sense. Such water bodies tend to evolve into marshland or peat bog, owing to accumulation of organic debris. Lakes in environments less favorable to vegetative growth may fill with inorganic sediment or be drained by stream erosion at their outlets. Obviously the career of a lake involves other factors in addition to organic productivity, but pollution with organic wastes can bring about extensive changes in properties of the water and in a relatively short time may convert a clear oligotrophic lake to a relatively turbid eutrophic one. The rates at which such changes might occur and the feasibility of reversing them represent areas that are receiving study by limnologists.

THE HYDROLOGIC CYCLE

A characteristic property of the free water of the earth is its continual motion, imparted primarily by the input of radiant energy from the sun. This energy input causes some liquid water to be converted to the vapor state and carried off by wind wherever a water surface is exposed to the atmosphere. When atmospheric conditions become favorable, the vapor returns to the liquid state with a release of energy, first to form the very small droplets of clouds and, if temperatures are low enough, tiny ice crystals. Rain or snow may be produced if the condensation proceeds under favorable conditions. The amounts of energy involved in water circulation in the atmosphere are very large in total and when concentrated, as in tropical storms, may have spectacular consequences. The water reaching the land surface by precipitation moves downslope in the general direction of the ocean or point of minimum gravitational energy.

A wide variety of representations of the hydrologic cycle exist in the literature (for one example, U.S. Department of Agriculture, 1955), and the hydrologist who is interested in detail can find many different paths through which continuous circulation can occur or places where water can be stored for very long periods. The cycle itself, however, deals only with pure H_2O and has indirect significance so far as natural-water chemistry is concerned. A systematic representation of a cycle of dissolved material which they called the solusphere was prepared by Rainwater and White (1958). Although the work of circulating water in transporting nonaqueous material is a major aspect of aqueous geochemistry, it is by no means the only one that must be considered. Livingstone (1963, p. 38) calculated that rivers of North America carry an average load of 85 metric tons per year in solution from each square mile of drainage basin.

SOURCES OF SOLUTES IN THE ATMOSPHERE

Table 4 shows the principal gaseous constituents of the atmosphere. Any liquid water in the atmosphere naturally would be expected to be saturated with respect to these gases, the amount in solution being proportional to the solubility and the partial pressure of each and to the temperature. Gases which enter into reactions with water in general are more soluble than those which do not. The effect of carbon dioxide is relatively great for this reason, even though it makes up only 0.03 percent by volume of normal air. Other gases such as H_2S, SO_2, NO_2, HCl, or NH_3 may be present in the air in some places, either as the result of air pollution by man, volcanic exhalations, or biological or chemical processes, and they influence rainfall composition.

Certain elements form solids or liquids with a significant vapor pressure at ordinary temperatures. Certain boron compounds, for example, (Gast and Thompson, 1959) tend to evaporate from the ocean to a significant extent for this reason. The elements iodine and mercury have appreciable vapor pressures at low temperature, but those elements are comparatively rare and do not influence air or rainfall composition appreciably.

Nuclides such as tritium and carbon-14, as noted earlier, are produced in the atmosphere by cosmic-ray bombardment. The atmosphere also contains extra-terrestrial material which is introduced from outer space. On the basis of the nickel content of snow in Antarctica, Brocas and Picciotto (1967) estimated from 3 to 10 million tons a year of such material falls on the earth's surface.

Most of the atmospheric particulate matter consists of terrestrial dust carried aloft by wind or propelled upward by volcanic eruptions and of sodium chloride or other salt picked up as a result of wind agitation of the ocean surface. The particulate matter is important in forming nuclei for condensation of water and as sources of solutes in precipitation.

The subject of atmospheric chemistry has a large literature of its own, some of which is summarized by Junge (1963). With increasing concern over air pollution, a considerable amount of research is being done in this field.

COMPOSITION OF ATMOSPHERIC PRECIPITATION

Studies of the composition of rainfall have been carried on for many years. In summarizing this subject, Clarke (1924a) quoted some 30 investigators who published data between 1880 and 1920. In more recent times, interest in this field seems to have increased, especially in northern Europe and the U.S.S.R. and in the United States. Continuing studies in Scandinavian countries have produced many data (Egner and Eriksson, 1955). Papers on the composition of rainfall and its effect on river and lake waters are plentiful.

Eriksson (1955, 1960) used rainfall-composition data to calculate the importance of atmospheric transport in the recirculation of ions to the ocean. Eriksson's estimates state that the average chloride content of rainfall on the land surface is about 10 kilograms per hectare per year, and the sulfur contribution is about the same, computed as S. In terms of SO_4^{-2}, the weight would be about three times as great.

Gorham (1955) made extensive observations of rainfall composition in the English Lake District, an area about 50 kilometers east of the Irish Sea. These data showed a resemblance to sea water in the ratios of sodium to chloride and magnesium to chloride. Gorham (1961) discussed the general significance of atmospheric factors on water quality in a later paper.

The best-known recent data on rainfall composition in the United States are those of Junge and his associates. These investigators operated about 60 rainfall-sampling stations distributed over the entire country (except Alaska and Hawaii) for a year, from July 1955 to July 1956. The results, described by Junge and Gustafson (1957) and by Junge and Werby (1958), showed that the average chloride concentration in rainfall rapidly decreases from several milligrams per liter near the oceans to a few tenths milligram per liter inland, whereas sulfate increases inland to values between 1 and 3 mg/l on the average. Nitrate and ammonia concentrations also were determined.

Feth, Rogers, and Roberson (1964) reported data for snow in the western United States, especially in the northern part of the Sierra Nevada, and Gambell and Fisher (1966) reported on composition of rain in North Carolina and Virginia. Mikey (1963) studied occurrence of fluoride in precipitation in certain areas of the U.S.S.R. Where appropriate, references will be made later to these and other studies in describing the chemical behavior of specific constituents.

The reported composition of rainfall is influenced by the methods used to obtain samples for analysis. The samples collected in some investigations represent only material present in rain or in snow, because particulate matter was filtered out before analysis and the sampling container was kept closed when rain was not falling. Other investigators desired to obtain total fallout and kept their sampling containers open at all times. The insoluble material, however, was generally filtered out of these samples also. For meteorologic purposes the composition of rainfall without any influence from antecedent or subsequent dry fallout is perhaps of primary interest. The geochemist, however, and most other users of such data may well need total values including dry fallout; this is the "bulk precipitation" defined by Whitehead and Feth (1964). The extent to which data of the two kinds may differ is uncertain Whitehead and Feth ascribed a considerable importance to the dry fallout factor, but Gambell and Fisher (1966) did not. It has also been pointed out by some investigators that aerosols may deposit particulate matter on vertical surfaces on which the wind impinges. Thus the foliage of trees near the seacoast may pick up salt particles from landward-blowing winds. The importance of this effect is not known.

The values given in table 6 for rainfall composition were taken from published sources. The table shows that rainfall composition is highly variable, not only from place to place but from storm to storm in a single area and within individual storm systems as well. A very large volume of air passes through a storm system. The very nature of the conditions that often produce rain, a mingling of air masses of different properties and origins, insures a high degree of vertical and horizontal nonhomogeneity. Analyses 3 and 4, table 6, represent samples of rain collected successively during a rainy period at Menlo Park, Calif. The later sample shows a considerably higher concentration of solutes than the earlier. Analysis 2 in table 6 represents bulk precipitation, and analysis 1 probably can be assumed to have been influenced by dry fallout. The other data in the table represent composition of rainfall unaffected by dry fallout.

TABLE 6.—*Composition, in milligrams per liter, of rain and snow*

Constituent	1	2	3	4	5
SiO_2	0.0		1.2	0.3	
Al	.01				
Fe	.00				
Ca	.0	0.65	1.2	8	1.41
Mg	.2	.14	.7	1.2	
Na	.6	.56	.0	9.4	.42
K	.6	.11	.0	.0	
NH_4	.0				
HCO_3	3		7	4	
SO_4	1.6	2.18	.7	7.6	2.14
Cl	.2	.57	.8	17	22
NO_2	.02		.00	.02	
NO_3	.1	.62	2	.0	
Total dissolved solids	4.8		8.2	38	
pH	5.6		6.4	5.5	

1. Snow, Spooner Summit, U S Highway 50, Nevada (east of Lake Tahoe), alt 7,100 ft, Nov 20, 1958 (Feth, Rodgers, and Roberson, 1964)
2. Average composition of rain August 1962 to July 1963 at 27 points in North Carolina and Virginia (Gambell and Fisher, 1966)
3 Rain, Menlo Park, Calif , 7 00 p m Jan 9 to 8 00 a m Jan 10, 1958 (Whitehead and Feth, 1964)
4 Rain, Menlo Park, Calif , 8 00 a m to 2 00 p m Jan 10, 1958 (Whitehead and Feth, 1964)
5 Average for inland sampling stations in the United States for 1 year Data from Junge and Werby (1958) as reported by Whitehead and Feth (1964)

INFLUENCE OF MAN

A major impact on the environmental factors influencing the composition of water results from the activities of man. The power of man to alter his environment is very great and is widely evident in the changes which he can bring about in water composition. Solutes may be directly added to water by disposal of wastes or directly removed in water treatment or recovery of minerals. The ecology of whole drainage basins may be profoundly altered by bringing forested land into cultivated agriculture. Water-movement rates and solute-circulation rates may be altered by water diversions and by structures and paved surfaces which replace open land as cities expand in population and area. Some aspects of these effects will be considered in more detail in other sections of this book.

A SYSTEMS APPROACH TO STUDY OF WATER CHEMISTRY

The concept of "systems" as applied here is an attempt to evaluate simultaneously all the factors of major significance which for example, might control the composition of a surface stream at a given point or perhaps, more simply, the water of a specified segment of an underground aquifer. The concept certainly is not a new one, for research has been directed for a long time at theoretical and practical evaluations of these factors. The principal objective here is to explore briefly the present state of development of this concept and its applicability.

OPEN VERSUS CLOSED SYSTEMS

In experimental studies of chemical equilibria, one approach commonly used is to confine all the reactants and products in an inert closed container and to allow reactions to proceed until no further change can be observed. A system of this kind is "closed" in the sense that definite, impervious boundaries isolate it from any outside influence that is not a controlled part of the experiment. If such a system reasonably can be shown to be at equilibrium, a rigorous application of the mass law can be made, and all factors affecting the system closely evaluated.

A completely closed system is not required for application of the mass law, if the variables acting on the system from outside are known and can be measured. For example, if the reaction vessel is open to the atmosphere at sea level, the pressure on the system will be maintained at 1 atmosphere, and if gaseous reaction products do not escape, the system could still be evaluated rigorously, by assuming any reactant introduced by contact with air was a controlling factor in the final equilibrium.

For example, exposing crystalline calcite to pure water in an inert container open to the air would give a constrained system if constant temperature and pressure were maintained. The concentration of solutes would remain constant once equilibrium was reached even if some of the water were lost by evaporation, but ultimately evaporation would alter the system by removing one of the phases. Natural systems cannot generally be considered to be confined by an inert container, but may still be treated as closed if the complicating factors can be satisfactorily evaluated.

An open system is one in which movement of reactants and products in and out is virtually unrestricted. Although some variables in such systems may change slowly, the system may not attain a true equilibrium. The chemical principles described earlier may be applied rigorously for some parts of such systems, and processes can be observed in them and described. It may be possible by simulation techniques or other means to devise models of the systems and observe effects of changed input or output.

Obviously, very few natural-water systems can be realistically viewed as being fully closed. Some, however, are reasonably fully constrained, or some parts of the system may be fully constrained and can be evaluated with some rigor by means of the mass law and related thermodynamic concepts. In some systems, the leakages are small enough to be neglected. The practical utility of considering groundwater chemistry as strongly controlled by chemical equilibria should hold promise in this field, comparable to the application of mathematical and physical concepts to water movement in aquifers, where a "homogeneous, isotropic" medium is assumed as a starting point.

THE PHASE RULE

Some indications of the number of things in a heterogeneous system that can be freely varied can be gained from the Gibbs phase rule. According to this principle, in any such system

$$\text{components} - \text{phases} + 2 = \text{degrees of freedom}.$$

For systems of interest here, phases are defined as those parts of the system which are homogeneous within themselves and have definable boundary surfaces; components are independent chemical constituents derived from (or going to make up) the phases; and the degrees of freedom are the number of variables such as temperature, total pressure, and the concentrations, or partial pressures, that must be fixed to define the system at equilibrium.

In the simple system $CaCO_3(c) + H_2O(l)$ (where no gas phase is present) the chemical equations involved are

$$CaCO_3(c) + H^+ = Ca^{+2} + HCO_3^-$$

and

$$H^+ + OH^- = H_2O.$$

The system considered may be defined as having only two phases, $CaCO_3(c)$ and $H_2O(l)$; the dissolved species include Ca^{+2}, HCO_3^-, H^+, and OH^-. The equations show, however, that these four dissolved species are not all *independent* constituents. For any value of $[H^+]$, the $[OH^-]$ would be fixed, and then any specified value for $[Ca^{+2}]$ would fix the value for $[HCO_3^-]$. Thus there are only two independent constituents which could be considered as components, and from the phase rule the degrees of freedom would be $2-2+2=2$. Two of these degrees of freedom represent the temperature and pressure, and if fixed values are assigned for them no possible degrees of freedom remain.

This means that in the system, as defined, the pH and activities of Ca^{+2} and HCO_3^- each have fixed values if temperature and pressure are specified. The result of applying the phase rule to this simple system also would be reached intuitively, but in more complicated systems it may be a useful approach to systematic evaluation (Sillén, 1967b). It is necessary, however, to be careful in defining the components.

Phase diagrams in which pressure and temperature are variables and other similar applications of phase-rule principles have been widely used in geochemistry in connection with rock melts and hydrothermal systems. The simpler approach using a specified constant pressure and temperature currently is predominant in considering natural-water chemistry, however, and has been followed in most diagrams in this book.

For systems where pressure and temperature are specified, an algebraic approach that involves reasoning similar to the phase rule can be used to see how many significant concentration variables a particular system might have. Mass-law equations for all equilibria involving solids or complexes known to be present are set down. Where other valid relationships among solutes can be expressed, such as balance of the molarities of cations and anions, these also are set down. The number of degrees of freedom, or potentially independent variables, can be determined by subtracting the number of equations from the total number of unknown variables.

From this kind of consideration one can determine how many things should be measured in a system to permit calculation of remaining components, always keeping in mind the limiting assumption for this approach, that the system is in a state of equilibrium. Measurement of as many variables as possible is generally desirable, as this may make it possible to test whether equilibrium actually does exist. In laboratory systems where conditions are carefully controlled, the array of equations may be used for calculation of equilibrium constants.

Calculations of the type described are given by Garrels and Christ (1964, p. 74-92) for carbonate systems and by Hem (1968) for solutions of aluminum complexes.

The mathematical analysis of natural-water systems suggests that, in general, the variability is largest where there are only a few phases. Increasing the number of phases which are at equilibrium with a given group of components decreases the possibilities of variation. This conclusion would be reached intuitively from simple chemical reasoning: the more equilibria there are in a system the fewer things will be left that can vary independently.

A ground-water system has a greater inherent likelihood for equilibria becoming established than a surface-water system, and as might be expected, one can generally observe much less variation with time in components in ground water than in surface water.

INFLUENCE OF WATER MOVEMENT

As the foregoing discussions have stated, all the water of interest to the hydrologist must be considered as moving. The rate of movement is rapid in rivers, but much slower in ground-water or soil-water environments. The movement also is slow and often relatively complex in lakes and reservoirs. Movement patterns and their effects must always be considered in natural-water systems. A solution moving through a porous solid where equilibrium is attained may display behavior like that of solutions moving through ion-exchange or chromatographic columns. Moving solutions in systems where equilibrium is not attained will have compositions influenced both by movement rates and by reaction rates.

SOME CHARACTERISTICS OF GROUND-WATER SYSTEMS

The application of equilibrium models to ground-water chemistry has had considerable appeal to theoretically minded investigators. Although the mineral composition of the solids in such systems commonly is heterogenous and poorly known, the activities of solute species can be determined completely, and because movement of water is slow, there is a considerable time span available for completion of slow reactions. Presumably, any reaction that reasonably could be expected to reach equilibrium would do so in the usual aquifer system.

Experience has shown that equilibrium models involving calcium carbonate and some other carbonates (Barnes, 1964; Barnes and Back, 1964a: Back, 1963) and oxidation and reduction of iron (Barnes and Back, 1964b) and probably of manganese (Hem, 1963a) can be significantly applied in ground-water systems. Some other kinds of equilibria, for example those of ion exchange, also may prove to be applicable in these systems. Some of the solute activities may be unstable, requiring extra care in sampling and analysis. Such problems, however, if they are recognized, usually can be overcome. A more difficult problem to evaluate can result from the pattern of movement of water through the system to be sampled. In ground-water systems, where strata of high and low permeability are interbedded, a flow pattern can occur in which water movement is largely confined to the more permeable layers. Differences between mineral composition in the layers may cause considerable variation in water composition with depth at any given site. Wells that penetrate several of these layers may yield water that is a mixture, enriched in the types of solutions present in the more permeable beds, but also influenced by the efficiency of well construction and development, rate of pumping, and related factors. A well influenced by factors such as these is unlikely to give usable chemical-quality information. Unfortunately there is no good way of evaluating the importance of these effects, nor sometimes of knowing for certain whether they are present or absent.

SURFACE WATER SYSTEMS—RIVERS

The water carried in streams is often considered to consist of a base-flow fraction made up of ground water that infiltrates into the channel and a direct runoff fraction which enters the drainage system during and soon after precipitation periods. The direct runoff presumably has had no residence time in the ground-water reservoir and only a short contact with soil or vegetation. Reactions in the soil zone, however, are commonly extensive enough that the direct runoff has a considerably higher dissolved-solids content than the original rain or snow. The base flow has a still greater dissolved-solids content. The solute concentration of river water thus tends to be inversely related to flow rate. At very high flow rates, the water may be very dilute.

It is usually not feasible to evaluate the composition of base flow exactly or, for most medium-sized and large streams, to separate the chemical effects of base flow completely from those of direct runoff. The quantity of base flow changes with time, the relative importance of different contributing sources changes, and the result is a complex fluctuation of solute concentration. Steele (1968a, p. 21) was able to develop a chemical means of distinguishing base flow from direct runoff for a small stream in northern California.

In addition to mixing of ground water and runoff, the natural factors that influence stream composition include reactions of water with mineral solids in the streambed and in suspension, reactions among solutes, losses of water by evaporation and by transpiration from plants growing in and near the stream, and effects of organic matter and water-dwelling biota. This latter set of natural factors results in fluctuations of composition that bear little relation to discharge rate.

Superimposed on all these factors are the influences of man—stream pollution and waste disposal by all kinds of activities within the river basin, and flow diversions and augmentation.

Chemical equilibria probably control a few properties of water in flowing streams. For example, the ion-exchange reactions of solutes with suspended sediment probably are rapid enough that they usually are at equilibrium. Kennedy and Brown (1964) found that sodium-saturated natural sediments from Brandywine Creek, Del., exchanged 90 percent of their adsorbed sodium for calcium in 3-7 minutes in laboratory experiments. Certain oxidations, ferrous to ferric iron for example, also normally may reach equilibrium. The equilibrium approach, however, seems inadequate for studies of most biologically mediated processes such as utilization and production of carbon dioxide and oxygen. A river is by nature a dynamic system, and kinetic principles would seem much better suited to stream chemistry than the steady-state equilibrium approach. For example, the processes whereby biota consume organic-pollution loads of streams often can be most effectively studied by application of kinetics.

Studies of river water require continuing, carefully planned observations and evaluations of the importance of the various effects. Ultimately, where sufficient critical data have been accumulated, a model of the basin, or part of the basin, may be constructed that will allow prediction of downstream results as basic variables change in upstream reaches.

LAKES AND RESERVOIRS

A lake which has a surface outlet represents a holding and mixing basin for streamflow. The detention time of water in a lake provides a potential opportunity for slow reactions to come closer to completion than they can in the rapidly moving water of a river. Mixing, however, may not be complete, so at any given time the water in one part of the lake may be greatly different in composition and properties from that in other parts of the lake. Closed-basin lakes become saline owing to evaporation of water and continued influx of solutes.

An important influence on lake composition is thermal stratification. During warm weather, an upper, heated layer of water of relatively low density may form at the surface, floating on the deeper, cooler water below and insulating the deeper layers from direct contact with atmospheric oxygen. In deep lakes, during the summer season, this stratification may persist for long periods and in time the deeper water becomes depleted in oxygen, owing to biochemical processes. In cooler seasons, the stratification disappears, and oxygen again becomes dispersed throughout the lake.

Hutchinson (1957) described the physical and chemical aspects of lakes in detail, and there is an extensive literature on stratification effects.

GEOCHEMICAL CYCLES

Before proceeding to more specialized discussion of water chemistry, there are certain geochemical concepts relating to behavior of the elements that should be considered. Geochemists sometimes speak of the cycle through which an element moves—from its initial incorporation in crystalline material deep within the earth, through processes by which it may be transported into or incorporated into other materials, and finally again to its original state by geologic processes. Those elements that are readily extracted from crystalline minerals and brought into solution are easily transported and have sometimes been referred to as "mobile."

For many years, geochemists have been interested in these circulation patterns, and some have tried to deduce the earth's erosional history from differences in distribution of elements between igneous rocks on one hand and the sedimentary rocks and the ocean on the other hand. If the premise is accepted that sediments were derived from igneous rocks having an average composition similar to the igneous rocks now available for sampling and if the surplus of eroded elements not needed to produce the observable volume of sediments was left in the ocean, it is possible to estimate how much erosion of the outer crust has taken place. Goldschmidt (1933; 1937; 1954) made a series of such estimates on the basis of the distribution of sodium among igneous and sedimentary rocks and the water of the oceans. His figure, which is widely known and quoted, is that the equivalent of 160 kilograms of igneous rock has been eroded from each square centimeter of the earth's surface.

If one calculates the balance of elements as the amount not accounted for in the total volume of sediments, using Goldschmidt's figure for the volume of eroded rock and the generally accepted abundance data for the elements, a figure in reasonable agreement with observed sea-water composition is obtained. Some of the elements, however, especially chlorine, are far too abundant in the ocean to fit this concept, and sources other than igneous rock, at least like that which can now be sampled, must be assumed for them.

Barth (1952) proposed a somewhat different view of the behavior of the elements in weathering. He suggested one might assume that a general balance has been attained between the rates of accretion of elements to the ocean (in solution in river water for the most part) and the rates at which these elements are returned to igneous and metamorphic rock. The average time in years for an atom of any element to remain in the ocean (T) can be computed from the formula

$$T = \frac{A}{dA/dt},$$

where A is total amount of the element dissolved in the ocean and dA/dt is the rate at which the element is added to the ocean. This rate can be obtained from observations on the total contributions of rivers as in the compilations by Livingstone (1963, p. 38–40) or Durum, Heidel, and Tison (1960). An earlier well-known set of figures is that prepared by Conway (1943), based in turn on the date of Clarke (1924a, p. 119). These values are corrected as necessary for the effect of recirculation in the atmosphere of ions from the ocean.

Amounts of most of the elements that are contributed annually to the ocean by rivers are poorly known. Goldberg and Arhennius (1958), however, prepared estimates of removal rates and residence times of a large number of elements by determining the composition and rate of accumulation of sediment in the ocean. This approach uses the same fundamental concept as Barth's, and residence times agreed reasonably well where they could be computed for some of the elements by both methods. The residence times of elements determined by Goldberg (1961) are given in table 7. These range from nearly 10^8 years for sodium to only 100 years for aluminum.

TABLE 7.—*Average residence time of elements in the ocean*

[After Goldberg (1963)]

Element	Residence time (yr)	Element	Residence time (yr)	Element	Residence time (yr)
Na	2.6×10^8	Rb	2.7×10^5	Sc	5.6×10^3
Mg	4.5×10^7	Zn	1.8×10^5	Pb	2.0×10^3
Li	2.0×10^7	Ba	8.4×10^4	Ga	1.4×10^3
Sr	1.9×10^7	Cu	5.0×10^4	Mn	1.4×10^3
K	1.1×10^7	Hg	4.2×10^4	W	1.0×10^3
Ca	8.0×10^6	Cs	4.0×10^4	Th	3.5×10^2
Ag	2.1×10^6	Co	1.8×10^4	Cr	3.5×10^2
Au	5.6×10^5	Ni	1.8×10^4	Nb	3.0×10^2
Cd	5.0×10^5	La	1.1×10^4	Ti	1.6×10^2
Mo	5.0×10^5	V	1.0×10^4	Be	1.5×10^2
Sn	5.0×10^5	Si	8.0×10^3	Fe	1.4×10^2
U	5.0×10^5	Y	7.5×10^3	Al	1.0×10^2
Bi	4.5×10^5	Ge	7.0×10^3		
Sb	3.5×10^5	Ce	6.1×10^3		

In most respects, the concept of residence time in the ocean is more satisfying to the chemist interested in aqueous behavior of the elements than is the Goldschmidt model with its dependence on average igneous and sedimentary rock composition. As Barth (1961) has pointed out, the igneous rocks that lie near the surface of the continents, which are the ones available for collection of rock samples for analysis, represent material that has in all probability been reworked many times and may, therefore, have a composition very different from its original one. The present compositions of rocks and the oceans represent the result of a long-continued process of fractionation, and the residence time of elements in the ocean is, therefore, useful as an index of their geochemical behavior. The elements whose chemistry definitely favors retention in aqueous species have long residence times, and those preferentially bound into crystal structures have short ones.

The general subject of geochemical cycles and circulation of elements in the ocean is extensively discussed in the geochemical literature. The aspects of the chemical behavior of the elements which will be considered here relate more specifically to what happens in dilute aqueous solutions. Ultimately the details of this part of the cycles of the elements will become well enough known so a firm scientific base will exist upon which the management practices needed to maintain water quality at desired levels can rest.

BUFFER CAPACITY

The concept of buffer capacity extended by Weber and Stumm (1963) to natural-water systems may be useful in developing water-quality models. According to this concept, the chemical contents of natural water are maintained at a certain level by buffer systems more or less analogous to the buffering of hydrogen-ion activity by a reserve supply of ions in the solution which react with H^+. The calcium and bicarbonate content of a water can thus be buffered by the existence of a reserve supply of calcium carbonate in the solid form and a consistent input of carbon dioxide species through the atmosphere and biological activity.

When pollutants are present in a water system, in amounts below the level which could overcome the natural buffer systems, presumably the natural condition will be approximately reestablished in time, and no long-term serious effects will be observed. When the capacity is greatly exceeded, however, problems are sure to result, and of course some pollutants may not be affected by natural buffering reactions.

Sillén (1967a) has used the Goldschmidt figures for volume of rock weathered and other considerations to construct an equilibrium model for sea water which may help explain the pH and composition with respect to many dissolved species. In this model the sediments in the ocean serve to add to the buffer capacity of the system. Sillén's proposals have aroused much interest among geochemists and oceanographers. Papers by Garrels (1965) and by MacKenzie and Garrels (1966) further explored the aluminosilicate equilibria that may exist in the oceans and their influence on the composition of sea water. This kind of buffering system might be capable of controlling the concentration of many elements in sea water and could perhaps maintain the composition nearly constant for very long periods of time.

EVALUATION OF WATER COMPOSITION

The composition of natural water must be determined by physical and chemical means, usually by collection and examination of samples. The standard practice of collection of samples and later analysis in the laboratory is changing somewhat in response to the growing automation trend in which automatic sampling and continuous-sensing devices are becoming more widely used. It is with the study and interpretation of water composition, however obtained, that we are principally concerned.

COLLECTION OF WATER SAMPLES

Sampling is a vital part of studies of natural-water composition and is perhaps the major source of error in the whole process of obtaining water-quality information. This fact is not well enough recognized, and some emphasis upon it seems desirable.

In any type of study where only small samples of the whole substance under consideration may be examined, there is inherent uncertainty because of possible sampling error. The extent to which a small sample may be reliably considered to be representative of a large volume of material depends on several factors. These include, first, the homogeneity of the material being sampled and, second, the number of samples, the manner of collection, and the size of the individual samples.

The sampling of a completely homogeneous body is a simple matter, and the sample can be very small. Because most materials are not homogeneous, obtaining truly representative samples depends to a great degree upon the sampling technique. A sample integrated by taking small portions of the material at systematically distributed points over the whole body represents the material better than a sample collected from a single point. The more portions taken, the more nearly the sample represents the original. The sample error would reach zero when the size of the sample became equal to the original volume of material being sampled, but for obvious reasons this method of decreasing sampling error is not practical.

One of the primary goals of a water-quality investigation may be to provide information by which the composition of the whole volume of water within or available to a region can be closely determined. The water may constitute a slowly circulating mass in a lake or reservoir, the water in an aquifer, or the quantity of water carried by a river within some finite time period. Also, information is required on the variations in composition from place to place within the water body and the variation of composition at a point, or over the whole water body, with passage of time. The design of a sampling program which will accomplish all these goals encounters different kinds of problems in surface- and ground-water systems, and rather careful attention to sample collection is required.

The purpose underlying a water-quality study largely governs the sampling procedures that will be followed. Commonly, the investigator wishes to know the composition of a cross-section of a river at a specific time. For some purposes, however, only the composition that would occur at a fixed water intake point is of interest, and here the procedure would be somewhat simpler to design.

SAMPLING OF RIVER WATER

To determine adequately the composition of a flowing stream, each sample, or set of samples taken simultaneously, must be representative of the entire flow at the sampling point at that instant. Furthermore, the sampling process must be repeated frequently enough to define changes with time that may occur in the water passing the sampling point. Changes occurring along the length of the stream can be evaluated by adding more sampling points.

The homogeneity of a stream at a cross section is determined by physical factors, such as proximity of inflows and turbulence in the channel. Locally, poor lateral or vertical mixing can be observed in most stream systems. Immediately below the confluence of a stream and a tributary there may be a distinct physical separation between the water of the tributary and that of the main stream, and, particularly in large rivers, this separation may persist for many miles downstream. The effect is more pronounced if the water of the tributary differs markedly from the water of the main stream in content of dissolved or suspended solids or in temperature. These effects may be of special interest in some studies, but if the average composition of the whole flow of a stream or its changes in composition over a period of time are the factors of principal significance, sampling locations where mixing is incomplete should be avoided.

An outstanding example of lack of mixing across a stream is afforded by the Susquehanna River at Harrisburg, Pa. The stream at the highway bridge where samples were collected is about half a mile wide and is split into two channels by an island. The composition of the water is indicated by six samples spaced across the stream and is given in figure 2. More than 20 years of observations by the U.S. Geological Survey (Anderson, 1963) show this pattern is always present in some degree, except at very high stages. The anthracite-mining region northeast of Harrisburg produces large volumes of drainage containing high sulfate concentrations and having a low pH. Tributaries entering the river from the west above Harrisburg, especially the

62 CHEMICAL CHARACTERISTICS OF NATURAL WATER

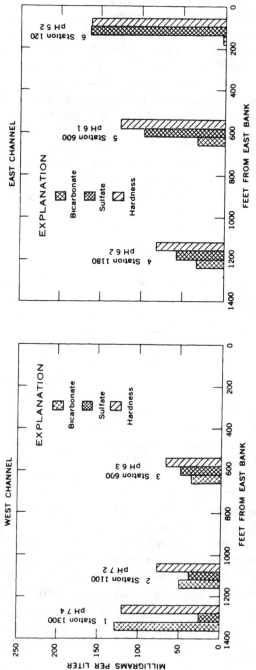

FIGURE 2.—Bicarbonate, sulfate, hardness, and pH of samples collected in cross section of Susquehanna River at Harrisburg, Pa., July 8, 1947.

Juniata River, carry alkaline water having much lower sulfate contents.

Obviously, it is difficult to characterize the whole flow of the stream at Harrisburg, although samples at one point would indicate what an intake located there would obtain.

In theory, a sample representing the instantaneous average composition of a stream should be obtained by combining depth-integrated samples of equal volume taken at places representing equal flow at several points across the stream. In practice, on small or medium-sized streams, it generally is possible to find a sampling section at which the composition of the water is uniform with depth and across the stream. The problem of obtaining adequate samples is thus simplified, and the cross section can be represented by a single grab sample taken at a convenient point. For larger streams, more than one sample may be required in the cross section. A portable conductivity meter may be useful in selecting a sampling site.

When a sampling point has been found and a procedure adopted which insures that each sample adequately represents the water flowing at that instant, there remains the problem of relating the sample to streamflow and the passage of time. The composition of all surface streams is subject to change with time. Long-term changes may result from long rainfall or runoff cycles or changes in land or in water utilization. Seasonal changes are to be expected from varying rates of runoff, evaporation, and transpiration typical of the seasons. Daily or even hourly changes of considerable magnitude may occur in some streams owing to flash floods, regulation of flow by man, dumping of wastes, or biochemical changes.

Stream discharge commonly is computed in terms of mean daily rates. A strictly comparable water-quality observation would be the daily mean of a continuously determined property. A single grab sample, however, ought to be considered only to represent the instantaneous discharge at the time of sampling.

To determine the water-quality regimen of a river at a fixed sampling point, samples should be collected or measurements made at such intervals that no important cycle of change in concentration could pass unnoticed between sampling times. For some streams, where flow is completely controlled by large storage reservoirs or is maintained at a nearly steady rate by large, constant ground-water inflows, a single sample or observation may represent the composition accurately for many days or weeks. For most streams, however, one sample cannot be safely assumed to represent the water composition closely for more than a day or two and for some streams not for more than a few hours. The U.S. Geological Survey began extensive investigations of chemical quality of river water shortly after 1900 (Dole, 1909; Stabler, 1911).

In these studies, samples were collected once a day by a local observer for a period of a year or more at each sampling site. Once-daily sampling schedules were followed in most subsequent work of this kind by the U.S. Geological Survey. Although this frequency of sampling may miss a few changes of significance, it generally has been thought to provide a reasonably complete record for most large rivers. The sample is assumed to represent all the water passing the sampling point on the day it was collected, and also it usually is assumed to represent a discharge rate equal to the daily mean. The descriptive text accompanying the river-water analyses for the early years of this century does not tell much about sampling methods and gives no reasons for the decision to obtain one sample each day. Continuous water-stage recorders were not yet in wide use at that time (Corbett and others, 1945, p. 191). It may be that investigators who were conditioned to accept once- or twice-daily gage readings as a basis for calculating mean daily water discharge felt that a once-daily sampling schedule was so obviously indicated that no alternative needed to be considered.

In recent years, equipment has been developed that can be installed on a streambank or bridge to obtain various measurements of water quality every few minutes and record them or transmit the results to a central point. Developments in this field have been rapid, and such equipment can obviously provide much more detailed information than could be obtained by the old sampling methods. Some limitations of sampling remain, however, as the water on which measurements are made has to be brought to the instrument through a fixed intake. The location of the intake represents a fixed sampling point.

Although it may seem somewhat surprising in retrospect that the technology of sample collection or onsite sensing of water quality was not improved for so long a time, the explanation is not difficult to find. The information obtained simply was not considered important enough to justify improvement of the reliability or accuracy when it appeared such improvement would be costly.

In the early studies mentioned above, the daily samples were combined into composite samples before the analysis was begun. The composites usually included 10 daily samples, and three composites were prepared for each month. In later investigations, a single determination, usually specific conductance, was made for each daily sample before making the composite with the remaining water. One of the principal reasons for combining individual daily samples into composites was the need for a large volume of water for the analysis. Another reason, of course, was economy. The composite samples usually have included from 10 to 30 daily samples, but shorter periods were used at times to avoid obscuring day-to-day changes and to study the composition of water at times of unusually high or low discharge rate. Samples of water that differed widely in conductance usually were not included in the same composite sample, nor were samples representing widely different discharge rates. Samples of those kinds were analyzed separately. In studies made on streams in the Missouri River basin beginning in 1946, the composite samples were prepared by using amounts of each daily sample proportional to the discharge rate observed at the time of sampling. That procedure yields discharge-weighted results. Similar procedures have been followed for some other streams since then, but for many streams, composites have continued to be made by using equal volumes of each daily sample. The publications in the U.S. Geological Survey series "Quality of Surface Waters of the United States," in which the results were released, describe compositing methods used.

Mean annual chemical composition of streamflow and total loads of dissolved solids carried can be obtained with reasonable accuracy from analytical records of composited daily samples. For many kinds of water-quality studies, however, compositing of samples is not desirable. Records obtained in this way may tend to obscure the relationship of water composition to discharge rate. The effect of compositing on computation of average composition will be considered later in discussing the use of averages.

The range between high and low extremes of dissolved-solids concentration at a sampling point on a stream is rarely as wide as that between high and low flow rates. Maximum dissolved-solids concentrations 20–40 times as great as the minimum have been observed over long periods of record in some nontidal streams in the United States, but for most of the larger rivers the range is much narrower. Usually the changes in dissolved-solids concentration are somewhat related to the rate of water discharge and the rate of change of discharge, but this relationship is complicated by other factors, as discussed under "Buffer capacity." For some streams, past records can be directly utilized as a means of estimating water quality from discharge. For other streams, such estimates have too low an accuracy to be of any value.

As automated or continuous-recording equipment for conductivity and other variables comes into wider use, properly designed supplementary sampling and chemical analyses will provide many details on water-quality regimens which are not attainable by sampling alone. Various combinations of continuous recording supplemented with periodic samples are now in use and are particularly useful in streams with tributaries that supply water that varies extensively in composition, in streams with large actual or potential inflows of waste, and in streams influenced by oceanic tidal inflow.

66 CHEMICAL CHARACTERISTICS OF NATURAL WATER

The emphasis in this discussion has been on the evaluation of the quality of water discharged by a stream at a fixed sampling point during a finite time period. The hydrologist may also be interested in using the quality record in other ways, which require a subtly different interpretation of the meaning of sample composition. For example, to correlate water composition with changes in flow rate one must consider the samples to represent only the water-discharge rate at the instant of sample collection and in effect dismiss the idea of extrapolating of the sample composition over a longer time period or of compositing samples taken at different times.

The Rio Grande at the San Acacia gaging station in central New Mexico is an example of a stream with considerable fluctuations of discharge and quality. Figures 3 and 4 show the way in which discharge fluctuates and conductivity of the water changes at different times of the year. Figure 3 covers the spring-runoff period in which melting

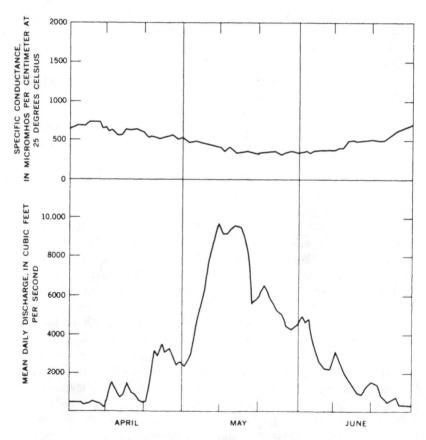

FIGURE 3 —Conductance of daily samples and mean daily discharge of Rio Grande at San Acacia, N. Mex., 1945

snow in the headwater region of the river caused the flow to increase from a few hundred to nearly 10,000 cubic feet per second during May. The flow decreased to low stages again in June. During this period, the conductance of the water declined and then rose again, but the day-to-day change was minor; the maximum for the 3 months was only about double the minimum. Daily, or less frequent, sampling will define such a period adequately. During the summer, much of the runoff results from flash floods in ephemeral tributaries in which both the quantity and the quality of the water vary widely. In figure 4 the results of samples collected from one to five times a day during part of the month of August were plotted with discharge rates observed at sampling times. On August 17th two samples collected a few hours apart showed a nearly threefold increase in concentration. Once-daily sampling can obviously give a distorted picture of such rapidly changing conditions.

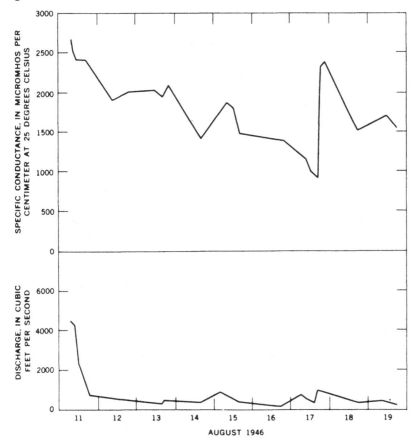

FIGURE 4—Conductance and discharge at times of sampling of Rio Grande at San Acacia, N. Mex under typical summer flow conditions

Samples collected once a week or once a month provide a reasonable indication of water composition for some streams. Systematic sampling in a stream basin during which many sources are sampled at about the same time are of considerable value for reconnaissance purposes as well as for general knowledge.

The implication given here that discharge or flow measurements should be available for sampling sites is intentional. Chemical analyses of river water generally require some sort of extrapolation, because the water sampled has long since passed on downstream by the time a laboratory analysis is completed. The discharge record provides a means of extrapolating the chemical record if the two are closely enough related. The discharge data also serve as a means of averaging the water analyses, give an idea of total solute discharges, and permit evaluating the composition of water that might be obtained from storage reservoirs.

LAKE AND RESERVOIR SAMPLES

Water stored in lakes and reservoirs commonly is poorly mixed. Thermal stratification and associated changes in water composition are among the most frequently observed effects. Single samples from lakes or reservoirs can be assumed to represent only the spot within the water body from which they come.

The effect of stratification on water composition is noticeable in concentrations of ions whose behavior is influenced by oxidation and reduction, the reduced species often increasing in concentration with depth below the surface. The components that are utilized by life forms in the water also are often considerably affected. The mechanics of stratification in lakes have been studied by many limnologists and will not be further reviewed here. Several studies suggest that an aeration technique may be useful in preventing stratification (Ford, 1963).

Reservoir outlets often are located in positions where they may intercept water that is depleted in dissolved oxygen. From a practical point of view, the water user is interested only in the composition of water available at the outlet, and most samples from storage reservoirs for which analyses are available come from released water.

GROUND-WATER SAMPLES

Most of the physical factors which promote mixing in surface waters are absent or much less effective in ground-water systems. Even in a thick sand of uniform permeability, the horizontal or downslope movement of water in the zone of saturation is slow, and mixing is relatively poor. In most sediments, the horizontal permeability is greater than the vertical permeability, and flow in a particular stratum may develop chemical characteristics that are substantially different from water in strata above or below.

Means whereby differing composition of water in different parts of the saturated zone can be studied are not entirely adequate. Wells commonly obtain water from a considerable thickness of saturated rock and often from several different strata. These components are mixed by the turbulent flow of water in the well before they reach the surface of the land and become available for sampling. Springs may obtain water from a lesser thickness of saturated material, but often the exact source is difficult to ascertain. Most techniques for well sampling and exploration are usable only in unfinished or nonoperating wells. Usually, the only means of evaluating the quality of water tapped by a well is an analysis of a pumped sample. The limitations of the well as a sampling device are obvious but unavoidable.

The differences of water quality with depth below the surface, and associated differences in lithology, are shown for three wells in the western part of Pinal County, Ariz., in figure 5. Water samples were obtained at several depths in each well during periods when the pumps were not operating, and the specific conductance of each sample is indicated on the diagram opposite the depth where the sample was obtained. Kister and Hardt (1966), in whose publication the illustration first appeared, reported a range in specific conductance in one well from 1,480 micromhos per centimeter at a depth of 300 feet below land surface to 29,400 micromhos at 550 feet. Water pumped from the well had a specific conductance of 5,960 micromhos per cm. As Kister and Hardt pointed out (p. 10), "Chemical analyses of water samples collected from the discharge pipe of a pumping well are not necessarily indicative of the quality of water throughout the sequence of sediments penetrated by the well."

Although the range of conductance is unusually great, the data show how water yielded by a well could change in quality in response to changes in pumping rate or regional drawdown of the water table. Many wells are influenced in some degree by water-quality stratification, and the interpretation of ground-water analyses must always consider the possibility of such effects. Electric logs may provide useful indications of the location of water of poor quality in the saturated material penetrated by wells.

Although one rarely can be certain that a sample from a well represents exactly the composition of all the water in the vertical section at that point, it is usually a useful indication of the average composition of available water at that point. Where a considerable number of wells reaching the same aquifer are available for sampling and show similar composition, the investigator usually is justified in drawing some conclusions about the chemistry of the ground water in the

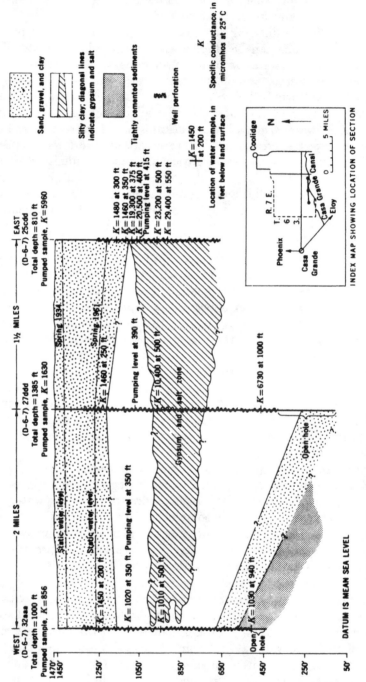

FIGURE 5.—Changes in conductance of ground water, Pinal County, Ariz. Modified from Kister and Hardt (1966).

aquifer. If a well penetrates a large relatively homogenous aquifer, the composition of water generally will not change much over long periods of time. Areal variations in ground-water quality occur in many aquifers.

In considering the variations of quality of ground water with time, it is probably better to think of aquifers as reservoirs, subject to slow changes in quality as the water circulates, than to consider them as "pipelines" or water courses through which water moves from one place to another. The latter concept, however, may be valid to some extent for water moving through the large openings of a cavernous limestone. Regardless of which view is held, the movement of the ground water in the aquifer is usually slow enough that changes in quality with time are satisfactorily shown by samples taken monthly, seasonally, or annually.

Some examples of month-to-month changes in very shallow ground water in the alluvium of the Gila River in Safford Valley, Ariz., (Gatewood and others, 1950, p. 73) are shown in figure 6. These fluctuations are more rapid than those usually observed in wells of greater depth, and because many factors such as changes in river discharge, rainfall, irrigation pumping, and return flow may influence the water composition, no well-defined pattern of quality fluctuation can be discerned.

A long-term trend in ground-water composition is shown in figure 7. The two wells indicated in the graph were used for irrigation in the Wellton-Mohawk area along the Gila River in southwestern Arizona (Babcock and others, 1947). The dissolved solids increased greatly over the period of record.

COMPLETENESS OF SAMPLE COVERAGE

In areas where hydrologic studies are being made, a decision is needed as to how many samples or other water-quality observations are required. Aside from administrative limitations in funds and manpower, this decision should be based on the conditions in the area to be studied. Factors to be weighed include the amount of information of this type already available, the hydrologic complexity of the area, the extent to which water of inferior quality is known or believed likely to occur, and other similar considerations. The aim of many water-quality investigations is to evaluate the resource as thoroughly as possible, and this usually requires many samples and field observations. The amount of laboratory work per sample often can be decreased if, by means of field determinations and laboratory determinations of certain key constituents, it can be shown that many of the samples have similar composition. The experienced water chemist thus can determine the water quality of an area by the most efficient combination of complete and partial chemical analyses. His place on the team of any hydrologic investigation is important beyond the actual performance of analytical determinations.

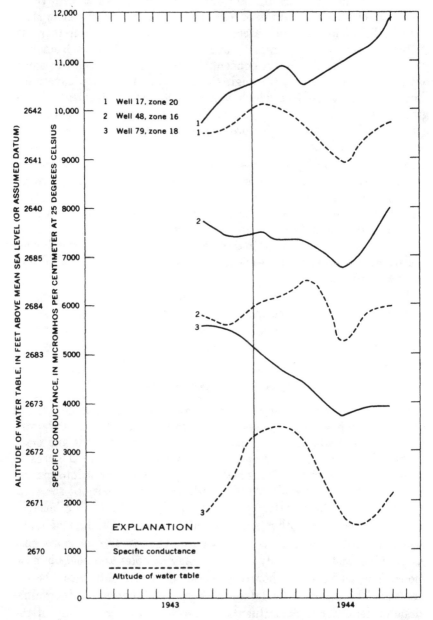

FIGURE 6.—Specific conductance and altitude of water table for three typical observation wells, Safford Valley, Ariz.

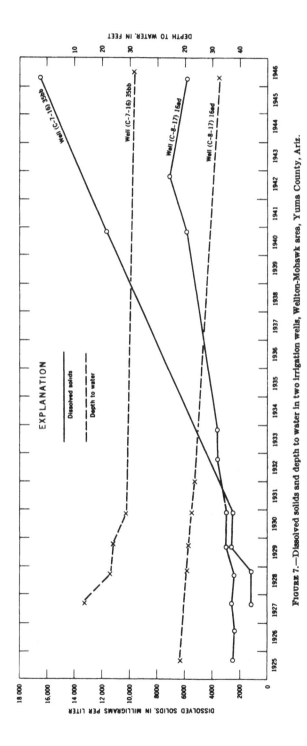

FIGURE 7.—Dissolved solids and depth to water in two irrigation wells, Wellton-Mohawk area, Yuma County, Ariz.

ANALYSIS OF WATER SAMPLES

The analysis of water for its dissolved components is a part of the work done by a large number of chemical laboratories, including many supported by State, Federal, and local governments, academic and research institutions, and private enterprise. The methods used in water analysis are fairly well standardized and will not be discussed here. There are certain procedures for field testing, however, which should be commented upon.

FIELD TESTING OF WATER

Examination of water in the field is an important part of hydrologic studies. Certain properties of water, especially its pH, are so closely related to the environment of the water that they are likely to be altered by sampling and storage, and a meaningful value can be obobtained only in the field. Other properties of water such as its specific conductance are easily determined in the field with simple equipment, and the results are useful in supplementing information obtained from analyses of samples and as a guide to which sources should be sampled for more intensive study. Commercial equipment is available or can be adapted to field use, ranging from pocket-size testing kits to trailer or bus-mounted mobile laboratories. In the more elaborate units, almost any kind of standard analysis can be made.

The early history of field testing shows that its importance was recognized as long ago as the early 1900's, but equipment available then generally was rather crude. As early as 1896, a portable Wheatstone bridge for measuring conductivity of water and saturated soil was being used by the U.S. Department of Agriculture (Scofield, 1932). The equipment for measuring water conductivity in the field has evolved into relatively sophisticated models which give readings of temperature and specific conductance of samples with null points indicated by a cathode-ray tube. The units are battery powered, light and easy to carry, and give results that are as accurate as those obtained in the usual laboratory installation. Sensing cells also can be incorporated into well-exploration equipment.

A rapidly expanding application has been the continuous measurement of conductivity or other characteristics of river water with equipment installed at the sampling site. These units can be made to record results in a form which can be directly fed into an electronic computer. Power for operating the installation usually is best obtained from a 115-volt alternating-current line.

Any determination which can be made by potentiometric methods can easily be accomplished in the field or can be built into an onsite sensor. The pH of a water, for example, generally is determined by means of a sensitive electrometer and suitable reference and glass electrodes. For most ground waters, the pH needs to be determined immediately after the water issues from the well or spring, if the value is to represent conditions within the system where the water occurs (Back and Barnes, 1961). Electrode systems for determining dissolved oxygen and many individual ions also are available.

Although almost any property or component of natural water can now be determined at a streambank location and probably could be automated if required, the cost of the most elaborate installations is high, and a real need must exist as justification.

ELECTRIC LOGS AS INDICATORS OF GROUND-WATER QUALITY

One of the most widely used means of geophysical exploration of subsurface conditions is electric logging of boreholes. One of the principal determinations made as part of the logging procedure is the resistance to passage of an electric current through the formations penetrated by the borehole. One form of resistivity logging uses a pair of electrodes which are spaced a specified distance apart and which are held against the side of the uncased hole. As the electrodes are moved up or down the hole, the electrical resistance observed between them changes in response to environmental changes. A recording device traces the resistance on a chart as the probe moves in or out of the hole. The result is a curve of resistance against depth below the surface.

The resistance of water-bearing material in place is a function of the resistance of the rock itself, the resistance of the interstitial water, and the length of the path through which current passes in the water contained in the interconnected openings in the rock. Resistance of most dry rocks is very high, and in effect the resistance measured by electric loggers is controlled by the water conductivity and the length of the current path. The resistance is expressed as resistivity (reciprocal of conductivity) usually in ohms per meter per square meter. Values obtained for resistivity of aquifers through electric logging are thus very closely related to conductivity of water and porosity of the rocks. The principal use of electric logs in the water-well industry has been as an aid to determining the physical properties of water-bearing formations and in correlating formations from well to well. Applications of logging equipment in hydrology were described by Patten and Bennett (1963).

Some investigators have proposed a modified interpretation with the conductivity of the water considered the unknown, the other features of the aquifer which influence the observed resisivity being computed on the basis of laboratory tests of drill cuttings from the formation or on the basis of previous experience with the same aquifer. Used in this way, the resistivity log provides an indication of water quality in place underground.

The reliability with which water conductivity can be determined from a resistivity log is highest where aquifer properties are well known. Some investigators have utilized resistivity logs to compute chemical analyses for water; however, circumstances under which this may be done satisfactorily occur only where a large number of chemical analyses are available for water from the formation in question and where these analyses show a well-defined relationship between conductivity and each of the constituents for which the computation is made. Jones and Buford (1951) described such computations for ground waters in Louisiana.

Another property generally determined in electric logging is the spontaneous potential which can be observed between an electrode and the land surface when no current is introduced. This potential is partly the result of electrochemical effects such as semipermeable-membrane effects, concentration cells, and changes in rock and water composition as the electrode is moved along the side of the hole.

Some work has been done in relating the observed potentials more specifically to the water-bearing materials and the composition of the water, but more would seem to be justified, as this technique, when carefully refined, might give a considerable insight into electrochemical relationships in ground-water systems that may influence both water quality and movement.

LABORATORY PROCEDURES

The procedures considered to be sufficiently accurate and most acceptable for general use in water analysis in the United States have been compiled in several publications. The most widely known of these is "Standard Methods for the Examination of Water, Sewage, and Industrial Waste" (American Public Health Association, 1965). It is revised every few years by the American Public Health Association, the American Water Works Association, and the Water Pollution Control Federation. Other compilations widely used are those of the American Society for Testing and Materials (1964), Association of Official Agricultural Chemists (1965), and U.S. Salinity Laboratory Staff (1954). The U.S. Geological Survey methods compilation by Rainwater and Thatcher (1960) is also well known. A review of periodical literature on water analysis is published every two years in the journal "Analytical Chemistry." The large volume of current literature in this field is indicated by the more than 600 articles referenced for the 2-year period October 1964 through September 1966 (Fishman and others, 1967).

EXPRESSION OF WATER ANALYSES

Various terms and units are commonly employed in the expression of data obtained in the chemical analysis of water. An understanding of those more frequently used is required for the interpretation of analyses.

HYPOTHETICAL COMBINATIONS

Water analyses published before 1900 generally are expressed in terms of concentrations of combined salts, such as sodium chloride or calcium sulfate. This kind of terminology probably was used in part as an attempt to describe the residue obtained when the water was evaporated, but it also predated the concept of dissociated ions in solution introduced in the late 19th century by Arrhenius.

Although water chemists often use terms such as "calcium bicarbonate water" to describe a solution in which Ca^{+2} and HCO_3^- are the principal ionic species, they recognize this as a form of shorthand or abbreviation for the much more clumsy expression "a water in which calcium and bicarbonate are the predominant cation and anion, respectively."

A few fairly recent books on geochemistry and occasional journal articles contain more misleading uses of this kind of shorthand. For example, the statement "Ferrous iron is transported in ground water as the bicarbonate" is still sometimes seen. The writers of this kind of statement usually do not seem to be referring to the ion pair $FeHCO_3^+$, which has never been chemically identified, but are speaking of a type of water that sometimes is termed "ferrous bicarbonate water." Such a water may not have Fe^{+2} as its predominant cation, but it has enough to be worth taking special note of. What these writers probably mean is "Ferrous iron may be relatively stable in ground water containing bicarbonate." More care in choice of terminology is needed.

Chemical calculations and standard gravimetric factors can be used to convert data reported in terms of combined salts into ionic form. Because the reporting of combined salts in solution can lead to confusion, this practice should not be used unless the species actually is known to occur in the form of a dissolved ion pair or complex.

ANALYSES REPORTED IN TERMS OF IONS

For many years it has been known that most inorganic salts in solution in water are dissociated into charged particles, or ions. A water analysis is intended to be a statement of the composition of a water solution, and it is, therefore, appropriate to use an ionic form of statement of the analysis. In accordance with the dissociation concept, water analyses are now generally expressed in concentrations of individual ions for those substances known to be dissociated in solution. Substances whose form in solution is unknown, or which may be suspected to occur in undissociated or even colloidal form, are commonly reported either as an oxide or as an uncombined element. For example, the sodium concentration is reported as the cation Na^+; iron is usually reported as the element Fe, and silicon as an equivalent concentration of the oxide SiO_2.

DETERMINATIONS INCLUDED IN ANALYSES

As more sophisticated equipment and analytical methods have come into use, the analysis of a water sample could include most of the elements in the periodic table, as well as a considerable suite of naturally or artificially produced isotopes of these elements and many specific organic compounds. The principal interest of the analyst and the great preponderance of data, however, deal with the major constituents which go to make up nearly all the dissolved inorganic material. Minor constituents are included when they pose actual or potential problems to water users or when the analysis is being made for special purposes.

The dissolved cations (positively charged ions) which constitute a major part of the dissolved-solids content usually are calcium, magnesium, sodium, and potassium; the major anions (negatively charged ions) are sulfate, chloride, fluoride, nitrate, and those contributing to alkalinity, most generally assumed to be bicarbonate and carbonate. The silicon present usually is nonionic and is reported in terms of an equivalent concentration of the oxide silica (SiO_2).

Some other dissolved constituents are included in many chemical analyses because they may be objectionable in water used for certain purposes. Sometimes these constituents attain concentrations comparable with those of major components. They include aluminum, boron, hydrogen ion or acidity, iron, manganese, phosphate, and forms of nitrogen other than nitrate.

Some chemical analyses include other constituents, most of which are reported in terms of the uncombined element, although specific ionic species are specified for some. These constituents include certain ones whose presence is objectionable in water to be used for domestic or industrial purposes and a few that may be major constituents in some kinds of solutions. Those for which there is enough information to be considered specifically here are the alkali metals, lithium, rubidium, and cesium: the alakaline earth metals, beryllium, strontium, and barium: the metallic elements, titanium, vanadium, chromium, molybdenum, cobalt, nickel, copper, silver, gold, zinc, cadmium, mercury, germanium, and lead; the nonmetals, arsenic, selenium, bromine, and iodine: and the radioactive elements, uranium, radium, radon, and thorium. Sometimes specific organic solutes are determined, and dissolved gases, especially oxygen, also are commonly reported.

Certain properties of water solutions besides the contents of specific ions have commonly been included in water analyses. Hardness in water is commonly expressed in terms of an equivalent quantity of calcium carbonate. Other properties often included in a water analysis are color, specific conductance, total dissolved solids, specific gravity, suspended matter, turbidity, biochemical or chemical oxygen demand, sodium-adsorption-ratio, and various forms of radioactivity. These constituents and properties will be discussed in more detail as appropriate in following sections of this report.

UNITS USED IN REPORTING ANALYSES

Over the years, a wide variety of units have been used in reporting water analysis. Considerable progress had been made toward standardization of these units, but to use the data available in published literature often requires a general understanding of the units and systems used in the past and how they compare with more modern units. The two most common types of concentration units are those which report weights of solute per weight of solution and those which report weights of solute per unit volume of solution.

WEIGHT-PER-WEIGHT UNITS

A concentration reported in weight-per-weight is a dimensionless ratio and is independent of the system of weights and measures used in determining it. For many years, the water analyses made by U.S. Geological Survey and many other laboratories in the United States were reported in parts per million. One part per million is equivalent to one milligram of solute per kilogram of solution. One percent, of course, is one part per hundred, or 10,000 parts per million.

Parts per thousand sometimes are used in reporting the composition of sea water. In this connection, the terms "chlorinity" and "salinity" have been defined in terms of parts per thousand (grams per kilogram) for use in studies of sea-water composition (Rankama and Sahama, 1950, p. 287).

At one time the unit "parts per hundred thousand" was in common use. "Parts per billion" or "parts per trillion" are sometimes used in reporting trace constituents.

WEIGHT-PER-VOLUME UNITS

Because water is a liquid, definite quantities for analyses are ordinarily measured in the laboratory by means of volumetric glassware. The laboratory results, therefore, are in terms of weights of solute per unit volume of water. These results must be converted to a weight basis to obtain parts per million values. The conversion usually is done by assuming that a liter of water weighs exactly 1 kilogram, hence, that milligrams per liter and parts per million are equivalent. This assumption is strictly true only for pure water at 3.89° C. The presence of dissolved mineral matter tends to increase the density, and at higher temperatures the density decreases. For practical purposes, however, the error introduced by assuming unit density does not reach a magnitude comparable to other anticipated analytical errors until the concentration of dissolved solids exceeds about 7,000 milligrams per liter. For highly mineralized waters a density correction should be used in computing parts per million from milligrams per liter. Volumetric glassware is calibrated for use at 20° C, and ordinary laboratory temperatures are usually close to this value. Concentrations expressed in milligrams per liter are strictly applicable only at the temperature at which the determination was made, but for most purposes for which the concentration values might be utilized, the effect of volume change caused by temperature changes of the solution is not important.

Many chemists, especially in countries where the metric system is the official standard, report water analyses in milligrams per liter. This procedure eliminates density corrections in reporting the analysis, and it probably will be more widely used in the future. It became standard practice within the U.S. Geological Survey in 1967.

In the United States and where the English system of units is used, analyses are sometimes expressed in grains per gallon. The particular gallon that is meant must be specified, as the United States and the Imperial, or British, gallons are not the same. The unit is still frequently seen in connection with hardness in water. Reports dealing with irrigation water often express concentrations in tons per acre-foot. The ton is 2,000 pounds, and an acre-foot is the amount of water needed to cover 1 acre to a depth of 1 foot.

Streamflow or discharge represents a rate quantity. Rate concepts may be significant in some studies of river-water composition, and the dissolved-solids load, a rate quantity, is generally expressed in tons per day.

Conversion factors which indicate the relationship of the various units to each other are given in table 8. Factors changing grains per gallon or tons per acre-foot to parts per million, or the reverse, would be the same as those shown for conversions to or from milligrams per liter if it is valid to assume unit density of the water. For highly mineralized water these factors must take into account the density of the water. Hardness values are sometimes expressed in degrees. Conver-

TABLE 8.—*Conversion factors for quality-of-water data*

[U.S. gallon is used for all units involving gallons]

To convert—	To—	Multiply by—
Grains per gallon	Milligrams per liter	17.12
Milligrams per liter	Grains per gallon	0.05841
Milligrams per liter	Tons per acre-foot	.001360
Milligrams per liter	Tons per day	second-feet × 0.002697
Tons per acre	Metric tons per hectare	2.2417
Parts per hundred thousand	Parts per million	10
Grams	Ounces (avoirdupois)	.03527
Ounces (avoirdupois)	Grams	28.35
Gallons (Imperial)	Gallons (U.S.)	1.2009
Liters	Quarts (U.S.)	1.057
Quarts (U.S.)	Liters	.9463
Second-foot days[a]	Acre-feet	1.983471
Second-feet[b]	Gallons per minute	448.8
Second-foot days	Gallons per day	646,317
Acre-feet[c]	Gallons	325,851
Acre-feet	Cubic feet	43,560
Cubic feet	Cubic meters	.028317
Cubic feet	Gallons	7.481
Ca^{+2}	$CaCO_3$	2.497
$CaCl_2$	$CaCO_3$	.9018
HCO_3^-	$CaCO_3$	.8202
HCO_3^-[d]	CO_3^{-2}	.4917
Mg^{+2}	$CaCO_3$	4.116
Na_2CO_3	$CaCO_3$	.9442
NO_3^-	N	.2259
N	NO_3^-	4.4266

[a] 1 sec-ft day = 1 cfs for 24 hr.
[b] 1 sec-ft = 1 cfs.
[c] 1 acre-ft = an area of 1 acre 1 ft deep.
[d] In reaction $2HCO_3 = CO_3^{-2} + H_2O + CO_2 g$.

sion factors for these units, which differ in different countries, are given in the discussion of hardness in water.

EQUIVALENT-WEIGHT UNITS

For manipulations that involve the chemical behavior of dissolved material, the chemist must express analytical results in units which recognize that ions of different species have different weights and electrical charges. For example, in the mass-law calculations discussed earlier, concentrations of ions and other dissolved species are given in moles per liter. A mole of a substance is its atomic or molecular weight in grams. A solution having a concentration of 1 mole per liter is a molar solution, and the molarity of a solution is thus its concentration in a weight per volume unit. A molal solution is one which contains 1 mole of solute per 1,000 grams of solvent. For dilute solutions, the two units are equal, within ordinary experimental error.

Concentrations in milligrams per liter are readily converted to moles per liter by dividing by the atomic or formula weight of the constituent, in milligrams. When parts per million values are treated in this way, the concentration unit obtained is one usually called formality, the number of formula weights per 1,000 grams of solution.

The concept of chemical equivalence can be introduced by taking into account the ionic charge. If the formula weight of the ion is divided by the charge, the result is termed the "combining weight." When a concentration value in milligrams per liter is divided by the combining weight of the species concerned, the result is an equivalent concentration that is useful for many purposes. Table 9 contains reciprocals of combining weights of cations and anions generally reported in water analyses. Milligrams-per-liter values may be converted to milliequivalents per liter by multiplying the milligrams per liter by the reciprocals of the combining weights of the appropriate ions.

The term "equivalents per million," which is used for the value obtained when parts per million is used instead of milligrams per liter as a starting point, is a contraction which has been generally adopted for the sake of convenience. In more exact language, the unit is "milligram-equivalents per kilogram" if derived from parts per million or "milligram-equivalents per liter" if derived from milligrams per liter. The term "milligram equivalents" is shortened by chemists to milliequivalents, abbreviated "meq."

In an analysis expressed in milliequivalents per liter, unit concentrations of all ions are chemically equivalent. This means that if all ions have been correctly determined, the total milliequivalents per liter of anions should exactly equal the total milliequivalents per liter of cations. The relation of water composition to solid-mineral composition is made more clearly evident when the analysis is expressed in milliequivalents per liter. There are disadvantages to the use of these units, however, in that they require knowledge or assumptions about the exact form and charge of dissolved species. Laboratory determinations do not always provide this kind of information. For species whose charge is zero, as for silica, an equivalent weight cannot be computed. A concentration, however, of such species in moles or millimoles per liter is generally equally useful.

Analyses in the U.S. Geological Survey report series "Quality of Surface Water for Irrigation" are mostly in terms of equivalents per million or milliequivalents per liter, and some organizations concerned with irrigation-water quality report most of their analyses in these units.

COMPOSITION OF ANHYDROUS RESIDUE

The means of expressing analytical results discussed to this point all utilize concentrations of solutes. These concentration values are generally the deciding factors in the evaluation of water quality.

TABLE 9.—*Conversion factors: Milligrams per liter* $\times F_1 =$ *milliequivalents per liter, milligrams per liter* $\times$ $F_2 =$ *millimoles per liter (based on 1961 atomic weights, referred to carbon-12)*

Element and reported species	F_1	F_2
Aluminum (Al^{+3})	0.11119	0.03715
Ammonium (NH_4^+)	.05544	.05544
Barium (Ba^{+2})	.01456	.00728
Beryllium (Be^{+3})	.33288	.11096
Bicarbonate (HCO_3^-)	.01639	.01639
Boron (B)		.09250
Bromide (Br^-)	.01251	.01251
Cadmium (Cd^{+2})	.01779	.00890
Calcium (Ca^{+2})	.04990	.02495
Carbonate (CO_3^{-2})	.03333	.01666
Chloride (Cl^-)	.02821	.02821
Chromium (Cr)		.01923
Cobalt (Co^{+2})	.03394	.01697
Copper (Cu^{+2})	.03148	.01574
Fluoride (F^-)	.05264	.05264
Germanium (Ge)		.01378
Gallium (Ga)		.01434
Gold (Au)		.00511
Hydrogen (H^+)	.99209	.99209
Hydroxide (OH^-)	.05880	.05880
Iodide (I^-)	.00788	.00788
Iron (Fe^{+2})	.03581	.01791
Iron (Fe^{+3})	.05372	.01791
Lead (Pb)		.00483
Lithium (Li^+)	.14411	.14411
Magnesium (Mg^{+2})	.08226	.04113
Manganese (Mn^{+2})	.03640	.01820
Molybdenum (Mo)		.01042
Nickel (Ni)		.01703
Nitrate (NO_3^-)	.01613	.01613
Nitrite (NO_2^-)	.02174	.02174
Phosphate (PO_4^{-3})	.03159	.01053
Phosphate (HPO_4^{-2})	.02084	.01042
Phosphate ($H_2PO_4^-$)	.01031	.01031
Potassium (K^+)	.02557	.02557
Rubidium (Rb^+)	.01170	.01170
Silica (SiO_2)		.01664
Silver (Ag)		.00927
Sodium (Na^+)	.04350	.04350
Strontium (Sr^{+2})	.02283	.01141
Sulfate (SO_4^{-2})	.02082	.01041
Sulfide (S^{-2})	.06238	.03119
Titanium (Ti)		.02088
Uranium (U)		.00420
Zinc (Zn^{+2})	.03060	.01530

Geochemists, however, have sometimes preferred to express analytical data for water in terms which they believed were more directly comparable to rock-composition data. To this end they have expressed analyses in terms of the percentage of each element or ion in the anhydrous residue remaining after evaporating the water. Clarke (1924 a, b) used this reporting procedure, usually with a value for total dissolved-solids concentration and percentages of the components which he termed "percentage composition of anhydrous residue."

The principal advantage of this method of expressing analyses is that it may demonstrate similarities among waters which have similar geochemical origin but whose analyses might appear dissimilar because of dilution effects. For example, the composition of water from a river usually will appear to change a great deal as a result of increases or decreases in flow rate when one examines water analyses expressed in concentrations. When the data are recalculated to percentage composition of dry residue, at least part of the fluctuation in composition disappears.

Although for some geochemical considerations the calculation of percentage composition of residue is useful, this is not a good way of expressing the chemical composition of a solution, and since Clarke's time, it has gradually been disappearing from the literature.

Instead of computing percentage composition from actual weights of constituents, some investigators have computed percentages based on the total anions or cations, in equivalents per million. The first extensive use of this kind of information was by Palmer (1911), and the procedure has been adapted extensively in more recent times in certain graphical methods of study of water quality. These methods will be considered later in this paper.

CONCENTRATION IN TERMS OF CALCIUM CARBONATE

The hardness of water is conventionally expressed in all water analyses made in the United States in terms of an equivalent quantity of calcium carbonate. Some such convention is needed for hardness because this is a property imparted by several different cations, which may be present in varying proportions. However, the actual presence of the indicated number of milligrams per liter in the form of $CaCO_3$ certainly should not be assumed.

Another convention which is followed by many water analysis laboratories is to express the results of the alkalinity titration in terms of an equivalent amount of calcium carbonate. Although the titrated alkalinity is sometimes difficult to assign exactly to one or more specific ionic species, such an assignment gives a much clearer indication of the composition of the solution. In any event, alkalinity (to methyl-orange end point) expressed as milligrams per liter of $CaCO_3$ can be converted to an equivalent concentration of HCO_3^- in milligrams per liter by dividing the former by 0.8202. (See table 8.)

The formula weight of $CaCO_3$ is very near 100. Thus hardness or alkalinity values in terms of milligrams per liter of $CaCO_3$ can be converted to milliequivalents per liter by dividing by 50. Analyses are occasionally seen in which other constituents are reported in terms of calcium carbonate. There would seem to be little benefit in this device for constituents that can readily be reported in terms of specific ions.

COMPARISON OF UNITS OF EXPRESSION

Table 10 shows a single water analysis expressed in milligrams per liter, millimoles per liter, milliequivalents per liter, percentage composition of dry residue, percentage of total cation and anion equivalents, and in grains per U.S. gallon. It is assumed the milligrams per liter values are equal to parts per million and milliequivalents per liter to equivalents per million for a water of this dissolved-solids concentration.

Although the U.S. Geological Survey records all analyses in both milligrams per liter (parts per million) and milliequivalents per liter, most of its publications contain only the milligrams per liter values. If other forms of expression are desired or required, the milligrams per liter values can be readily recomputed to other forms.

As previously noted, the practice of expressing water analyses in terms of hypothetical combined salts has nearly disappeared. An analysis in this form can be converted to an ionic statement by application of gravimetric factors (ratios of atomic and molecular weights). For example, to convert a reported value of milligrams per liter of $CaCl_2$ to milligrams per liter of Ca, the formula used is

$$\text{milligrams per liter Ca} = \text{milligrams per liter } CaCl_2 \left(\frac{\text{atomic weight Ca}}{\text{molecular weight } CaCl_2} \right).$$

TABLE 10.—*Chemical analysis of a water sample expressed in six ways*

[Sample from flowing well 488 ft. deep. Water from Lance Formation. NW¼ sec 30, T. 57 N., R. 85 W. Sheridan County, Wyo Collected Aug. 3, 1946]

Constituent	mg/l or ppm	meq/l or epm	mM/l	Gravimetric percent	Percentage of epm	Grains per U.S. gallon
Silica (SiO_2)	7.9		0.131	0.40		0.46
Iron (Fe)	.17		.003	.01		.01
Calcium (Ca)	37	1.85	.925	1.87	6.1	2.16
Magnesium (Mg)	24	1.97	.985	1.21	6.5	1.40
Sodium (Na) } Potassium (K) }	611	26.58	26.58	30.80	87.4	35.69
Bicarbonate (HCO_3)	429	7.03	7.03	[1] 10.63	23.1	25.06
Sulfate (SO_4)	1,010	21.03	10.52	50.90	69.2	59.00
Chloride (Cl)	82	2.31	2.31	4.14	7.6	4.79
Fluoride (F)	.6	.03	.032	.03	.1	.04
Nitrate (NO_3)	.0	.00	.000	.00	.0	.00
Boron (B)	.2		.019	.01		.01
Dissolved solids Calculated	1,980	60.80	48.535	100.00	200.0	115.65
Hardness as $CaCO_3$: Total	191		1.91			11.16
Noncarbonate	0		.00			0
Specific conductance (micromhos per cm at 25° C)	2,880	2,880	2,880	2,880	2,880	2,880
pH	7.3	7.3	7.3	7.3	7.3	7.3

[1] As carbonate (CO_3)

The quantity in parentheses is the gravimetric factor which can be computed or obtained from tables in standard chemical handbooks. The milligrams per liter value thus obtained should be added to those obtained for calcium from similar calculations for other combined salts containing calcium. Totals for other ions are obtained by similar calculations.

SIGNIFICANCE OF PROPERTIES AND CONSTITUENTS REPORTED IN WATER ANALYSES

The properties and constituents that are determined in water analyses are discussed individually in the following sections. For most constituents, the subjects considered are the form of dissolved species, solubility in relation to other dissolved substances, the source of the material, the range of concentration which may be expected, and factors that may influence the accuracy or precision of analytical determinations. Appropriate environmental influences are considered, and chemical analyses are used to illustrate the discussions.

The chemistry of some solutes in natural water is reasonably well understood. For others, much remains to be learned. New research results will tend to make many statements made here subject to rather rapid obsolescence. This is one reason for keeping the discussions of individual constituents generalized and rather brief.

NATURE OF THE DISSOLVED STATE

The discussions so far have presupposed that the dissolved state was well enough understood that it did not need to be defined. An exact definition of the term, however, does pose some difficulties. Usually solutions are described as completely uniform mixtures in which each dissolved particle is surrounded by solvent molecules and has no direct contact with other like particles except as permitted by migration through the solution. The particles may carry an electrical charge, or they may be electrically neutral. There can be no doubt that individual ions, or pairs of ions, or even complexes made up of several ions are in the dissolved state. If the particles unite to form aggregates that will settle out of the solvent by gravity, they certainly can no longer be said to be in the dissolved state.

Between these extremes lies a wide range of particle sizes which may, under favorable conditions, be retained in suspension indefinitely. The particles in this size range generally are considered as colloidal, and the smaller colloidal particles merge into the dissolved state. Some textbooks suggest size ranges, for example, Glasstone and Lewis (1960, p. 571) stated that colloidal particles range in diameter from about 5 millimicrons up to about 0.2 micron. A micron is 10^{-4} centimeter, or 10^4 angstrom units. Therefore, the lower limit of diameter of colloidal particles is 50 angstroms, by this definition. A considerable number of organic compounds that may occur in natural water, such as those that impart a brown color, have molecules that are more than 50 angstroms in diameter; hence they can occur only in the colloidal state even though all particles may be single molecules.

Natural surface waters and some ground waters carry both dissolved and suspended particles. The amounts of the latter present in ground water before it is brought to the land surface generally are small. But in river water the concentration of suspended material may be large, and at high stages of flow in many streams it greatly exceeds the dissolved-solids concentration. This suspended material poses something of a dilemma for the water chemist. In the usual water-chemistry study, the suspended solids are removed before analysis either by filtration or by settling. Either procedure removes particles more than a few tenths of a micron in diameter, those which are large enough to produce light scattering. The suspended solids are then discarded, and the clear solution is analyzed. Some kinds of suspended material, however, may have an important bearing on the sanitary condition of the water, and analyses for the biochemical oxygen demand or bacterial counts must be made on unfiltered samples. In some of the studies of the U.S. Geological Survey, the quantity of suspended load of streams is determined, but except for separation into size ranges, very little has been done to determine the chemical composition of suspended loads of streams. The water user can justifiably say he is not interested in the composition of material he would filter out of the water before he used it in any event; hence for many practical water-use problems the practice of filtering a sample and analyzing only the filtrate can be justified.

Concentrations reported in analyses made by the U.S. Geological Survey represent amounts "in solution" at the time the sample was collected unless some special notation is used to identify constituents to which this does not apply. "In solution" is taken to mean material not removed by filtration. This procedure does not apply, obviously, to material present partly or wholly in particulate form for which concentration data might be obtained (for example biochemical oxygen demand).

For the geochemist or the chemist interested in processes in stream-water chemistry, the reactions among water, solutes, and suspended matter are significant. Kennedy (1965) found that the cation-exchange capacity of stream sediment can be large and that ions carried on exchange positions of suspended-sediment particles could be a significant part of the total ion load. On reaching the ocean, the adsorbed ions are probably exchanged for sodium. The adsorbed ions also may be exchanged for solute ions in the stream water if the composition of the water changes owing to inflows of fresh or saline water.

Certain minor constituents may be held very strongly by metal-hydroxide particles, rock minerals, or living or dead organic matter, all of which could be present in river water as suspended particles.

Since so little is known of the significance of adsorption reactions on suspended sediment, the subject cannot be usefully considered further here. The behavior of many of the minor constituents in river water, however, cannot be fully explained without consideration of the sediment fraction.

When samples are filtered before analysis, the common practice is to use a plastic membrane filter with pores 0.45 micron in diameter. Particles nearly 100 times as large as the lower limit of colloidal particles size would pass through this filter. It is not valid, therefore, to assume that only dissolved species are present in filtered samples. Certain metals tend to form oxide or hydroxide particles with a diameter 0.10 micron and may occur in this form in natural water. On the other hand, if some particulate material large enough to be caught by the filter is present, the pores may effectively be blocked to passage of particles considerably smaller than the nominal pore diameter. Filtration, therefore, provides a rather inexact means of separating suspended from dissolved constituents, but nothing better seems available.

HYDROGEN-ION ACTIVITY (pH)

The effective concentration (activity) of hydrogen ions could be expressed in the same kinds of units as other dissolved species, but H^+ concentrations in milligrams per liter or moles per liter are very low for water solutions that are not strongly acid. The activity of hydrogen ions can be most conveniently expressed in logarithmic units, and the abbreviation "pH" represents the negative base-10 log of the hydrogen-ion activity in moles per liter.

Theoretical concepts and various practical aspects of pH determination have been discussed at length in the literature. Bates (1964) has prepared an excellent and authoritative summary of these topics. The notation "pH" is now generally taken to mean hydrogen-ion activity rather than concentration, although the distinction between these concepts was not understood at the time Sorensen proposed the use of the pH notation in 1909. Throughout this discussion the term "hydrogen ion" is used with the reservation that such species exist in hydrated form in aqueous solution.

Pure water is itself dissociated to a slight extent into the species H^+ and OH^-. This dissociation may be written

$$H_2O(l) = H^+ + OH^-.$$

In mass law form, the equilibrium can be expressed as the equation

$$\frac{[H^+][OH^-]}{[H_2O]} = K_w.$$

By convention, the activity of the liquid water is taken to be unity in this very dilute solution, and the constant K_w is then equal to the product of the activities of H^+ and OH^-. This ion-activity product for water at 25°C is in exponential terms $10^{-14.000}$ (Covington and others, 1966). The two-place log of K_w is -14.00. At neutrality, by definition $[H^+] = [OH^-]$ and therefore pH = 7.00. At higher temperatures K_w increases, and the neutral value for pH becomes smaller; the value 30°C given by Covington, Robinson, and Bates (1966) is $10^{-13.837}$. Neutral pH at 30°C therefore would be 6.92.

The hydrogen-ion content of a natural water is computed in moles per liter (or milligrams per liter, which for H^+ is nearly the same as millimoles per liter) and is usually in the trace constituent range. At pH 7 only 1×10^{-7} moles per liter of hydrogen ion is present, for example. The major constituents of most waters are in the concentration range from 10^{-4} moles per liter and up. Thus the hydrogen-ion content does not begin to approach the status of a major component of the solution until the pH goes below 4.0.

The dissociation equilibrium for water is always applicable to any aqueous solution, but many other equilibria and many nonequlibrium reactions which occur in natural water among solute, solid and gaseous, or other liquid species also involve hydrogen ions. The pH of a natural water is related in one way or another to all these reactions. For example, the reaction of dissolved carbon dioxide in water, which is one of the most important in establishing pH, is represented by the three steps

$$CO_2(aq) + H_2O(l) = H_2CO_3(aq),$$

$$H_2CO_3(aq) = H^+ + HCO_3^-,$$

and

$$HCO_3^- = H^+ + CO_3^{-2}.$$

Both the second and third steps produce H^+ and influence the pH of the solution. Other reactions involving dissociations of acidic solutes include

$$H_2PO_4^- = HPO_4^{-2} + H^+,$$

$$H_2S(aq) = HS^- + H^+,$$

and

$$HSO_4^- = SO_4^{-2} + H^+.$$

Many of the reactions between water and solid species use up H^+, for example,
$$CaCO_3(c) + H^+ = Ca^{+2} + HCO_3^-$$
and
$$\underset{\text{(albite)}}{2NaAlSi_3O_8(c)} + 2H^+ + 9H_2O(l) = \underset{\text{(kaolinite)}}{Al_2Si_2O_5(OH)_4(c)} + 4H_4SiO_4(aq) + 2Na^+.$$

The second of these involves alteration of water, and the first could also be written as
$$CaCO_3(c) + H_2O(l) = Ca^{+2} + HCO_3^- + OH^-.$$

These are commonly called hydrolysis reactions, and all such reactions influence or are influenced by pH.

Certain reactions in which precipitates are formed such as
$$Fe^{+3} + 3H_2O = Fe(OH)_3(c) + 3H^+$$
influence pH, and most oxidation reactions are pH related,
$$FeS_2(c) + 8H_2O = Fe^{+2} + 2SO_4^{-2} + 16H^+ + 14e^-.$$

This reaction requires something to take up the electrons it produces—that is something must be reduced as the sulfur is oxidized. The usual oxidizing agent probably is atmospheric oxygen. The rest of the complete equation then would be
$$7/2\ O_2 + 14e^- + 14H^+ = 7H_2O.$$

Adding these together gives
$$FeS_2 + 7/2\ O_2 + H_2O = Fe^{+2} + 2SO_4^{-2} + 2H^+.$$

If a particular reaction or set of reactions involving water, hydrogen ions, other solutes, and solids or gases has attained equilibrium, the pH that should be attained can be computed by the mass law and a set of simultaneous equations. The equilibrium constants or standard free energies of the reacting species participating also must be available and their concentrations determined by analysis. The degree to which the measured pH agrees with the computed value may serve as indication of the existence of a particular chemical equilibrium.

The pH of a natural water under equilibrium conditions is effectively controlled by a single dominant chemical equilibrium or set of interrelated equilibria. The controlling species are generally those present in the system or available to it in the largest quantity or those whose reaction rate is fastest. Reactants which are present in lesser quantity or which react more slowly assume a subordinate role, with the pH of the system controlling their behavior.

To illustrate by a simplified example, when pure water is in contact with a constant supply of gas containing CO_2 such as the atmosphere, carbon dioxide will be dissolved up to a specific solubility limit depending on temperature and pressure. At 25°C and 1 atmosphere of pressure the chemical equilibria involved are those for the dissociation of dissolved carbon dioxide and of water that already have been given. Mass-law equations for these are

$$\frac{[H_2CO_3]}{pCO_2} = K_s = 10^{-1.43},$$

$$\frac{[HCO_3^-][H^+]}{[H_2CO_3]} = K_1 = 10^{-6.35},$$

$$\frac{[H^+][CO_3^{-2}]}{[HCO_3^-]} = K_2 = 10^{-10.33},$$

and

$$[H^+][OH^-] = K_w = 10^{-14.00}.$$

If no other solutes are present, the total concentrations of cations must balance those of anions, and we can also write

$$[H^+] = [HCO_3^-] + 2[CO_3^{-2}] + [OH^-].$$

Because this will be a very dilute solution, ion activities and actual concentrations are the same within measurable limits. The quantity pCO_2 is the partial pressure of carbon dioxide in the gas phase—that is, the volume percent of CO_2 multiplied by the total pressure in atmospheres. The bracketed quantities are activities of solute species, in moles per liter.

There are five equations governing this system and a total of six variables—the pCO_2 and five solute activities. Therefore, if one variable is specified, all the others will be fixed. The average value of pCO_2 for the atmosphere is $10^{-3.53}$ and is nearly constant; therefore, pure water in contact with air having the average CO_2 content will have a fixed pH and content of the other ions. The pH that can be calculated for these conditions is 5.65.

Although natural water cam be expected to contain other ions, rainwater might sometimes be expected to approach the conditions specified in this example, and a pH in the vicinity of 5.6 is frequently observed. The pH of ordinary laboratory distilled water often is near this value also.

If it is further specified that the system contains an excess of solid calcite, $CaCO_3c$, the above four equilibria still apply, along with a fifth representing the solution of calcite:

$$\frac{[Ca^{+2}][HCO_3^-]}{[CaCO_3][H^+]} = K_{eq}.$$

The activity of the solid is unity, by convention, and can be eliminated from the equation. One more solute species is added to the ion balance equation giving

$$2[Ca^{+2}] + [H^+] = [HCO_3^-] + 2[CO_3^{-2}] + [OH^-].$$

The system now has seven variables related by six equations, and again, if one is specified all the others will be fixed. For example, at equilibrium in this system, according to Garrels and Christ (1964, p. 83), if the gas phase is ordinary air the pH will be 8.4.

Garrels and Christ (1964, p. 83–91) gave calculations for other conditions involving carbonates. It should be evident from the examples how the pH enters into the equilibrium calculations. Furthermore, it should be apparent that this way of studying natural-water composition has many interesting possibilities. It is unsafe, however, to assume that pH in natural water is always at an equilibrium value.

The pH of solutions that are not at equilibrium may be related to the rates of reactions that are occurring and may tend to change considerably with the passage of time; many natural waters display such effects.

BUFFERED SOLUTIONS

A solution is said to be buffered if its pH is not greatly changed by the addition of moderate quantities of acid or base. Buffering effects occur in systems where equilibria involving hydrogen ions occur, and the range of pH over which buffering is effective depends on the nature of the solute species. Most natural waters are buffered to some extent by reactions which involve dissolved carbon dioxide species. The most effective buffering action by these species is within the pH range of most natural waters, from near 6.0 to about 8.5.

Weber and Stumm (1963) have described the nature of buffering effects in natural water and defined the term "buffer capacity" as the amount of acid or base required to produce a unit change in pH.

In the dissociation of H_2CO_3,

$$H_2CO_3 = HCO_3^- + H^+,$$

the pH is finally a function of the ratio of $[HCO_3^-]$ to $[H_2CO_3]$:

$$\frac{[HCO_3^-]}{[H_2CO_3]} = \frac{K_1}{[H^+]}.$$

If strong acid is added to this system, the reaction will be driven to the left, but a considerable amount of hydrogen ion may be absorbed before the ratio changes enough to alter the pH noticeably. This buffering effect is strongest in solutions that contain considerable amounts of dissolved CO_2 species. For example, if the total activity of such species is 2×10^{-2} moles per liter, when their ratio is unity, each will be present in an activity of 1×10^{-2} moles per liter (610 mg/l HCO_3), and the pH will be 6.35. To produce enough change in the ratio to cause a noticeable change in pH (about 0.01 pH unit) would require changing at least 10^{-4} moles per liter of hydrogen ions from their positions in the HCO_3^- and H_2CO_3 species. The free hydrogen ions in the system, as indicated by pH, would be less than 10^{-6} moles per liter. Thus the quantity of reacted H^+ would be some 100 times as great as the activity of free H^+.

Other reactions within the system also are affected when H^+ is added or removed, and redistribution of the H^+ among all these other reactions adds to buffering effects. The observed pH of a natural water is in this sense a residual effect and a remainder after the requirements of the system are satisfied. The pH of some waters is held at levels considerably below the buffering range of dissolved carbon dioxide species by redox reactions, such as the oxidation of pyrite or sulfur in other forms.

RANGE OF pH OF NATURAL WATER

As stated previously, the pH of pure water at 25°C is 7.00. Most ground waters found in the United States have pH values ranging from around 6.0 to 8.5, but water having lower pH is not uncommon in thermal springs. Water with a pH much over 9.0 is unusual but by no means unknown. Values as high as 11.6 (Feth, Rogers, and Roberson, 1961) and 11.7 (Barnes and others, 1967) have been observed in springs.

River water in areas not influenced by pollution generally has a pH between about 6.5 and 8.5. Where photosynthesis by aquatic organisms takes up dissolved carbon dioxide during the daylight hours, a diurnal pH fluctuation may occur and the maximum pH value may sometimes reach as high as 9.0. Livingstone (1963, p. G9) gave an example of diurnal pH fluctuations in what was evidently a poorly buffered lake water in which the maximum pH exceeded 12.

Natural waters whose pH is below 4.5 contain a low proportion of dissociated to undissociated carbon dioxide species, and many such waters have other sources of hydrogen ions. Most commonly, the source is a reaction that involves oxidation of some form of sulfur.

Table 11 contains three analyses of waters whose pH is controlled by equilibria involving carbon dioxide species and solid calcium

TABLE 11.—*Analyses of waters whose pH is unstable*

[Analyses by U.S. Geological Survey. Date under sample number is date of collection. Source of data: unpub. data, U.S. Geol Survey files]

Constituent	(1) Jan. 5, 1967		(2) Jan. 5, 1967		(3) June 14, 1950	
	mg/l	meq/l	mg/l	meq/l	mg/l	meq/l
Silica (SiO$_2$)	22		22		19	
Iron (Fe)	.02		.00		.01	
Manganese (Mn)	.01					
Zinc (Zn)	.02					
Calcium (Ca)	93	4.64	22	1.10	264	13 17
Magnesium (Mg)	58	4.77	54	4.44	79	6.50
Sodium (Na)	15	.65	15	.65	513	22.31
Potassium (K)	1.2	.03	1.1	.03	23	.59
Strontium (Sr)	.50	.01				
Carbonate (CO$_3$)					0	
Bicarbonate (HCO$_3$)	556	9.02	297	4.87	964	15.80
Sulfate (SO$_4$)	50	1.04	55	1.15	147	3.06
Chloride (Cl)	16	.45	17	.48	815	22.99
Fluoride (F)	.3	.02			.2	.01
Nitrate (NO$_3$)	.1	.00			3.2	.05
Boron (B)	.02					
Dissolved solids:						
Calculated	530		332		2,340	
Residue at 180°C						
Hardness as CaCO$_3$	470		277		984	
Noncarbonate	19		34		194	
Specific conductance (micromhos per cm at 25°C)					3,940	
pH	7.54		8.40		6.5	

1. Spring on headwaters of Blackbird Creek, SE¼ sec. 6, T 6 S., R. 5 E., San Mateo County, Calif. Temp, 7°C. pH measured at time of collection Calcium determined on acidified sample.
2. Reanalysis of unacidified fraction of sample 1 after 6 months of storage in laboratory.
3. Spring discharging into Little Colorado River, 13 miles above mouth, Coconino County, Ariz. One of a group known as Blue Springs Water-bearing formation Redwall Limestone. Water contains CO$_2$ and deposits travertine. Temp, 20.6°C. Discharge of group about 200 cfs.

carbonate. Analysis 1 is for a spring whose water was rather highly charged with carbon dioxide. The pH measured in the field at the time of sampling was 7.54. The water precipitated calcium carbonate in the sample bottle, and a second analysis of the sample made 7 months after the first shows a major change in both calcium and bicarbonate concentrations, accompanied by an increase in pH of a full unit. Analysis 3 represents a spring issuing from limestone near the mouth of the Little Colorado River in Arizona. This water deposits travertine in the riverbed, and it is likely that a pH value considerably higher than the 6.5 reported in analysis 3 would be observed in the water where precipitation was nearing completion. The pH of 9.4 reported for analysis 1, table 18, may be associated with hydrolysis reactions involving silicates in a system containing little carbon dioxide. The pH of the sodium carbonate brine from Wyoming (analysis 2, table 18) probably is much higher than any of those tabulated, but it was not reported in the published analysis. The low pH value for analysis 3, table 18, results from the oxidation of sulfur species. Fumaroles in the vicinity yield sulfur dioxide.

The pH of a water sample can also be affected by oxidation of ferrous iron. The data below represent successive determinations of ferrous iron and pH on a sample of water collected from the overflow of the Seneca anthracite mine at Duryea, Pa., on July 16, 1963, (unpub. analyses by U.S. Geol. Survey):

Date (1963)	Time after sampling (days)	Fe^{+2} (mg/l)	pH
July 16	0	--------	4.98
July 30	14	135	3.98
Aug. 20	35	87	3.05
Sept. 24	70	41	2.81
Oct. 29	105	2.2	2.69

The sample contained a very high concentration of iron, and its pH decreased by more than 2 units as the ferrous iron was oxidized and precipitated as ferric hydroxide.

MEASUREMENT AND INTERPRETATION OF pH

The pH of natural water is ordinarily determined by measuring the potential between a glass and a reference electrode immersed in the solution. The potential must be measured with a sensitive electrometer or similar device that does not permit a significant flow of current. The design of pH meters has been greatly improved in recent years, and equipment now available will measure pH either in the field or laboratory to the nearest 0.01 pH unit with excellent stability and reproducibility. Because the pH is a logarithm, measurements to two decimal places may still be imprecise compared to the usual measurements of concentrations of the other solute species.

The literature of published water analyses contains large numbers of pH determinations whose significance can be misinterpreted. The pH of a water represents the interrelated result of a number of chemical equilibria. The equilibria in a ground-water system are altered when the water is taken into a well and pumped to the surface. A pH measurement taken at the moment of sampling may represent the original equilibrium conditions in the aquifer satisfactorily, but if the water is put into a sample bottle and the pH is not determined until the sample is taken out for analysis some days, weeks, or months later, the measured pH may have no relation to the original conditions. Besides gains or losses of carbon dioxide, the solution may be influenced by reactions such as oxidation of ferrous iron, and the laboratory pH can be a full unit different from the value at the time of sampling. A laboratory determination of pH can be considered as applicable only to the solution in the sample bottle at the time the determination is made.

Accurate measurement of pH in the field should be standard practice for all ground-water samples. Methods and precautions for measurements required to obtain good results are given by Barnes (1964).

SPECIFIC ELECTRICAL CONDUCTANCE

Electrical conductance, or conductivity, is the ability of a substance to conduct an electric current. Specific electrical conductance is the conductance of a body of unit length and unit cross section at a specified temperature. This term is synonymous with volume conductivity and is the reciprocal of volume resistivity (Weast, 1968). The American Society for Testing and Materials (1964, p, 383) has defined electrical conductivity of water as "the reciprocal of the resistance in ohms measured between opposite faces of a centimeter cube of an aqueous solution at a specified temperature." This definition further specifies that units for reporting conductivity shall be "micromhos per centimeter at $t°C$." Because the definition already specifies the dimensions of the cube to which the measurement applies, the added precaution of including the length in the unit may not be essential and is often omitted in practice. Geophysical measurements of resistivity, however, commonly are expressed in ohm-meters, referring to a cube 1 meter on a side, so it may be well to emphasize that conductances of water refer to a centimeter cube. The standard temperature for laboratory measurements is 25°C, but some values taken at other temperatures exist; thus it is important that the temperature be specified.

UNITS FOR REPORTING CONDUCTANCE

Because conductance is the reciprocal of resistance, the units in which specific conductance is reported are reciprocal ohms, or mhos. Natural waters have specific conductances much less than 1 mho, and to avoid inconvenient decimals, data are reported in micromhos—that is, the observed value in mhos is multiplied by 10^6. Before October 1, 1947, the specific conductance values reported by the U.S. Geological Survey were mhos $\times 10^5$. To convert these older values to micromhos, they should be multiplied by 10.

PHYSICAL BASIS OF CONDUCTANCE

Pure liquid water has a very low electrical conductance: a few hundredths of a micromho per centimeter at 25°C. This value has only theoretical significance as water this pure is very difficult to produce. The presence of charged ionic species in solution makes the solution conductive. As ion concentrations increase, conductance of the solution increases; therefore, the conductance measurement provides an indication of ion concentration.

The relationship between ionic concentration and specific conductance is fairly simple and direct in dilute solutions of single salts. Figure 8 is a graph of the specific conductance at 25°C of solutions

SIGNIFICANCE OF WATER PROPERTIES AND CONSTITUENTS

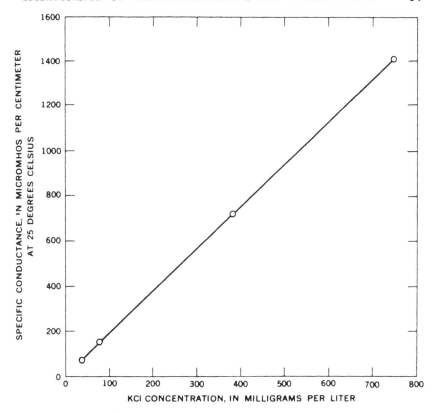

FIGURE 8.—Specific conductance of potassium chloride solutions.

of potassium chloride whose concentrations are as much as about 0.01 molar (746 mg/l). The relationship is a straight line over this range of concentration. As the concentration is increased, however, the slope decreases slightly; so for a concentration of 7,460 mg/l the conductance is 12,880 micromhos per centimeter rather than near 14,000 as an extrapolation of the slope in figure 9 would predict. This general behavior is typical of all salts, but the slope of the straight portion of the curve and the degree to which it flattens with increasing concentration is different for different salts.

Figure 9 shows the change in conductance of a solution containing 746 mg/l of KCl between 0° and 35°C. Over this temperature range the conductance of the solution more than doubles. This demonstrates the need for referring specific-conductance measurements to a definite temperature. Most values for natural waters are referred to 25°C (77°F). The response of the conductance value to temperature change is somewhat different for different salts and different concentrations, but in dilute solutions for most ions an increase of 1°C increases conductance by about 2 percent.

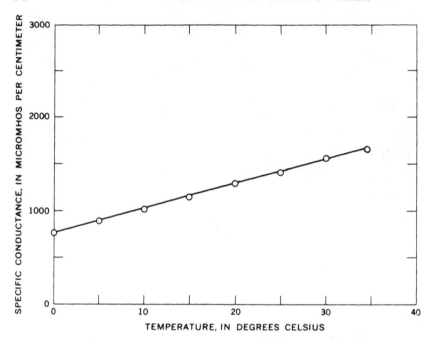

FIGURE 9.—Specific electrical conductance of a 0.01 molar solution of potassium chloride at various temperatures.

To conduct a current, solute ions actually must move through the solution to transfer charges, and the effectiveness of a particular ion in this process depends upon its charge, size, the way it interacts with the solvent, and other factors. The property called ionic mobility represents the velocity of an ion in a potential gradient of 1 volt per centimeter. (It should be noted that this is not the same property as the geochemical mobility of an element, a concept relating to the ease of transport of elements in geochemical cycles.) Ionic mobility is decreased by increasing concentration owing to interferences and interactions among the ions. Limiting values attained in solutions of very low concentration for the major ions in natural water are in the general vicinity of 6×10^{-4} centimeter per second at 25°C (Glasstone and Lewis, 1960, p. 445).

Very accurate and precise measurements of conductance have been used in research studies of the charge and composition of complex ions and for other purposes in solutions that did not contain too many ionic species.

It is apparent that even in relatively simple solutions the relationships that affect conductance may be complicated. More complete discussions of theory and application of conductance such as those in textbooks of physical chemistry (Glasstone and Lewis, 1960, p 415–

451) further emphasize this fact. Natural waters are not simple solutions. They contain a variety of both ionic and undissociated species, and the amounts and proportions of each may range widely. When applied to natural water, therefore, the conductance determination cannot be expected to be simply related to ion concentrations or to total dissolved solids, and a rigorous theoretical development of the meaning of conductivity values is rarely justifiable. The determination is easily made, however, and gives results which are very useful in a practical way as general indications of dissolved-solids concentration or as a base for extrapolating other analytical data when the general characteristics of the solution already are known from previous work.

Figure 10 is a plot of the dissolved solids contained in composite samples of water from the Gila River at Bylas, Ariz., for a year (U.S. Geological Survey, 1947) against the specific conductance of the samples. A reasonably well-defined relationship is indicated for the range; so for any given conductance value a dissolved-solids value can be estimated with an uncertainty of only about ± 100 mg/l.

The regression line has some curvature; hence the straight-line formula

$$KA = S,$$

where K is conductance in micromhos, S is dissolved solids in milligrams per liter, and A the conversion factor, does not fit exactly. This kind of formula is often used, however, in calculating approximate dissolved-solids values from conductance determinations. For the analyses of natural waters given in this report, the range of A is about from 0.54 to 0.96, which represents nearly the full range to be expected. Usually A is between 0.55 and 0.75, the higher values generally being associated with waters high in sulfate concentration.

Figure 11 shows the relationship of specific conductance to hardness and ion concentrations for the same set of chemical analyses used in preparing figure 10. Rather well-defined relationships again exist for chloride and sulfate, and almost as good a relationship for hardness (calcium+magnesium) is indicated. The bicarbonate content of these solutions (not shown in fig. 11) is less closely related to conductance, however.

The data used in figures 10 and 11 were obtained from a river which has a rather saline base flow maintained by irrigation drainage and ground-water inflows. The chemical characteristics of the base flow are relatively constant and are subject to dilution by runoff. It seems evident that a record of conductivity at this station could be used to compute the other chemical characteristics of the water with a good

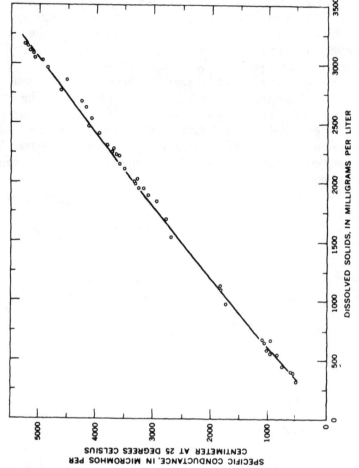

FIGURE 10.—Dissolved solids and specific conductance of composites of daily samples, Gila River at Bylas, Ariz., October 1, 1943, to September 30, 1944.

SIGNIFICANCE OF WATER PROPERTIES AND CONSTITUENTS

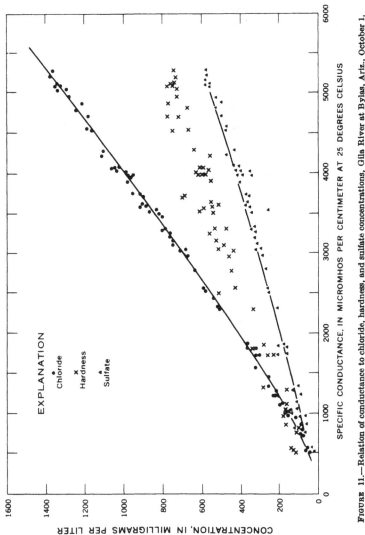

FIGURE 11.—Relation of conductance to chloride, hardness, and sulfate concentrations, Gila River at Bylas, Ariz., October 1, 1943, to September 30, 1944.

level of accuracy for major ions, except at high flow when the relationships would not be as well defined.

Where a satisfactory set of relationships between conductance and ion concentrations can be developed from a few years of intensive sampling, a record of major-ion concentrations, suitable for most uses, could be obtained by measuring conductivity continuously and feeding the output into a computer which would be set up to convert the data to concentrations, averaged for whatever periods might be desired. The sampling and analysis could be directed toward determination of constituents which did not correlate with conductance. The extent to which such uses of conductance data can be made is, of course, dependent on ascertaining relationships like those in figures 10 and 11, however, this cannot be done for all streams or all dissolved species of interest.

Conductance determinations are also useful in extrapolating ground-water analyses in areas where comprehensive analyses are available for a part of the sampled points. Conductivity probes are of value in exploring wells to determine differences in quality with depth.

RANGE OF CONDUCTANCE VALUES

The specific conductance of the purest water that can be made would approach 0.05 micromho per centimeter, but ordinary single-distilled water or water passed through a deionizing exchange unit will normally have a conductance of at least 1.0 micromho per centimeter. Carbon dioxide from the air in the laboratory dissolves in distilled water that is open to the air, and the resulting bicarbonate and hydrogen ions impart most of the observed conductivity.

Rainwater has ample opportunity before touching the earth to dissolve gases from the air and also may dissolve particles of dust or other airborne material. As a result, rain may have a conductance much higher than distilled water, especially near the ocean or near sources of atmospheric pollution. Feth, Rogers, and Roberson (1964) reported conductivities of melted snow in the western United States ranging from about 2 to 42 micromhos per centimeter. Whitehead and and Feth (1964) observed values greater than 100 in serveral rainstorms at Menlo Park, Calif.

The conductance of surface and ground waters has a wide range, of course, and in some areas may be as low as 50 micromhos, where precipitation is low in solutes and rocks are resistant to attack. In other areas, conductances of 50,000 or more may be reached; this is the approximate condutance of sea water. When concentrations of ions become so high, the relation between concentration and conductance becomes ill defined.

ACCURACY AND REPRODUCIBILITY

The equipment which is usually used to obtain conductance values of water, if carefully operated, may produce accuracy and precision from ± 2 to ± 5 percent.

The accuracy of field measurements of conductivity with good equipment compensated for temperature effects should be equivalent to laboratory measurements. Permanently installed conductivity measuring devices, however, require periodic maintenance to prevent electrode fouling and (or) other interferences which may cause erroneous readings and loss of record.

SILICA

The element silicon, as noted earlier, is second only to oxygen in abundance in the earth's crust. The chemical bond between silicon and oxygen is very strong, and the Si^{+4} ion is the right size to fit closely within the space at the center of a group of four closely packed oxygen ions. Silicon thus located is said to be tetrahedrally coordinated with respect to oxygen. The same structure occurs also with hydroxide ions, which are nearly the same size as the O^{-2} ion. The SiO_4^{-4} tetrahedron is a fundamental building unit of most of the minerals making up igneous and metamorphic rocks and is present in some form in most other rocks and soils, as well as in natural water. The term "silica," meaning the oxide SiO_2, is widely used in referring to silicon in natural water, but it should be understood that the actual form is hydrated and is generally represented as H_4SiO_4 or $Si(OH)_4$.

The structure and composition of silicate minerals cannot be considered in detail here, but some knowledge of the subject is useful in understanding the behavior of silicon in natural water. There are six principal patterns in which the SiO_4 tetrahedra are joined to build up the framework of silicate minerals. The kind of pattern which occurs is a function of the relative abundance of oxygen in the rock as compared to the abundance of silicon. In systems where oxygen is abundant relative to silica, the predominant pattern is one in which adjacent tetrahedra are joined by sharing a divalent cation such as magnesium, for example, in the magnesian olivine, forsterite (Mg_2SiO_4). This pattern extends in three dimensions, and silicates of this type are called nesosilicates. The second structural pattern is made up of pairs of tetrahedra sharing one oxygen between them, the sorosilicates. Few natural minerals have this structural pattern. A third pattern consists of rings in which three or more tetrahedra each share two oxygens. A six-membered ring structure occurs in the mineral beryl ($Be_3Al_2Si_6O_{18}$). The silicates having isolated rings in their structure are called cyclosilicates.

A structural pattern in which each tetrahedron shares two oxygens with neighbors can produce a long single chain as in the pyroxenes ($R_2(SiO_3)_2$). If two adjacent chains are cross linked by sharing some additional oxygens, the structure of the amphiboles is formed. For those amphiboles in which no aluminum is substituted for silicon and no univalent cations are present, the composition may be expressed as $R_7(OH)_2Si_8O_{22}$. The symbol R in these formulae represents various divalent cations; most commonly these are Ca^{+2}, Mg^{+2}, and Fe^{+2}. The chain silicates are called inosilicates.

The tetrahedra also may form planar structures in which three oxygens of each tetrahedron are shared with adjacent tetrahedra, all lying in a plane. The resulting sheet structure appears in such mineral species as micas and clays, for example, kaolinite ($Al_2Si_2O_5(OH)_4$). The sheet structures are called phyllosilicates The tectosilicates are made up of tetrahedra in which all oxygens are shared among adjacent tetrahedra in three dimensions. The structure of quartz $(SiO_2)_n$ follows this pattern. Feldspars also partially display it, but contain some aluminum and other cations in place of silicon, as in potassium feldspar ($KAlSi_3O_8$).

The structural features of silicate minerals are described in greater detail by Hückel (1950, p. 740–755) and in many more recent texts on inorganic chemistry and mineralogy. The structural pattern is rather closely related to the stability of the various mineral species toward attack by water. The strength of the silicon-oxygen bond is greater than that for the metal-oxygen bonds which occur in the silicate minerals. Thus, the resistance to chemical attack, which involves breaking the bonds holding the structure together, is greatest in those mineral structures where a larger proportion of the bonds are between silicon and oxygen.

The nesosilicates and inosilicates represent structures in which a relatively high proportion of the bonding is the linking of metal cations to oxygen. These bonds represent zones of weakness which can be disrupted relatively easily as compared to silicon-oxygen or aluminum-oxygen bonds. The ferromagnesian minerals, which belong largely to these two classes of silicate structures, are less resistant to weathering attack than structures like the tectosilicates where silicon-oxygen bonding predominates to a greater degree.

Crystalline SiO_2 as quartz is a major constituent of many igneous rocks and also constitutes the bulk of the grains in most sandstones. Quartz is one of the most resistant of all rock minerals to attack by water. The cryptocrystalline and amorphous forms of silica such as chert and opal are more soluble. It seems probable, however, that most of the dissolved silica observed in natural water results originally from the chemical breakdown of silicate minerals in processes of weathering. The amounts of silica which can be retained in solution are probably influenced by side reactions among the substances released when the silicate structures are destroyed.

The conditions of temperature and pressure under which most of the silicate minerals are formed are very different from the comditions of normal weathering. The weathering reactions must, therefore, for the most part be treated as irreversible processes. Certain reactions among the products of the initial breakdown of silicates can be reversed and may therefore, be examined by equilibrium and mass-law calculations. The irreversible steps, however, may be more suitably studied by consideration of rates and mechanisms of reaction.

FORMS OF DISSOLVED SILICA

The practice of representing dissolved silica as the oxide SiO_2 has been followed by all water analysts. Some of the older literature implied that the material was present in colloidal form, but occasionally a charged ionic species, usually SiO_3^{-2}, was specified. In most waters, it is very evident that the dissolved silicon does not behave like a charged ion. Nor does it have behavior typical of a colloid in most waters.

The dissociation of silicic acid, which for this discussion may be assumed to have the formula $Si(OH)_4$ or H_4SiO_4, begins with the reaction

$$H_4SiO_4(aq) = H^+ + H_3SiO_4^-.$$

Values collected by Sillén and Martell (1964, p. 145) for the equilibrium constant for this reaction at 25°C range from $10^{-9.41}$ to $10^{-9.91}$. These values indicate that the first dissociation step is half completed at a pH value somewhere between 9.41 and 9.91, and silicate ions ($H_3SiO_4^-$) might constitute no more than 10 percent of the total dissolved silica species at a pH between 8.41 and 8.91. Any silicate ions which might be present are converted to silicic acid in the alkalinity titration and appear in the analysis as an equivalent amount of carbonate or bicarbonate. As a result, the ionic balance is maintained in the analysis, and there is no obvious indication that the ionic species are incorrectly identified.

Analysis 1, table 12, is for a water which has a pH of 9.2. By using the dissociation constant for silicic acid determined by Greenberg and Price (1957) of $10^{-9.71}$, it can be calculated that about 0.389 milliequivalents per liter of dissociated silicate ions are present. The analysis reports 0.800 milliequivalents per liter of carbonate alkalinity, of which almost half actually must have been caused by silicate.

TABLE 12.—*Analysis of waters containing high proportions of silica*

[Analyses by U.S. Geol. Survey. Date below sample number is date of collection. Source of data: 1, 2, 4, 5, and 9, unpub. data, U.S. Geol. Survey files; 3, White, Hem, and Waring (1963, p. F40); 6, U.S. Geol. Survey Water-Supply Paper 1022 (p. 266); 7, Lohr and Love (1954b) (p. 286); 8, U.S. Geol. Survey Water-Supply Paper 1102. (p. 400)]

Constituent	(1) Nov. 23, 1953		(2) Aug. 1, 1947		(3) Oct. 16, 1957		(4) June 13, 1952		(5) Apr. 18, 1952		(6) Apr. 11-19, 1944		(7) Oct. 12, 1951		(8) May 6, 1947		(9) Mar. 22, 1952	
	mg/l	meq/l	mg/l	meq/l	mg/l	meq/l	mg/l	meq/l	mg/l	meq/l	mg/l	meq/l	mg/l	meq/l	mg/l	meq/l	mg/l	meq/l
Silica (SiO_2)	99		103		363		49		38		48		71		62		29	
Aluminum (Al)	.04				.2													
Iron (Fe)			.0		.06		.01		.24		.03		0		.01		.33	
Manganese (Mn)							.00		.00								.00	
Calcium (Ca)	2.4	0.120	6.5	0.324	.8	0.04	32	5.2	12	0.559	43	2.15	32	8.8	22	1.098	17	0.848
Magnesium (Mg)	1.4	.115	1.1	.090	.0	.00	12	0	6.6	.543	20	1.64	8.8		4.4	.362	1.7	.140
Sodium (Na)	100	4.348	40	1.729	352	15.30	30		7.2	.313	28	1.22	42		20	.870	7.4	.323
Potassium (K)	2.9	.074			24	.61	5.2		3.1	.079	4.6	.12	0		0	0	0	0
Carbonate (CO_3)	24	.800	0	0	(?)	3.86	0		0	0	0	0	0		0	0	69	1.131
Bicarbonate (HCO_3)	111	1.819	77	1.262	23	.48	220		85	1.393	254	4.15	161		104	1.704	6.9	.144
Sulfate (SO_4)	30	.625	15	.312	405	11.42	11		4.4	.092	8.3	.17	54		29	.604	1.1	.031
Chloride (Cl)	10	.282	17	.479	25	1.32	7.9		1.2	.034	28	.79	12		0	.000	1.0	.005
Fluoride (F)	22	1.158	1.6	.084	1.8	.03	7.2		1.3	.016	.5	.03	.7		.3	.016		
Nitrate (NO_3)	.5	.008	.4	.006	4.4		2.9	.08	.2	.003	.1	.00	.6		1.4	.023	0	.000
Boron (B)	.61		.0															
Dissolved solids:																		
Calculated	348		222		1,310		259		115		306		310		190		98	
Residue on evaporation							257		114						158		98	
Hardness as $CaCO_3$	12		20		2		129		57		190		116		73		49	
Noncarbonate	0		0		0		0		0		0		0		0		0	
Specific conductance (micromhos at 25° C)	449		167		1,790		358		136		463		404		198		130	
pH	9.2		6.7		9.6		7.8		7.9				7.9		8.1		7.1	
Color							10		8				2				5	

[1] Includes some silicate ion, probably $H_3SiO_4^-$, calculated as 0.389 meq/l.
[2] Original analysis reports carbonate and hydroxide alkalinity of 3.86 meq/l total; probably this is mostly attributable to silicate.

1. Flowing well 7S 6E-9ba2, Owyhee County, Idaho. Depth, 800 ft. Temp, 50.0° C.
2. Spring on Rio San Antonio, SW¼ sec. 7, T. 20 N., R. 4 E. (unsurveyed) Sandoval County, N. Mex. Temp. 38.3° C. Flow, about 25 gpm. Water from rhyolite.
3. Spring, 650 ft south of Three Sisters Springs in Upper Geyser Basin, Yellowstone National Park, Wyo. Temp. 94° C. Also reported 1.5 mg/l As, 5.2 mg/l Li, 1.5 mg/l Br, 0.3 mg/l I, 1.3 mg/l PO_4^{-3}, and 2.6 mg/l H_2S.
4. Drilled well, NW¼ sec. 10, T. 2 N., R. 32 E., Umatilla County, Oreg. Water from basalt of the Columbia River Group. Depth, 761 ft.
5. Flowing well, SE¼ sec. 16, T. 12 N., R. 17 E., Yakima County, Wash. Water from basalt. Depth, 1,078 ft. Temp, 17.2° C.
6. Eagle Creek at P-D pumping plant near Morenci, Greenlee County, Ariz. Mean discharge for composite period, 9.9 cfs. Drainage basin about 600 sq. mi. in which rocks exposed are virtually all extrusive volcanic rocks.
7. Main pump station, Albuquerque, N. Mex. Public supply. Seven wells 250-716 ft. deep. Water from Santa Fe Formation (valley fill).
8. Middle Loup River at Dunning, Nebr. Discharge, 394 cfs. Drainage from Nebraska sand hills, flow maintained by ground water.
9. Well at Valdese General Hospital, Rutherford College, Burke County, N.C. Depth, 400 ft. Water from mica schist. Temp, 15.0° C.

The method generally used for determining silica concentration in water requires that the silica form a colored molybdate complex in solution. The response of dissolved silica to this procedure is normally rapid and apparently complete, which suggests that silica must be present in units of very small size, probably approaching, if not actually having, the dimensions of single molecules. Some hot-spring waters are known, however, in which the total silica is several hundred milligrams per liter and in which the silica tends to polymerize on standing at 25°C to form colloidal particles. White, Brannock, and Murata (1956) observed that the polymerized silica thus formed reacted very slowly with complexing reagents. Analysis 3, table 12 represents water from a hot spring which precipitates silica on cooling. Polymerized silica is not determined accurately by the complexing procedure.

In reviewing the available literature on the dissolution of silica, Krauskopf (1956a) concluded that silica in natural water was mostly in the form of monomolecular silicic acid, H_4SiO_4aq. This is equivalent to a silicon ion tetrahedrally coordinated with four hydroxide ions, and by analogy with mineral structures it seems the most likely form. Some investigators quoted in Sillén and Martell (1964, p. 145) suggest polynuclear species such as $Si_4O_4(OH)_{12}^{-4}$, and Iler (1955, p. 19) proposed that in solution silica might be six coordinated with OH^- to form ions of the same type as fluosilicate, SiF_6^{-2}. In the absence of more definite evidence of the existence of more complicated species, the simplest form, $Si(OH)_4$, is probably the best postulate. This may also be considered equivalent to SiO_2 combined with two water molecules.

Data in Sillén and Martell (1964, 265) suggest that silica in acid water that contains fluoride would tend to form the fluosilicate complex ion, SiF_6^{-2}. This species is not likely to be significant above a pH of about 4.

SOLUBILITY CONTROLS

The solubility of quartz has been determined by Morey, Fournier, and Rowe (1962) as 6.0 mg/l at 25°C and 26 mg/l at 84°C (as SiO_2). Fournier and Rowe (1962) reported the solubility of cristobalite to be 27 mg/l at 25°C and 94 mg/l at 84°C. Morey, Fournier, and Rowe (1964) reported the solubility of amorphous silica to be 115 mg/l at 25°C, and Akabane and Kurosawa (1958) reported the solubility of this material at 100°C to be 370 mg/l.

These results show the strong effect of temperature on solubility of silica. It also should be noted that at 25°C the solid that is first to form is amorphous silica and that this material is not readily converted to the better crystallized forms of lower solubility. Morey, Fournier, and Rowe (1962) reported one experiment in which quartz grains were rotated rapidly in water for 386 days at 25°C during which time the SiO_2 concentration in the water was 80 mg/l. This concentration dropped suddenly to 6 mg/l and held that value for a subsequent period of about 5 months. It was assumed this was the reversible solubility for quartz.

Natural water normally has more dissolved silica than the quartz equilibrium value, but less than values shown for amorphous silica, which is the likely upper equilibrium limit. The general tendency for silica contents of natural water to fall within a relatively narrow range suggests some other type of solubility control may exist, but it has not yet been determined what this might be.

REACTION MECHANISMS

The dissolution of a silicate can be typified by the reaction of the sodium feldspar albite with water containing H^+, forming kaolinite and soluble products:

$$2NaAlSi_3O_8(c) + 2H^+ + 9H_2O(l) = Al_2Si_2O_5(OH)_4(c) + 4H_4SiO_4(aq) + 2Na^+.$$

Although this reaction is shown as only a single step, it actually must be a rather complicated process. The kaolinite produced by this reaction may dissolve reversibly under some conditions (Polzer and Hem, 1965):

$$Al_2Si_2O_5(OH)_4 + 6H^+ = 2Al^{+3} + 2H_4SiO_4 + H_2O;$$

dissolved silica may be precipitated if high concentrations are reached:

$$H_4SiO_4(aq) = SiO_2(c) + 2H_2O.$$

Both reactions may exert some control over silica concentrations in water. An extensive review of the state of knowledge of silica solubility was published by Siever (1962).

The kinetics of dissolution of feldspar have not been studied extensively, although Wollast (1967) obtained some data on potassium feldspar. Van Lier, de Bruyn, and Overbeek (1960) reported that the rate of solution of quartz was controlled by the surface area exposed to the solution and gave, as the zero order rate constant at 25°C, the value 10^{-18} gram centimeter^{-2} second^{-1}.

OCCURRENCE IN NATURAL WATER

Analyses in table 12 were selected to show several types of water which have rather high concentrations of dissolved silica. Some of these show the increased solubility attained as water temperatures rise.

Sea water near the surface is very low in silica (often less than 1 mg/l) apparently because marine organisms extract and utilize silica in their shells and skeletons. Similar depletion effects are noticeable in the surface layers of some lakes and reservoirs.

The range of concentrations of silica most commonly observed in natural water is from 1 to around 30 mg/l. Concentrations up to 100 mg/l, however, are not infrequent in some areas. Davis (1964) quoted a median value of silica for surface waters of 14 mg/l and for ground water of 17 mg/l. The higher concentrations generally are found in ground waters and are related to rock type and to temperature of the water. A general review of silica occurrence in natural water has been published by Ginzburg and Kabanova (1960).

The silica concentration of water samples collected and stored in some types of glass bottles may increase owing to solution of the glass. Bottles made of Pyrex and other resistant formulations of glass probably are not a serious source of contamination for most natural waters, however. Polyethylene and polypropylene bottles are superior to glass in this respect.

ALUMINUM

Although aluminum is the third most abundant of the elements in the earth's outer crust, it rarely occurs in natural water in concentrations greater than a few tenths of a milligram per liter. The exceptions are mostly waters of very low pH. The chemical behavior of aluminum in water is complicated and still not fully understood. Recent research by the U.S. Geological Survey and other organizations, however, has established some important facts regarding species likely to occur in solution and the most probable solubility controls. Some of these findings are briefly noted here.

Because aluminum is so abundant and widely distributed, most natural waters have ample opportunity to dissolve it. The low concentrations common to water at near-neutral pH must, therefore, be a result of the chemistry of the element. This behavior is also indicated by the short oceanic residence time (table 7).

SOURCES OF ALUMINUM IN WATER

Aluminum occurs in many silicate rock minerals such as the feldspars, feldspathoids, micas, and many amphiboles. The aluminum ion is small enough to fit approximately into fourfold coordination with oxygen and therefore can substitute, in a sense, for silicon. It also is commonly six coordinated, in which case it occupies crystal sites similar to those occupied by magnesium and iron.

Aluminum hydroxide in the form of gibbsite is a fairly common mineral, and less common species are the fluoride of aluminum and sodium, cryolite, and the basic sulfate alunite. Aluminum also is present in zeolites. The most common of all the sedimentary aluminum-bearing minerals are the clays. The clay minerals have a layer structure in which aluminum octahedrally coordinated with six oxide or hydroxide ions forms one type of layer, and silicon tetrahedrally coordinated with oxygen forms a second type of layer. These layers alternate in various ways, forming the various clay structures. Diagrams of the structures of the common clay minerals have been given by many investigators, including Robinson (1962. p. 31-58). Clays are present in most natural-water environments.

SPECIES IN SOLUTION

The cation Al^{+3} predominates in many solutions where pH is less than 4.0. The actual ion is probably an octahedron of six water molecules with the aluminum ion at the center (fig. 12). One of these water molecules may become an OH^- ion if the pH is raised slightly, and at pH 4.5-6.5 a process of polymerization occurs which results in units of various sizes with the structural pattern of gibbsite (Hsu and Bates, 1964; Hem and Roberson, 1967). Under some conditions, the polymerized species may be small enough to qualify as ions, and various polynuclear aluminum hydroxide complexes are decsribed in the literature. In the solutions studied by Hem and Roberson, however, aging of these polynuclear ions resulted in crystalline particles a few hundredths to a few tenths of a micron in diameter. Above neutral pH, the predominant dissolved form of aluminum is the anion $Al(OH)_4^-$.

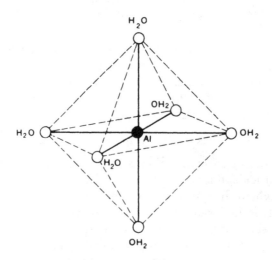

FIGURE 12.—Schematic diagram of hydrated aluminum ion, $Al(H_2O)_6^{+3}$ (From Hem 1968a, in "Trace Inorganics in Water," fig 1.)

The polymerization of aluminum hydroxide species proceeds in a different way in the presence of dissolved silica than when silica is absent. When sufficient silica is present, the aluminum appears to be rapidly precipitated as rather poorly crystallized clay-mineral species (Folzer and others, 1967). In the presence of fluoride, strong complexes of aluminum and fluoride are formed. The forms AlF^{+2} and AlF_2^+ appear to be most likely in natural water containing from a few tenths milligram per liter to a few milligrams per liter of fluoride (Hem, 1968). Soluble phosphate complexes of aluminum have been reported (Sillén and Martell, 1964, p. 186), and the sulfate complex $AlSO_4^+$ may predominate in acid solutions where much sulfate is present (Hem, 1968).

Obviously, the solubility of aluminum is strongly influenced by complexing, which means that the chemical factors that control aluminum concentrations are not easily defined.

SOLUBILITY CONTROLS

Not many of the reactions by which aluminum is dissolved or precipitated can be treated reliably by equilibrium methods. In solutions that contain fluoride, sulfate, or other complexing agents, the form of dissolved species can be determined from the ionic equilibria (Hem, 1968), and if silica concentrations are very low, the solubility product for aluminum hydroxide as bayerite gives a reasonable basis for calculating solubility of aluminum in alkaline solutions (Hem and Roberson, 1967). In solutions whose pH is below 4.0, the solubility can be reasonably calculated from the solubility product for gibbsite. Between the high and low pH regions, aluminum solubility reaches a minimum. The minimal solubility calculated by Roberson and Hem (1969) is a little less than 0.01 mg/l, near a pH of 6.0. This would represent a solution with a low dissolved-solids concentration containing less than 0.10 mg/l of fluoride. Higher fluoride concentrations increase the solubility of aluminum. Figure 13 is a graph of solubility of aluminum as a function of pH in a system where complexing species other than OH are absent. The polymeric forms of aluminum and hydroxide cannot be represented in a solubility diagram, as they are unstable and are converted to solid particulate matter on aging (Hem and Roberson, 1967). The dashed line labeled "microcrystalline gibbsite" represents the activity of uncomplexed aluminum, and the other two dashed lines represent the activity of $Al(OH)_4^-$. The solid line shows the effect of the monomeric complex $AlOH^{+2}$ on the solubility.

Although the influence of silica cannot be fully evaluated by equilibrium calculations, aluminum solubility is generally considerably decreased when silica is present owing to formation of clay-mineral

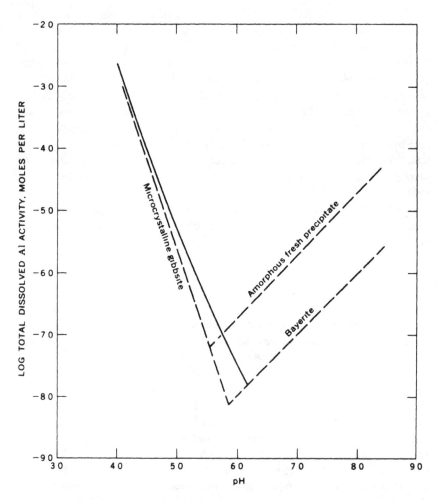

FIGURE 13.—Equilibrium activities of free aluminum for three forms of aluminum hydroxide (dashed lines) and calculated activity of $Al^{+3}+AlOH^{+2}$ (solid line).

species. The equilibrium activity of aluminum in uncomplexed form can be calculated for various silica concentrations and pH values by using equations given by Polzer and Hem (1965) and may be less than 0.001 mg/l in some natural waters. This concentration, however, is not equivalent to the total possible dissolved aluminum which could be much greater if complexing anions were present.

REACTION MECHANISMS AND THEIR IMPLICATIONS

The reaction shown previously for dissolution of albite was written as a single-step process. It is obvious, however, that the change in structure from the tectosilicate pattern of feldspar to the phyllosilicate structure of kaolinite requires extensive disruption and rearrangement of silicon-oxygen and aluminum-oxygen bonds. The process is generally slow, especially at pH levels controlled by dissolved carbon dioxide at concentrations available to most natural water. The aluminum and silicon released probably rearrange themselves fast enough to keep up with the rate of dissolution of the feldspar. Whether any polymerized aluminum hydroxide can be expected to be carried off in runoff water from areas where feldspars are being actively eroded remains to be learned. Many investigations of weathering processes, however, have assumed the aluminum is not removed.

The alteration of kaolinite to gibbsite by removal of silica also presents some problems in mechanisms and rates. The reaction could be shown as a desilicification:

$$Al_2Si_2O_5(OH)_4(c) + 5H_2O = 2Al(OH)_3(c) + 2H_4SiO_4(aq).$$

This reaction would require only splitting off silica layers from aluminum hydroxide layers and is a plausible single-step process. The aluminum hydroxide also would tend to be attacked, however, and some aluminum would be brought into solution, especially at low pH.

OCCURRENCE OF ALUMINUM IN WATER

Only a rather small amount of information is available regarding actual aluminum concentrations in water. Occasional reported concentrations of 1.0 mg/l or more in water with near neutral pH and no unusual concentrations of complexing ions probably represent particulate material. Whether this particulate material is aluminum hydroxide or an aluminosilicate is not presently known. The writer's work (Hem and Roberson, 1967) has shown, however, that gibbsite crystals near 0.10 micron in diameter have a considerable physical and chemical stability. Particles of this size will pass through most filter media and may need to be considered in water-quality evaluations. Carefully determined aluminum values for runoff from granitic terrane show only a few hundredths milligram per liter at most (Feth, Roberson, and Polzer, 1964). These determinations probably did not include particulate material, however.

The addition of alum in water treatment to flocculate suspended particles may leave a residue of aluminum in the finished water probably as small suspended hydroxide particles. For a study of quality of water supplies of major United States cities (Durfor and Becker, 1964), spectrographic analyses were made for various constituents including aluminum. The aluminum contents generally were higher in finished water than in raw water where alum was used as a flocculating agent.

Water with a pH below 4.0 may contain several hundred or even several thousand milligrams per liter of aluminum. Such water occurs in some springs and in drainage from mines. Table 13 contains analyses for some waters that are high in aluminum. Some of the commonly used analytical procedures are not sensitive to complexed or particulate aluminum species, and published information on aluminum concentrations in natural water may not all be reliable.

IRON

The element iron is an abundant and widespread constituent of rocks and soils. Concentrations of only a few tenths of a milligram per liter of iron in a water can make it unsuitable for some uses. For this reason, a determination of iron is frequently included in water analyses, and the information on iron contents of water is voluminous, even though the amounts present in most waters are small.

Research on chemistry of iron in water in recent years has clearly established the general principles and has demonstrated that iron concentrations in water are responsive to chemical equilibria. These equilibria involve processes of oxidation and reduction; precipitation and solution of hydroxides, carbonates, and sulfides; complex formation especially with organic material; and the metabolism of plants and animals.

SOURCES OF IRON

In igneous rocks the principal minerals containing iron as an essential component include the pyroxenes, amphiboles, biotite, magnetite, and the nesosilicates such as olivine. The composition of olivine ranges from Mg_2SiO_4 to Fe_2SiO_4 (forsterite to fayalite) with ferrous iron substituting freely for magnesium. Most commonly, the iron in igneous rocks is in the ferrous form, but may be mixed with ferric iron as in magnetite (Fe_3O_4).

In sediments, iron occurs in the ferrous form in such species as the polysulfides pyrite or marcasite (FeS_2), the carbonate siderite, and in the mixed oxide magnetite which also is an igneous or metamorphic mineral. Ferric iron is commonly mixed with ferrous in glauconite (greensand), and ferrous and ferric iron are widespread minor components of most sediments.

The ferric oxides and hydroxides are very important iron-bearing minerals. Hematite, Fe_2O_3, is a dehydrated form. The hydrous species $Fe(OH)_3$ is generally without a well-defined crystal structure and may contain less than the indicated amount of water. The hydrous ferric oxide content of rocks and soils is commonly responsible for their red or yellow color.

SIGNIFICANCE OF WATER PROPERTIES AND CONSTITUENTS 115

TABLE 13.—Analyses of waters high in dissolved aluminum or manganese

[Analyses by U S Geological Survey Date under sample number is date of collection. Source of data: Numbers 1 and 2, Scott and Barker (1962 p 23; 111), 3, unpub. data, U S Geol Survey files, 4, U S. Geol Survey Water-Supply Paper 1948 (p 38), 5, White, Hem, and Waring (1963, p. F26)]

Constituent	(1) Dec. 13, 1955		(2) Dec. 3, 1955		(3) Aug 31, 1958		(4) Mar. 1-6, 8-10, 1963		(5) Mar. 25, 1953	
	mg/l	meq/l	mg/l	meq/l	mg/l	meq/l	mg/l	meq/l	mg/l	meq/l
Silica (SiO$_2$)	98	------	10	------	92	------	9.7	------	31	------
Aluminum (Al)	28	------	.1	------	.35	------	3.5	------	.2	------
Iron (Fe)	.88	------	.04	------	.02	------	.10	------	2.7	------
Manganese (Mn)	9.6	------	1.3	------	.31	------	2.5	------	.22	------
Calcium (Ca)	424	21.16	58	2.89	67	3.34	32	1.60	28	1.40
Magnesium (Mg)	194	15.96	13	1.07	0	.00	11	.90	1.9	.16
Sodium (Na)	416	18.10	23	1.00	477	20.75	12	.52	6.8	.30
Potassium (K)	11	.28	2.8	.07	40	1.02	3.7	.09	4.2	.11
Hydrogen (H)	------	.10	------	------	------	------	------	.16	------	------
Carbonate (CO$_3$)	0	------	0	------	0	.00	0	------	------	------
Bicarbonate (HCO$_3$)	0	------	101	1.66	1,020	16.72	0	------	121	1.98
Sulfate (SO$_4$)	2,420	50.38	116	2.42	169	3.52	171	3.56	14	.03
Chloride (Cl)	380	10.72	39	1.10	206	5.81	5.0	.14	10	.03
Fluoride (F)	1.8	.09	0	------	6.8	.36	.1	.01	1	.01
Nitrate (NO$_3$)	3.1	.05	.6	.01	1.8	.03	5.3	.09	.2	.00
Orthophosphate (PO$_4$)	.0	------	.1	------	.11	------	------	------	0	------
Boron (B)	------	------	------	------	2.8	------	------	------	------	------
Dissolved solids										
Calculated	3,990	------	314	------	1,570	------	256	------	137	------
Residue at 180°C	4,190	------	338	------	1,560	------	260	------	------	------
Hardness as CaCO$_3$	1,860	------	198	------	168	------	125	------	78	------
Noncarbonate	1,860	------	115	------	0	------	125	------	0	------
Specific conductance (micromhos at 25°C)	4,570	------	517	------	2,430	------	507	------	192	------
pH	4.0	------	7.0	------	6.7	------	3.8	------	6.9	------

1 Well, 7 miles northeast of Montecello, Drew County, Ark. Depth, 22 ft. Water-bearing formation, shale, sand, and marl of the Jackson Group. Also contained radium (Ra), 1.7 μμc/l and uranium (U), 17 μg/l
2 Composite from two radial collector wells at Parkersburg, Kanawha County, W. Va Depth, 52 ft. Water from sand and gravel. Also contained copper (Cu), 0.01 mg/l and zinc (Zn), 0.01 mg/l.
3 Wagon Wheel Gap hot spring, Mineral County, Colo. Discharge, 20 gpm. Temp. 62.2°C. Associated with vein of Wagon Wheel Gap fluorite mine. Also contained 2.3 mg/l Li, 0.9 mg/l NH$_4$, 0.3 mg/l Br, and 0.3 mg/l I.
4. Kiskiminitas River at Leechburg (Vandergrift), Pa. Composite of nine daily samples. Mean discharge for period, 10,880 cfs
5. Well, 167 ft. deep, Baltimore County, Md Water-bearing formation, Port Deposit granitic gneiss Also contained 0.01 mg/l copper (Cu).

The solution of iron from silicate minerals is a slow process normally, but near-surface weathering of iron-bearing silicates may produce an accumulation of ferric oxide or hydroxide. The oxide and sulfide species of iron minerals are usually the principal sources from which the dissolved iron of ground water is derived.

Iron is an essential element in both plant and animal metabolism. Iron, therefore, is to be expected in organic wastes and in plant debris in soils, and the activities in the biosphere may have a strong influence on the occurrence of iron in water.

SPECIES OF IRON IN NATURAL WATER

The most common form of iron in solution in ground water is the ferrous ion Fe^{+2}. The complex $FeOH^+$ may occur in solutions very low in dissolved carbon dioxide species. An ion pair $FeSO_4(aq)$ reported in Sillén and Martell (1964, p. 240) could be important in solutions which have more than a few hundred milligrams per liter of sulfate. Ferrous complexes are formed by many organic molecules, but information concerning the exact species which might be expected in natural water is not available. Above a pH of 11, the anion $HFeO_2^-$ can exist in appreciable quantity, but this high pH is rarely attained in natural systems.

Ferric iron can occur in acid solutions as Fe^{+3}, $FeOH^{+2}$, $Fe(OH)_2^+$, and polymeric forms, the predominant form and concentration depending on the pH. Above a pH of 4.8, however, the solubility of ferric species is less than 0.01 mg/l (as Fe). The more concentrated solutions of ferric iron may contain partly polymerized forms of ferric hydroxide, and the species $Fe(OH)_3(aq)$ has been reported by Lamb and Jacques (1938). The solubility of this form, however, was reported by them to be less than 1 µg/l (microgram per liter) as Fe. Colloidal ferric hydroxide is commonly present in surface water, and small quantities may persist even in water that appears to be clear. Anionic ferric species such as FeO_4^{-2} are unlikely to be found in natural water because they require stronger oxidizing conditions and higher pH than are likely to occur. Ferric iron forms stronger complexes than ferrous iron with various organic substances and also can form complexes with chloride, fluoride, sulfate, and phosphate. The available evidence, however, seems to suggest that only the organic complexes have any practical significance in natural-water systems. The exact nature of the association of iron with organic matter remains uncertain also. Some investigators have suggested the most important action of the organic matter is reduction of ferric to ferrous iron. Others emphasize the peptizing action of organic matter on ferric hydroxide suspensions (W. L. Lamar written commun., 1967).

SIGNIFICANCE OF WATER PROPERTIES AND CONSTITUENTS

SOLUBILITY CONTROLS

Most of the reactions by which iron is dissolved in or precipitated from water are relatively rapid. They present an opportunity for the application of chemical equilibrium calculations to predict the concentrations of iron to be expected in solution. Calculations of this type by Hem and Cropper (1959) have been verified by field studies of ground water by Barnes and Back (1964b). The details of the calculations will not be considered here, as they are amply covered in other publications (Hem, 1961b). The equilibria involved include oxidation and reduction of dissolved-iron species and solution and precipitation of hydroxide, carbonate, and sulfide solids containing oxidized or reduced iron. They can be used to formulate equations based on the mass law and the Nernst equation.

The variables of principal importance which influence iron solubility include the pH and redox potential (Eh) and the dissolved carbon dioxide and sulfur species. If the dissolved carbon dioxide and sulfur species and dissolved iron are assigned fixed total activity values, a diagram in which pH is plotted as abscissa and Eh as ordinate can be prepared, and it will show the conditions where hydroxide, carbonate, or sulfide iron minerals will be stable at 25°C and 1 atmosphere of pressure. This stability field diagram, or Eh-pH diagram, is a useful device for summarizing the aqueous chemistry of many elements.

Figure 14 is an Eh-pH diagram for a system in which the total activity of iron is fixed at 10^{-7} molar, equivalent to 5.6 µg/l. The other conditions specified are a total activity of dissolved carbon dioxide species equivalent to 1,000 mg/l as HCO_3^- and a total activity of dissolved-sulfur species of 96 mg/l as SO_4^{-2}. The temperature is 25°C, and the pressure on the system is 1 atmosphere.

Stability regions for the solids are shaded, and the predominant dissolved forms are indicated in the various unshaded areas. The sloping upper and lower boundaries within the diagram are the limits within which water itself is chemically stable. Above the upper boundary water is oxidized to yield oxygen gas, and below the lower boundary reduced to yield hydrogen gas. The water-stability region is the only area within which solution equilibria can be maintained.

Besides indicating stabilities of solids, Eh-pH diagrams may be used to show the solubility of iron. The conditions chosen for the solubility diagram (fig. 15) are similar to the ones for the stability-field diagram, except that for figure 15 the total activity of dissolved carbon dioxide species is lower, equivalent to 61 mg/l as HCO_3^-. The contour lines on the diagram represent activities of dissolved iron in moles per liter from 10^{-7} to 10^{-3} (equivalent to 0.0056 to 56 mg/l of Fe). The lines are located by calculating the successive positions of the boundaries of

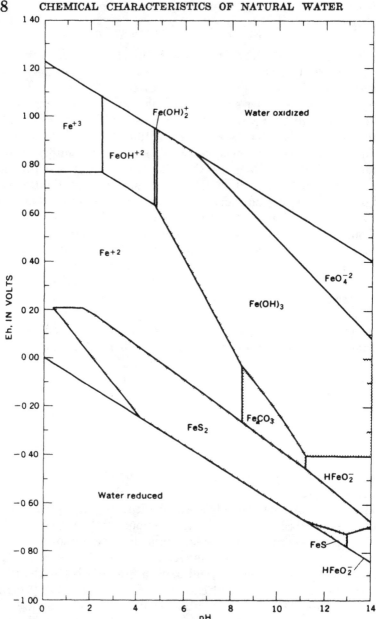

FIGURE 14.—Fields of stability for solid and dissolved forms of iron as function of Eh and pH at 25°C and 1 atmosphere of pressure. Activity of sulfur species 96 mg/l as SO_4^{-2}, carbon dioxide species 1,000 mg/l as HCO_3^-, and dissolved iron 0.0056 mg/l.

solid species attained as iron activity is changed by a factor 10. When ferrous iron activity is 10^{-7} molar at the activity of HCO_3^- specified, ferrous carbonate would not be formed, and the solubility boundaries at the lower edge of the ferric hydroxide field reflect this fact.

SIGNIFICANCE OF WATER PROPERTIES AND CONSTITUENTS 119

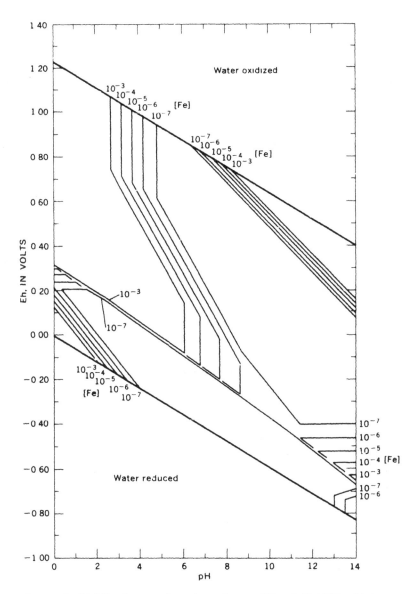

FIGURE 15 —Solubility of iron in moles per liter as function of Eh and pH at 25°C and 1 atmosphere of pressure Activity of sulfur species 96 mg/l as SO_4^{-2}, and carbon dioxide species 61 mg/l as HCO_3^-.

As figure 15 demonstrates, there are two general Eh-pH conditions under which iron solubility is very low. One of these is a condition of strong reduction and covers a wide pH range, within the field of stability for pyrite. The second is a condition of moderate oxidation above a pH of 5 and is in the ferric hydroxide stability region. Between these regions, and especially at low pH, iron is relatively soluble. The high solubility regions at high pH and high Eh are outside the Eh-pH range that is common in natural systems.

It is evident from the solubility diagram that relatively small shifts in Eh or pH can cause great changes in iron solubility. Thus, when pyrite is exposed to oxygenated water or ferric hydroxide is in contact with reducing materials, iron will tend to go into solution. It also is evident that if measurements of pH and of dissolved iron are made, the equilibrium Eh of the system can be calculated, or at least a range of possible values given.

Within the usual pH range of natural water (pH 5–9), the maintenance of an Eh below 0.20 and above -0.10 volt can maintain a considerable ferrous iron concentration in equilibrium. This goes a long way toward explaining the behavior of iron in underground water, where a relatively low Eh can be maintained readily. It is notable also that in a considerable part of this range of Eh, the solubility of iron may be controlled by precipitation of ferrous carbonate. In this part of the system, iron solubility is a function of pH and dissolved bicarbonate species, but is independent of Eh. As soon as an iron-bearing ground water dissolves oxygen from the air, however, the Eh goes up, and iron is oxidized to the ferric form which precipitates as ferric hydroxide.

The accuracy with which the Eh-pH diagram can be used in solubility calculations and applied to real systems to compute Eh has limits, although the more qualitative demonstration of principles afforded by the diagrams is so obvious it needs no further comment. The conditions which must be met when the diagrams are applied to real systems include at least three requirements:

1. That the system is at equilibrium. Reactions involving oxidation or reduction of sulfur seem to require presence of biota to attain equilibrium.
2. That solid species in the system are the same as those specified by the diagram and are reasonably pure.
3. That complexes not allowed for in the computation are absent, or negligible.

Serious problems may be encountered in actually measuring the variables of interest in the system itself and relating them to computed values. Although measuring potentials by means of a platinum or gold electrode, a reference electrode, and a pH meter or similar device appears to be relatively simple, the results may have questionable significance. The influence of oxygen dissolved from the atmosphere is particularly strong, and even slight contact of the system with air will often give potentials related to dissolved oxygen rather than to other components of the system.

The redox potential, which should be observed in a water saturated with oxygen at the partial pressure prevailing in the atmosphere, can be readily calculated. Measurements in such conditions, which are commonly made with a reference electrode and a platinum or gold electrode, give a potential near +0.45 volt at a pH of 7.0. This is considerably below the theoretical value. The meaning of the observed potential in this system has been extensively studied. Some investigators have thought the complicated mechanism of oxidation by dissolved oxygen was responsible. Others have suggested the potential was caused by reactions of oxygen at the electrode surface. Morris and Stumm (1967) pointed out some of the difficulties involved in measuring Eh in dilute solutions and concluded that measured potentials frequently may not be closely related to theoretical Eh values in natural systems because of lack of reversibility of the electrochemical reactions and other factors.

Even though the measurement of Eh may be difficult, or even impossible in some systems, this does not prevent using the concept in describing the general behavior of dissolved species and solids in natural systems. In a general way, it seems evident also that measured potentials in systems, where iron species are dominant, are of practical value in predicting and evaluating tendencies of ground waters to corrode or encrust well casings and screens (Barnes and Clarke, 1969).

The chemical processes involved in corrosion of iron and steel can be understood more easily by utilizing the Eh-pH diagram. It is readily possible to show a stability field for metallic iron on a diagram like figure 14, but this stability region is entirely in the part of the diagram where water is unstable and thus tends to be reduced. This implies that corrosion of iron could be expected in almost any environment where it is exposed to water. However, where a metallic iron surface is in contact with moving water, the corrosion rate may be slowed by the formation of a protective coating of rather highly insoluble material. For example, Eh-pH conditions within the $Fe(OH)_3$ stability region are favorable for formation of solid ferric hydroxide, but unfavorable for much iron going into solution. Regions in figure 15 where iron is the most highly soluble tend to be conditions where severe corrosion of iron may take place.

OCCURRENCE OF IRON IN WATER

The water of a flowing stream at near-neutral pH, even when rather grossly polluted, cannot contain at equilibrium significant concentrations of uncomplexed dissolved ferrous iron. Iron normally present in such water occurs either as particulate ferric hydroxide or as some form of organic complex. The latter also may be particulate Water that is naturally colored often is rather high in iron. An example is analysis 4, table 14. There usually is no direct correlation between iron concentration and intensity of color, however.

Hem and Cropper (1959) were able to decrease iron contents of uncolored surface-water samples to less than 0.01 mg/l by filtration and centrifugation. Most natural streams contain particulate ferric hydroxide, which may influence the behavior of some other constituents, especially metals that are present in minor concentrations (Jenne, 1968). Laboratory experiments by Hem and Skougstad (1960) showed that ferric hydroxide in alkaline solution decreased copper concentrations to very low levels.

Although the studies of colored organic material mentioned earlier in this paper have not identified the material completely, the suggestion is frequently made that the organic material is a dispersed colloid and that the particles can carry adsorbed metal oxides or hydroxides, including iron.

In lakes and reservoirs where a stratified condition becomes established, water at and near the bottom may become depleted in oxygen and attain a low Eh. Ferrous iron can be retained in solution in water of this type to the extent of many milligrams per liter (Livingstone, 1963, p. G11). The iron content of lake water also can be influenced by aquatic vegetation, both rooted and free-floating forms (Oborn and Hem, 1962).

Water of very low pH, which may result from discharge of industrial wastes, drainage from mines (analysis 7, table 14), or sometimes inflows of thermal springs, can, of course, carry high concentrations of ferric and ferrous iron. Zelenov (1958) noted the large amounts of iron which may be discharged by hot springs in volcanic areas.

Ground water with a pH between 6 and 8 can be sufficiently reducing to carry as much as 50 mg/l of ferrous iron at equilibrium, where bicarbonate activity does not exceed 61 mg/l. In many areas, the occurrence of 1.0–10 mg/l of iron in ground water is common. This type of water is clear when first drawn from the well, but soon becomes cloudy and then brown from precipitating ferric hydroxide. Wells that yield water of this type may appear to be eratically distributed around the area, and some exhibit changes in composition of their water with time that are difficult for hydrologists to explain. Analyses 1–3, 8, and 9, table 14, represent iron-bearing ground water. Brines, such as represented by analysis 5, table 14, may contain more than 1,000 mg/l of iron.

The chemical principles, as represented in the Eh-pH diagram, exert a well-defined control over all such occurrences of iron in ground water. The complications lie mostly in the water-circulation system. Recharge reaching the water table is generally oxygenated owing to

SIGNIFICANCE OF WATER PROPERTIES AND CONSTITUENTS 123

TABLE 14.—*Analyses of waters containing iron*

[Analyses by U.S. Geological Survey. Date below sample number is date of collection. Source of data: 1, 5, and 6, unpub data U S Geol. Survey files; 2, 3, and 9, Scott and Barker (1962, p 63, 101); 4, U S Geol Survey Water-Supply Paper 1948 (p 297); 7, U.S. Geol Survey Water-Supply Paper 1022 (p. 21), 8, Simpson (1929, p. 298)]

Constituent	(1) May 28, 1952		(2) Mar. 8, 1952		(3) Feb. 27, 1952		(4) Oct 1–31, 1962		(5) Mar. 11, 1952		(6) Jan. 30, 1952		(7) Aug. 8, 1944		(8) June 24, 1921		(9) Oct 26, 1954	
	mg/l	meq/l	mg/l	meq/l	mg/l	meq/l	mg/l	meq/l	ppm	epm	mg/l	meq/l	mg/l	meq/l	mg/l	meq/l	mg/l	meq/l
Silica (SiO$_2$)	20		12		26		11		9.1		8.1		21		23		7.9	
Aluminum (Al)			1.2		1.2								29				.6	
Iron (Fe)	2.3		2.9		10		1.4		32		.31		15	3.23	4.8		11	.32
Manganese (Mn)	.00										.34		10	.36				
Calcium (Ca)	126	6.29	2.7	0.135	8.8	0.439	18	0.898	7,470	372.75	264	13.17	119	5.94	136	6.79	8.4	0.419
Magnesium (Mg)	43	3.54	2.0	.164	8.4	.691	8.0	.658	1,330	109.37	17	1.40	68	5.59	35	2.88	1.5	.123
Sodium (Na)	13	.56	35	1.522	34	1.478	9.3	.40 }	43,800	1,940.59	52	2.26 }	17	.74	960	41.74	1.5	.065
Potassium (K)	2	.05	1.7	.044	2.9	.074			129	3.30	31	.80					3.6	.092
Bicarbonate (HCO$_3$)	440	7.21	100	1.639	65	1.065	69	1.131	76	1.25	61	1.00	0	.00	249	4.08	30	.492
Sulfate (SO$_4$)	139	2.89	5.6	.117	71	1.478	29	.604	47	.98	757	15.76	817	17.01	1,260	26.23	5.9	.123
Chloride (Cl)	8	.23	2.0	.056	2.0	.056	6.4	.181	83,800	2,363.43	24	.68	22	.62	734	20.70	1.8	.051
Fluoride (F)	.7	.04	.1	.005	.3	.016			0		.8	.04		.01			.1	.005
Nitrate (NO$_3$)	2	.00	.6	.010	.0	.000	2.9	.046			.0	.00	.4	.01	7.5	.12	.4	.006
Dissolved solids																		
Calculated	594		113		187				137,000		1,280		1,260		3,280		47	
Residue on evaporation	571		101		180		156		140,000		1,180				3,450		44	
Hardness as CaCO$_3$	490		15		56		78		24,200		730		845		484		27	
Noncarbonate	131		0		3		21		24,100		679		845		280		2	
Specific conductance (micromhos at 25°C)	885		162		264		188		146,000		1,460		1,780				68.8	
pH	7.6		7.4		6.4		6.9		7.4		7.5		3.0				6.3	
Color	1		23		7		140		15		2		8				3	
Acidity (total) as H$_2$SO$_4$													342					

1 Well 3, Nelson Rd , Water Works, Columbus, Ohio. Depth, 117 ft. Water from glacial outwash Temp, 13.3°C.
2 Well 79 8-50 Public supply, Memphis, Tenn. Depth, 1,310. Water from sand of the Wilcox Formation. Temp, 22.2°C
3 Well 5 290-1, 6 miles southeast of Maryville, Blount County, Tenn. Depth, 66 ft. Water from Chattanooga Shale. Temp, 14.4°C.
4 Partridge River near Aurora, Minn. Composite sample. Mean discharge, 30.8 cfs.
5 Brine produced with oil from well in NW¼ sec 3, T. 11 N., R. 12 E., Okmulgee County, Okla. Depth, 2,394 ft. Water from the Gilcrease sand of drillers, Atoka Formation.
6 Drainage at collar, drill hole 89, 7th level Mather A iron mine, Ishpeming, Mich. Temp, 15.1°C.
7 Shamokin Creek at Weighscale, Pa Discharge, 64.2 cfs, affected by drainage from coal mines
8 Flowing well, Minneapolis, St Paul, and Sault Ste Marie R.R., Enderlin, Ransom County, N Dak Depth, 613 ft. Water from Dakota Sandstone.
9 City well 4, Fulton, Miss. Depth, 210 ft. Water from the Tuscaloosa Formation. Temp, 17.2°C.

contact with air, and any reduced iron minerals, especially pyrite, which the solution contacts will be attacked to yield ferrous iron and sulfate. The oxygen in the circulating water is ultimately depleted by this and other reactions, but a considerable amount of ferrous iron can be dissolved by the time the water has passed for some distance through the aquifer. If oxidizable matter is scarce, the oxygen may persist in the ground water for a long time. Reduced iron minerals below the water table will tend to disapper in time where the water circulates freely, but the processes of oxidation or cation exchange that cause the disappearance may not be completed until after the water has moved a long distance.

Ground water that is high in dissolved iron can be associated with the oxidation of reduced iron minerals at a regional or local contact between oxidizing and reducing conditions. Broom (1966) identified such a zone in the interbedded sands of the Mississippi embayment in northeastern Texas. In eastern Maryland and at other localities on the Atlantic coastal plain, the permeability of some coastal sedimentary beds is much greater than others, and a well may encounter solutions with different oxidation-reduction potentials at different depths. Mingling of these waters at contacts between strata and particularly where the strata are locally short circuited either naturally or by a well that penetrates them can cause deposition or solution of iron minerals. In many wells corrosion of the iron casing adds iron to the water pumped out, and iron may be precipitated in various forms within the well under some conditions.

Water-bearing strata which contain oxidized iron minerals and organic debris may provide an environment favorable for reduction of ferric iron and give rise to rather high concentrations of ferrous iron in the circulating ground water.

Studies of iron occurrence in ground water by field measurements of pH and Eh have been made by several investigators in this country and abroad (Back and Barnes, 1965; Germanov and others, 1959). The studies of Barnes and Clarke (1969) show how these measurements can be related to the corrosion and encrustation of well screens and casing. Measurements of Eh of ground water require careful exclusion of air from the equipment and proper shielding and operation of measuring devices (Back and Barnes, 1961) and are subject to limitations noted earlier in this discussion.

Much uncertainty and confusion seem to exist regarding the role and importance of bacteria in the occurrence of iron in ground water. The ways in which bacteria may be involved in the behavior of iron in water include the following:

1. Processes in which bacteria exert a catalytic effect to speed re-

actions that are thermodynamically favorable but which occur rather slowly in the absence of bacteria.
2. Processes which require a contribution of energy and can be promoted by bacteria that consume some other substance as a source of energy.

Processes of the first type are involved in the oxidation of ferrous iron by *Crenothrix* and *Leptothrix*. Both genera require oxygen; hence they live in environments where ferrous iron is unstable. The bacteria may become established in wells and remove some iron by precipitation of ferric hydroxide before the water gets to the surface of the ground. Sulfur-oxidizing bacteria may exert an indirect effect on iron behavior in catalyzing such reactions as the oxidation of pyrite. Processes of the second type typically involve iron- and sulfur-reducing species which require an oxidizable substance, usually organic, as a source of food and energy. Reduction or oxidation of iron may occur incidentally, and iron may play no essential role in the life processes of the bacteria. Some species of bacteria that have been shown to influence the behavior of iron are discussed by Clark, Scott, and Bone (1967).

The effect of bacteria may thus be either to increase or to decrease dissolved-iron concentrations in water. Bacterial colonies in wells and pipelines may be partly dislodged by moving water from time to time and thus give rise to occasional releases of accumulated ferric hydroxide. The occurrence of iron in water from wells, however, is primarily a chemical phenomenon and cannot be ascribed solely to the bacteria. The biota associated with iron solution and deposition are often responsible for acceleration of chemical reactions, but are more in the nature of symptoms of the iron problem rather than causes.

Because of the unstable nature of ferrous iron in samples of alkaline ground water, the iron originally present may have been oxidized and precipitated by the time the analysis is made. Concentrations of iron reported in analyses by the U.S. Geological Survey represent the amounts in solution at the time of sampling. Waters suspected of containing iron generally are filtered at the time of sampling by using membranes with average pore diameters of 0.45 micron and an aliquot acidified to prevent preciptiation. The iron in untreated samples represented by analyses 1-3, 8, and 9, table 14, had largely been lost from solution before analysis, and amounts reported are the totals originally present in a filtered acidified fraction of the sample. The oxidation and precipitation reactions decreased the pH and altered the alkalinity originally present in the unacidified solutions used for the rest of the analysis. Although it might be possible to reconstruct these data to give the original composition of the water at the time of sampling, it is not common practice to do so. The effect of precipitation of iron may be readily noticeable in dilute solutions such as the one represented by analysis 9, table 14.

MANGANESE

Small concentrations of manganese in water may constitute an objectionable impurity. For this reason, many water analyses include manganese determinations although the concentrations measured are usually comparatively low. The water-quality standards proposed by the U.S. Public Health Service (1962) recommend an upper limit of 0.05 mg/l of manganese.

SOURCES OF MANGANESE

Water chemists often tend to think of manganese and iron as closely related in behavior, but there are some very important differences between the aqueous chemistries of the two elements. Although manganese is much less abundant than iron in the earth's crust, it is one of the most common elements and is widely distributed in rocks and soils. In igneous-rock minerals, manganese is usually in the reduced form (Mn^{+2}) substituting for some other divalent ion of similar size. The ferro-magnesian minerals such as biotite and hornblende commonly contain some manganese. The most common forms of manganese in rocks and soils, however, are oxides and hydroxides in which the oxidation state of the element is $+2$, $+3$, or $+4$. These oxides tend to adsorb other metallic cations very strongly, and the common naturally occurring species of the manganese oxides usually have many impurities. Manganese carbonate (rhodochrosite) is a fairly common manganese-bearing mineral.

Manganese is an essential element in plant metabolism, and it is to be expected that organic circulation of manganese can influence its occurrence in natural water. Specific mention of manganese accumulation in tree leaves appears in published literature. Some species of trees, however, are much more effective accumulators than others. Bloss and Steiner (1960) found considerable quantities in the leaves of the chestnut oak, and Ljunggren (1951) reported similar findings for the needles of spruce trees in southern Sweden. Aquatic plants were noted by Oborn (1964) to be accumulators of manganese. Manganese in plant parts that die back or are shed, as for example in leaves, becomes available for solution in runoff and soil moisture. The importance of this source of manganese in river water is not completely known, although some preliminary studies by Slack and Feltz (1968) on the effects of fallen leaves on water quality of a small stream in Virginia showed it could be important at times.

Deposits of manganese oxides occur as crusts on the surfaces of rocks in the beds of streams in certain areas of the world. Observations in the literature concern streams in Sweden (Ljunggren, 1953), Colorado (Theobald and others, 1963), and Maine (Canney and Post, 1964). The mechanisms which produce these deposits are probably different, but none was studied in enough detail to make possible a complete explanation. The deposits studied by Ljunggren seemed to be associated with an aquatic moss. In the other two areas, however, the deposition seemed to occur without any organic intervention.

Manganese oxide deposits also occur in fresh-water lakes apparently as a result of processes similar to those causing deposits of bog iron ore (Ljunggren, 1953; 1955). In reservoirs and lakes, when thermal stratification occurs, the water near the bottom may be able to dissolve considerable concentrations of manganese from deposits left there under oxidizing conditions prevailing at an earlier time. Water withdrawn from many public supply reservoirs contains excessive concentration of manganese at times. An extensive literature describing such occurrences exists. Ingols and Wilroy (1963), in studies of occurrences of manganese in reservoirs of the southeastern United States, noted that reduction by organic material and bacteria may bring into solution considerable amounts of manganese from the oxides present in soil and rock covered by the water of the reservoir.

Manganese oxide nodules are scattered on the floor of the oceans in many localities. The nodules occur in shallow water as well as at great depths. Goldberg (1963) attributed the deposition of the nodules to a slow oxidation of manganese dissolved in water in contact with the oxide surface. Mannheim (1965) suggested the nodules found at the shallower depths at least were partly the result of local high concentrations of manganese in water at the ocean bottom, resulting from the squeezing out of solutions associated with buried sediment. Many writers have commented on the possible importance of submarine volcanic activity as a source of the manganese.

The possible importance of suspended manganese oxides in streams as controls over the behavior of minor cations has been pointed out by Jenne (1968). There is not much information on the amounts of manganese oxides which are present in suspension in river water. Turekian and Scott (1967) in a study of 18 rivers mostly in the Southeastern United States found a manganese concentration of 1.2 percent in the sediment of the Susquehanna River in Pennsylvania and manganese concentrations over 0.1 percent in most of the other streams. The sediment concentration in the streams was low at the time of sampling, however, and the chemical composition of sediment at times of higher sediment loads may not be adequately shown by these results. One of the notable properties of manganese oxide is a

128 CHEMICAL CHARACTERISTICS OF NATURAL WATER

tendency to form coatings on other mineral surfaces, and this is a further complicating factor in determining the amounts present in ordinary streamflow.

MANGANESE-SOLUBILITY CONTROLS

The solubility of manganese in natural water and the predominant dissolved forms to be expected can be effectively summarized by means of Eh-pH diagrams. Figure 16 is drawn for a system in which

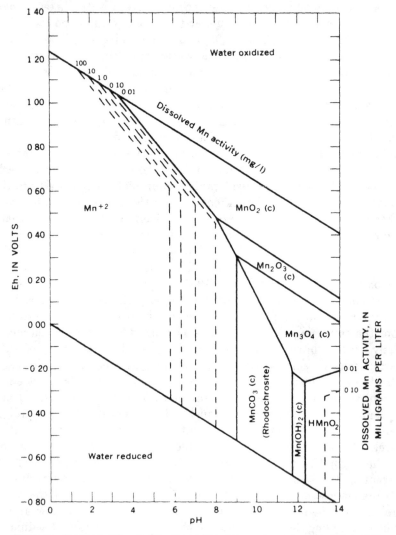

FIGURE 16 —Fields of stability of solids and solubility of manganese as a function of Eh and pH at 25°C and 1 atmosphere of pressure Activity of dissolved carbon dioxide species 100 mg/l as HCO_3^- Sulfur species absent

the activity of dissolved carbon dioxide species is 100 mg/l as HCO_3^-. The manganese carbonate assumed to be present in this system is well-crystallized rhodochrosite; a freshly precipitated form of $MnCO_3$ is known to be more soluble (Hem, 1963a).

At a pH near neutrality in this system, the predominant dissolved form of manganese would be the divalent ion Mn^{+2}, and a concentration between 1.0 and 10 mg/l would be stable at the Eh one might find in surface water exposed to the air. A small increase in pH, however, would shift the system into the oxide-stability regions where both pH and Eh control the solubility. The oxides shown in the diagram are pure forms approximately equivalent to pyrolusite (MnO_2), in which manganese has a valence of $+4$; braunite, in which the valence of manganese is nominally $+3$; and hausmannite (Mn_3O_4), which contains a mixture of Mn^{+2} and the more oxidized species.

Some writers have stated that dissolved Mn^{+3} forms are present in natural water. Most published thermodynamic data, however, suggest that these forms would not be stable at equilibrium. The Eh-pH diagram is applicable only to equilibrium conditions. Manganese oxidation and reduction reactions probably are less likely to reach equilibrium quickly than are similar reactions of iron. Organic complexes of manganese also may affect the behavior of the element.

The complex ion $MnHCO_3^+$ has been shown to be present in solutions containing bicarbonate (Hem, 1963b) and could represent half the manganese if bicarbonate concentration were near 1,000 mg/l. The ion pair $MnSO_4^\circ$ was also reported by Nair and Nancollas (1959) and can be important in water containing more than 1,000 mg/l of SO_4^{-2}. The Eh-pH diagram does not specify the presence of sulfur, as manganese sulfide minerals are rare and have no large influence on the chemical behavior of manganese in aqueous systems.

Although the assumptions inherent in Eh-pH diagrams must be kept in mind, figure 16 does agree reasonably well with much of the observed behavior of manganese in natural water. Manganese concentrations of the order of 1 mg/l are relatively stable in oxygenated water unless the pH is rather high. The diagram thus correctly shows that manganese is difficult to remove from water by aeration alone, unless the pH is maintained a a high level. The stability of some forms of manganese oxides toward attack by water, their adsorption of other ions, and some other important aspects of the chemistry of manganese, however, do not seem to be adequately represented by equilibrium considerations alone.

Although manganese forms complexes with many kinds of organic material, the divalent Mn^{+2} ion is relatively stable in uncomplexed form in the conditions normally to be expected in river water. No correlation has been established between organic color and manganese concentrations in river water.

MANGANESE-REACTION RATES AND MECHANISMS

Studies by Morgan (1967) and by Hem (1963a; 1964) show that the rate of oxidation of Mn^{+2} and precipitation of the oxidized form as $MnO_2(c)$, is greatly increased by increases in pH and that the reaction goes on more rapidly as surface area increases. The nature of surface-catalysis effects such as those shown by manganese oxides is not fully understood. Probably the reaction pathways which can be followed at such a surface are different from the ones which must be followed by reactions in the bulk of the solution. Manganese oxidation also can be catalyzed by feldspar surfaces (Hem, 1964) and probably by other mineral surfaces.

Bacteria may influence the rates of manganese oxidation, probably in a manner similar to the effects of bacteria in oxidation of ferrous iron. Tyler and Marshall (1967) found *Hyphomicrobium* in deposits of manganese oxide in pipelines, and Schweisfurth (1963) discussed the role of bacteria in manganese oxidation.

OCCURRENCE OF MANGANESE IN WATER

Manganese is often present to the extent of more than 1 mg/l in streams that have received acid drainage from coal mines. Manganese usually persists in the river water for greater distances downstream from the pollution source than the iron contained in the drainage inflows. As the acidity is gradually neutralized, ferric hydroxide precipitates first. Manganese, however, also disappears from solution after a longer time. These effects can be noted for streams in the Ohio River basin (U.S. Geological Survey Water-Supply Paper 1948). Analysis 4, table 13, represents water from a river influenced by acid mine drainage from which most of the iron has been precipitated. Analysis 7, table 14, is for water affected by mine drainage in which much iron was still present. The water of the upper Mississippi River contains appreciable amounts of manganese in solution at times (Wiebe, 1930). The source in this instance is not mine drainage, but probably is related to the abundant vegetation in the basin and to the organic debris in alluvium along the stream. The usual practice has been to omit manganese determinations on water from streams not subject to extensive pollution, and there is little information on normal loads of manganese in river water and the way these might vary seasonally.

Some ground waters that contain objectionable concentrations of iron also have considerable amounts of manganese. but ground waters that contain more manganese than iron are rather unusual. Analysis 2, table 13, represents water from two radial collector wells located along the Ohio River at Parkersburg, W. Va. Manganese concentration in this water was 9.6 mg/l compared to 0.88 mg/l from iron. It seems likely the manganese is dissolved as water moves toward the wells from the bottom of the river. The occurrence of manganese in these wells was discussed by Gidley (1952). Analysis 5, table 13, represents a rather common type of water where both iron and manganese are present in objectionable concentrations, although the water has a low dissolved-solids concentration.

Many of the ground waters reported to carry large manganese concentrations are from thermal springs. White, Hem, and Waring (1963, p. F50) reported some examples of these. In many places these springs seem to be closely associated with manganese oxide deposits. Analysis 3, table 13, represents water from a thermal spring which contains 2.5 mg/l of manganese.

CALCIUM

In most natural fresh water, calcium is the principal cation. The element is very widely distributed in the common minerals of rocks and of soil. The chemical controls on the occurrence of calcium in water are perhaps somewhat simpler than the controls on most multivalent elements, but are intricate enough to provide an ample number of problems for future research.

SOURCES OF CALCIUM

Calcium is an essential constituent of many igneous-rock minerals, especially of the chain silicates, pyroxene and amphibole, and the feldspars. The plagioclase feldspar group of minerals represents mixtures in various proportions of the end members albite, $NaAlSi_3O_8$, and anorthite, $CaAl_2Si_2O_8$. Calcium also occurs in other silicate minerals which are produced in metamorphism. Some calcium is, therefore, to be expected in water which has been in contact with igneous and metamorphic rock. The concentration generally is low, however, mainly because the rate of decomposition of most igneous-rock minerals is slow. The decomposition of anorthite can be represented as

$$CaAl_2Si_2O_8 + H_2O + 2H^+ = Al_2Si_2O_5(OH)_4 + Ca^{+2}.$$

The normal composition of plagioclase feldspar lies between the pure sodium and pure calcium forms, and decomposition will, therefore, generally yield both calcium and sodium and some soluble silica. Given a long enough contact time, pH may rise to a point where calcium carbonate precipitates from the solution.

The most common forms of calcium in sedimentary rock are the carbonates. The two crystalline forms, calcite and aragonite, both have the formula $CaCO_3$, and the mineral dolomite can be represented as $CaMg(CO_3)_2$. Limestone consists mostly of calcite with admixtures of magnesium and other impurities. A rock is commonly termed dolomite if the magnesium is present in amounts approaching the theoretical 1:1 mole ratio with calcium. Other calcium minerals common in sediments include the sulfates, gypsum ($CaSO_4 \cdot 2H_2O$) and anhydrite ($CaSO_4$), and the fluoride, fluorite (CaF_2). Calcium is also a component of some types of zeolite and montmorillonite.

In sandstone and other detrital rock, calcium carbonate commonly is present as a cement between particles or a partial filling of the interstices. Calcium also is present in the form of adsorbed ions on negatively charged mineral surfaces in soils and in rocks.

SOLUTE SPECIES

Calcium ions are relatively large, having an ionic radius of 0.99 angstrom. The charged field around the ion is therefore not as intense as those of smaller divalent ions. Calcium ions do not have a strongly retained shell of oriented water molecules surrounding them in solution. The usual dissolved form can be simply represented as the ion Ca^{+2}. Calcium does form complexes with some organic anions, but the influence of such species on the concentrations of calcium in natural water is probably not important. In some solutions, ion pairs such as $CaHCO_3^+$ can exist. Data published by Greenwald (1941) show that about 10 percent of the calcium might be in this form if the bicarbonate concentration were near 1,000 mg/l. The ion pair $CaSO_4(aq)$ is more important. In solutions where sulfate concentrations exceed 1,000 mg/l, more than half the calcium could be present in the form of the $CaSO_4$ ion pair. Both generalizations assume calcium concentration is small compared to the bicarbonate or sulfate contents.

Garrels and Christ (1964, p. 96) also mentioned hydroxide and carbonate ion pairs with calcium. These species could be present in appreciable concentrations in strongly alkaline solutions. Other calcium ion pairs such as those with phosphate are known, but are not likely to be significant in natural water.

CHEMICAL CONTROLS OF CALCIUM CONCENTRATION

Equilibria involving carbonates are the major factor in limiting the solubility of calcium in most natural water. Barnes (1965) has shown that some departure from equilibrium is to be expected in surface streams, because some of the variables in the system may be almost continuously changing. For the less rapidly changing conditions prevailing in ground water, a closer approach to the equilibrium condition is to be expected. From his studies of the rates of calcite solution, Weyl (1958) concluded that saturation with respect to calcite should prevail for all water in limestone, below the water table. Various investigators have found extensive departures from his generalization. Back, Cherry, and Hanshaw (1966) observed both supersaturation and undersaturation in water from wells penetrating the Ocala Limestone in Florida and suggested that solution of limestone might continue for hundreds of feet below the water table.

When natural-water systems are examined with reference to carbonate equilibria, it must be remembered that pH measurements, even when made in the most careful way, must be taken on samples that have not changed since they were within the actual system one is trying to evaluate. For practical purposes, this means the determination must be made immediately when the source is sampled. The measurement of pH to two decimal places is generally about all that can be accomplished, and because of the exponential nature of the quantity, the precision may be low compared to that obtained by a chemical analysis for dissolved species. Also, it should be remembered that a ground-water body may not be sampled representatively by a pumping or flowing well and that standard conditions of temperature and pressure and purity of solids are commonly assumed when it may be that such assumptions do not apply closely.

The solubility equilibrium for calcite, which is more stable than aragonite at 25°C and 1 atmosphere of pressure, is

$$CaCO_3(c) = Ca^{+2} + CO_3^{-2}.$$

This can be used directly in calculating the saturation status of a particular solution. The concentration of carbonate ions, however, usually is not directly determinable to the desired accuracy. The equilibrium

$$CO_3^{-2} + H^+ = HCO_3^-$$

can be added to the above to give

$$CaCO_3(c) + H^+ = Ca^{+2} + HCO_3^-,$$

for which the equilibrium constant is 97 (Hem, 1961a). Analytical data for concentrations of calcium and bicarbonate can be converted to activities by techniques outlined by the writer (Hem, 1961a). Some aspects of this equilibrium already have been discussed in this paper, and computations will not be repeated here.

The graphical representation of the above relationship given in figure 17 enables a quick estimate of whether a particular water is supersaturated or unsaturated with respect to calcite. For many years

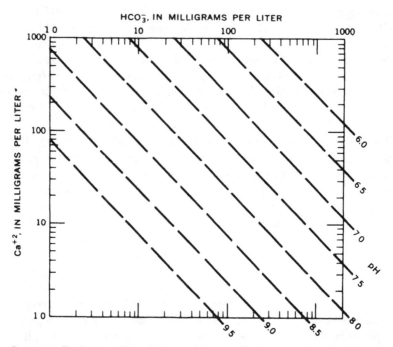

FIGURE 17.—Equilibrium pH in relation to calcium and bicarbonate activities in solutions in contact with calcite. Total pressure 1 atmosphere; temperature 25°C.

water-treatment plant operators have used somewhat similar calculations to evaluate the stability of a water toward precipitation or solution of calcite. The most desirable condition is for some degree of supersaturation in the water as it enters the distribution system, as this causes precipitation of a protective film of calcite in pipelines and retards corrosion. The Langelier index, named for the man who described these calculations, is a number indicating the extent of supersaturation or unsaturation in terms of logarithmic units. Thus an index of $+1.0$ describes a water supersaturated by a factor of 10, or whose pH is 1 unit above the equilibrium value. Langelier's (1936) computations did not utilize ion-activity coefficients.

The input which most likely will control the solubility of calcite is the hydrogen-ion supply. For the most part, the availability of H^+ is a function of the availability of carbon dioxide. The amount of carbon dioxide present in normal air is 0.03 percent, or in terms of partial pressure, 0.0003 atmosphere.

The relationship of calcite solubility to partial pressure of carbon dioxide in an associated gas phase is represented in figure 18. The data used for this graph were obtained experimentally by Frear and Johnston (1929) and represent a system at 25°C and 1 atmosphere of

SIGNIFICANCE OF WATER PROPERTIES AND CONSTITUENTS 135

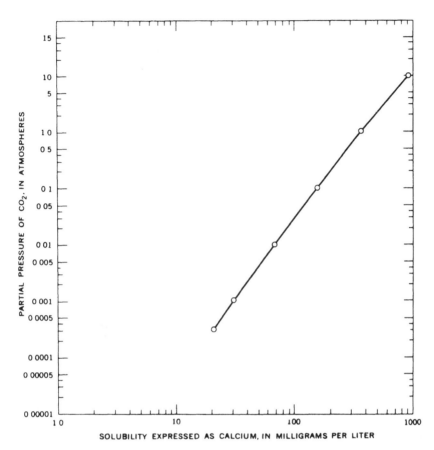

FIGURE 18 —Solubility of calcium carbonate (calcite) in water at 25°C in the presence of carbon dioxide.

total pressure, with no other solutes. The solubility indicated for calcium in water in contact with air is about 20 mg/l. Garrels and Christ (1964, p. 83), using other published thermodynamic data, calculated a solubility of 16 mg/l under these conditions.

If the input of H^+ is solely from the water itself and carbon dioxide is absent, the solubility of calcite is 5.4 mg/l of calcium (Askew, 1923). It is of interest to note that this solution would have a pH between 9.9 and 10.0. Garrels and Christ (1964), p. 81) observed this experimentally. They further calculated (p. 88) that rainwater which had dissolved all the carbon dioxide possible through equilibrium with air and then was equilibrated with calcite in the absence of a gas phase could dissolve hardly any more calcium than pure water and would also reach a high pH.

From figure 18 it is evident that the concentrations of calcium, which are common in waters from regions where carbonate rocks occur, are well above the level specified for the partial pressure of carbon dioxide in the atmosphere. Water percolating through the soil and the unsaturated zone above the water table, however, is exposed to air in pore spaces in the soil which is greatly enriched in carbon dioxide. Partial pressures of carbon dioxide in soil air are commonly 10–100 times the levels reached in the atmosphere (Buckman and Brady, 1960, p. 246). The carbon dioxide content of soil air for the most part results from plant respiration and decay of dead plant material and can be expected to be greatest in environments which support dense of vegetation.

In environments where hydrogen ions are supplied for rock weathering by processes other than dissociation of dissolved carbon dioxide species, for example by oxidation of sulfur or sulfides, calcium may be brought into solution in amounts greater than the stoichiometric equivalent of bicarbonate. In such a system, or where water is in contact with solid gypsum or anhydrite, the maximum calcium concentration that could be reached would generally be determined by equilibria in which gypsum is the stable solid.

The equilibrium solubility of gypsum can be computed by taking into account the solubility product, the effect of the ion pair $CaSO_4(aq)$, and the effects of increasing ionic strength. If no ions besides calcium and sulfate are present, saturation is reached when the calcium concentration is about 600 mg/l. Analysis 3, table 15, which represents a spring issuing from gypsum, approaches saturation, and the effect of other ions is minor, as this water has 636 mg/l of calcium. Increasing the ionic strength increases the equilibrium concentration of calcium. In a solution containing 2,500 mg/l of chloride and about 1,500 mg/l of sodium, the equilibrium concentration of calcium would be near 700 mg/l. These calculations presuppose nearly equivalent concentrations of calcium and sulfate. Frequently this does not occur in natural water. When the solution is in equilibrium with gypsum, however, any increase in sulfate activity would be matched by a decrease in calcium. (See also fig. 21.)

Natural brines are fairly common in which the predominant dissolved ions are calcium and chloride. An example is represented by analysis 4, table 15. There is no general agreement as to how such a composition is reached. Water that has been trapped underground for a long time could be altered from its original composition by selective permeability of strata for different solutes, the bacterial reduction of sulfate and other dissolved ions, and adsorption or desorption of dissolved ions, as well as by chemical solution or precipitation of minerals. The formation of calcium chloride brine was ascribed by Valyashko and Vlasova (1965) to a combination of concentration and ion-exchange processes.

Cation-exchange equilibria have a considerable influence on the calcium concentrations in most natural water. The exchangeable cations on mineral surfaces in rocks and in soils provide a major reservoir of ions, and the composition of the water coming in contact with these reflects the results of exchanges between the water and solid surfaces. Divalent cations are generally held more strongly than monovalent ones. Exchange reactions can alter individual ion concentrations, but do not change the total milliequivalents per liter of cations that are present. The suspended-sediment concentration of river water generally shows a sharp increase when discharge increases. The work of Kennedy (1965) shows that when sediment concentrations are high, the proportion of adsorbed to solute cations in a given volume of water can exceed 1. The influence of sediment on the ratio of calcium to sodium in solution could thus be strong at times. The influence is strongest during flood stages when the water is low in dissolved-solids concentration, tailing off, or perhaps almost disappearing, as flow and sediment concentrations decrease.

The process of cation exchange in ground-water bodies has been extensively observed since the early studies of Renick (1924) called attention to natural softening of ground water by cation exchange in sediments underlying the northern Great Plains area of the United States. Ground water which has exchanged calcium for sodium is common in many areas (see analyses 1 and 2, table 17). The reverse effect, exchange of sodium for calcium, also can be expected, but is less well documented. In localities near the ocean where sea water has entered fresh-water aquifers, the advancing salt-water front commonly carries higher proportions of calcium to sodium than are characteristic of sea water (Poland and others, 1959, p. 193), owing to release of calcium from exchange positions by the sodium brought in by the advancing salt water.

In irrigated areas, the exchange of calcium for sodium in soil moisture may proceed forward or in reverse at different times and in any specific spot might fluctuate extensively. The residual salts from the irrigation water used by the crop or evaporated from the soil, however, will partly reappear in drainage water. The water of the Salt River, which is extensively used for irrigation in the Phoenix area, Arizona, is shown by analysis 5, table 17, to contain nearly twice as many milliequivalents per liter of Na^+ as of Ca^{+2} and Mg^{+2}. The ion ratio in the effluent, principally drainage of used irrigation water, is represented by analysis 6, table 17; it has a slightly higher proportion of the divalent ions, and thus suggests that the exchange reactions are occurring, even though a good deal of the calcium present in the initial water probably is precipitated as carbonate or sulfate in the irrigated area. The effect is more strongly shown by analysis 6, table

CHEMICAL CHARACTERISTICS OF NATURAL WATER

TABLE 15.—*Analyses of waters in*

[Analyses by U.S. Geological Survey Date below sample number is date of collection Source of data Survey Water-Supply Paper 1163 (p 360), 1, 4,

Constituent	(1) Mar 28, 1952		(2) May 20, 1950		(3) Nov 25, 1949		(4) Mar 26, 1952	
	mg/l	meq/l	mg/l	meq/l	mg/l	meq/l	mg/l	meq/l
Silica (SiO$_2$)	8.4		22		29		130	
Iron (Fe)	.04						41	
Manganese (Mn)								
Calcium (Ca)	46	.295	144	7.19	636	31 74	93,500	4,665 65
Magnesium (Mg)	4 2	.345	55	4.52	43	3 54	12,100	995 35
Sodium (Na)	1.5	.065	29	1 24	17	74	28,100	1,222.35
Potassium (K)	.8	.020					11,700	299.17
Bicarbonate (HCO$_3$)	146	2.393	622	10.19	143	2 34	0	.00
Sulfate (SO$_4$)	4.0	.083	60	1.25	1,570	32 69	17	.35
Chloride (Cl)	3.5	.099	53	1.49	24	.68	255,000	7,193.55
Fluoride (F)	.0	.000	.4	.02				
Nitrate (NO$_3$)	7.3	.118	.3	.00	18	.29		
Dissolved solids								
Calculated	[1] 149		670		2,410		[2] 408,000	
Residue on evaporation	139						415,000	
Hardness as CaCO$_3$	132		586		1,760			
Noncarbonate	12		76		1,650			
Specific conductance (micromhos at 25°C)	250		1,120		2,510		[4]	
pH	7.0						5 29	
Color	5							

[1] Includes 1.0 mg/l aluminum (Al)
[2] Includes strontium (Sr) 3,480 mg/l 79 45 meq/l, bromide (Br) 3,720 mg/l, 46 54 meq/l, iodide (I) 48 mg/l, 0.38 meq/l
[3] Includes 0 70 mg/l lithium (Li) and 0 03 mg/l zinc (Zn).
[4] Density at 46°C, 1 275 g/ml zinc (Zn)

1. Big Spring, Huntsville, Ala Water-bearing formation, Tuscumbia Limestone Temp, 16 1°C
2. Spring on Havasu Creek, near Grand Canyon, Ariz Water-bearing formation, limestone in Supai Formation, 100 gpm Temp, 19 4°C Water deposits travertine
3. Jumping Springs, SE¼ sec 17, T 26 S, R. 26 E, Eddy County, N Mex Water-bearing formation, gypsum in Castile Formation, 5 gpm.

15. This analysis represents water from a well in the area irrigated with Salt River water, where recharge is composed of irrigation drainage water. The milliequivalents per liter content of sodium in the water is far lower than that of calcium.

OCCURRENCE OF CALCIUM IN WATER

Commonly, the flow of a river which drains a humid area contains more calcium than any other cation. The general relationships among the constituents are much like those in analysis 7, table 15. This is for the Cumberland River at Smithland, Ky., a short distance above its mouth. The water is far below saturation with respect to calcite at the pH reported in the analysis. A river water may have a higher pH during summer days, when photosynthesis by aquatic vegetation is actively occurring, than in winter, when photosynthesis is less active. This process depletes the dissolved carbon dioxide. The calcium content of the Cumberland River water, 25 mg/l, is not greatly above the amount which could occur in equilibrium with calcite in air (20 mg/l as shown by fig. 18). It is interesting to note that the average concentration of calcium in river water, shown as 15 mg/l in table 3, is also

SIGNIFICANCE OF WATER PROPERTIES AND CONSTITUENTS 139

which calcium is a major constitutent
2, 6, and 10, unpub. data, U.S. Geol. Survey files; 3 and 5, Hendrickson and Jones (1952); 8, U.S. Geol. and 9, Scott and Barker (1962, p. 19, 47, 59)]

(5)		(6)		(7)		(8)		(9)		(10)	
Jan. 26, 1948		June 23, 1949		May 19, 1952		1948 and 1949		June 6, 1954		Dec. 8, 1954	
mg/l	meq/l	mg/l	meq/l	mg/l	meq/l	mg/l	meq/l	mg/l	meq/l	mg/l	meq/l
-----	-----	74	-----	4.9	-----	17	-----	24	-----	15	-----
-----	-----	-----	-----	-----	.04	-----	-----	-----	.02	1.0	-----
-----	-----	-----	-----	-----	-----	-----	-----	-----	.00	.01	-----
99	4.94	277	13.82	25	1.248	394	19.66	88	4.39	96	4.790
28	2.30	64	5.26	3.9	.321	93	7.65	7.3	.60	19	1.562
4.1	.18	53	2.29	4.5	.196	333	14.48	19	.83	18	.783
				1.4	.036			2.8	.07	1.5	.038
287	4.70	85	1.39	90	1.475	157	2.57	320	5.24	133	2.180
120	2.50	113	2.35	12	.250	1,150	23.94	6.7	.14	208	4.330
6	.17	605	17.06	2.2	.062	538	15.17	13	.37	25	.705
-----	-----	-----	.2	-----	.1	.005	-----	-----	.3	.02	.021
2.8	.05	35	.56	1.9	.031	5.0	.08	4.6	.07	.4	.006
401	-----	1,260	-----	100	-----	2,610	-----	323	-----	³449	-----
-----	-----	-----	-----	99	-----	-----	-----	322	-----	468	-----
362	-----	954	-----	78	-----	1,370	-----	250	-----	318	-----
127	-----	884	-----	5	-----	1,240	-----	0	-----	209	-----
651	-----	2,340	-----	172	-----	3,540	-----	543	-----	690	-----
				6.7				7.5		7.8	
				5				2		3	

4. Brine well 3 Monroe, SE¼ sec. 27, T. 14 N., R. 2 E., Midland, Mich. Depth, 5,150 ft. Water-bearing formation, Sylvania Sandstone. Temp, 46.1°C.
5. Rattlesnake Spring, sec. 25, T. 24 S., R. 23 E., Eddy County, N. Mex. Water-bearing formation; alluvium, probably fed by Capitan Limestone; 2,500 gpm.
6. Irrigation well NE¼ sec. 35, T. 1 S., R. 6 E., Maricopa County, Ariz. Water-bearing formation, alluvium. Temp, 25.6°C.
7. Cumberland River at Smithland, Ky. Discharge, 17,100 cfs.
8. Pecos River near Artesia, N. Mex. Discharge-weighted average 1949 water year; mean discharge 298 cfs.
9. City well at Bushton, Rice County, Kans. Depth, 99 ft. Water-bearing formation, Dakota Sandstone.
10. Industrial well, Williamanset, Mass. Depth, 120 ft. Water-bearing formation, Portland Arkose. Temp, 12.2°C.

not far from an equilibrium value for a system containing calcite in contact with air.

Rivers in more arid regions, and especially where some of the more soluble rock types are exposed, tend to have much higher dissolved-calcium concentrations. Analysis 8, table 15, is a discharge-weighted average of the composition observed for a year in the Pecos River at Artesia in southeastern New Mexico. It shows clearly the influence of gypsum, which is very abundant in the Pecos drainage basin.

When river water is impounded in a storage reservoir, changes may occur in calcium content as a result of calcium carbonate precipitation. The increased pH near the water surface, caused by algae and plankton, may bring about supersaturation, and precipitation can occur around the edges of the water body. A conspicuous white deposit has developed around the edges of Lake Mead on the Colorado River between Arizona and Nevada. In part, the deposition in that reservoir is also related to inflows that are high in calcium and to the solution of gypsum beds exposed in the reservoir; solution also tends to increase calcium activity (Howard, 1960, p. 119--124). The fresh-water deposit called marl, which is formed in lakes, partly is made up of calcium carbonate.

In closed basins, water can escape only by evaporation, and the residual water can be expected to change in composition after a fairly well-defined evolutionary pattern. The less soluble substances, including calcium carbonate, are lost first, followed by calcium sulfate (Swenson and Colby, 1955, p. 24–27). The Salton Sea in Imperial Valley, Calif., is fed by irrigation drainage and since its origin, during flooding by the Colorado River some 60 years ago, has become rather highly mineralized. Hely, Hughes, and Irelan (1966) concluded water of the Salton Sea in 1965 was at saturation with respect to gypsum.

Ground water associated with limestone would commonly be expected to be near saturation with respect to calcite. Unfortunately, most available analytical data do not include pH determinations that are satisfactory for testing this expectation rigorously. Analyses 1, 2, and 5, table 15, all represent springs issuing from limestone, but only for one is the pH given, and even in this instance the determination was made in the laboratory long after the sample was obtained. The observed value is within about 0.5 pH unit of what would represent saturation for the measured concentrations of calcium and bicarbonate. The spring from the Grand Canyon region (analysis 2) yields water highly supersaturated with respect to calcite at the CO_2 pressure of the atmosphere and deposits travertine. Other travertine-depositing springs are represented by analyses given by White, Hem, and Waring (1963, p. F54).

Analysis 5, table 15, represents water from a spring near Carlsbad Caverns, N. Mex. The analysis shows some influence of the gypsiferous rocks in this region. The properties of analysis 3 have been mentioned as showing a water at saturation with respect to gypsum.

Analysis 9 represents water from a sandstone with calcareous cement that has given the solution a character typical of saturation with calcite. Analysis 10 is for a water from calcareous and gypsiferous shale.

MAGNESIUM

Water chemists often have a tendency to think of the behavior of magnesium as being very similar to that of calcium, perhaps because the two elements are the principal causes of the property of hardness. The geochemical behavior of magnesium, however, is substantially different from that of calcium. Magnesium ions are smaller than sodium or calcium ions and, therefore, have a stronger charge density and a greater attraction for water molecules. In solution in water, the magnesium ion probably is surrounded by six water molecules in an octahedral arrangement, similar to that described for aluminum.

The hydration shell of the magnesium ion is not as strongly held as that of aluminum ions, but the effect of hydration is much greater for magnesium than for the larger ions calcium and sodium. The tendency for crystalline magnesium compounds to contain water or hydroxide is probably related to this hydration tendency.

SOURCES OF MAGNESIUM

In igneous rock, magnesium is typically a constituent of the dark-colored ferromagnesian minerals. Specifically, these include olivine, pyroxenes, amphiboles, and dark-colored micas, along with various less common species. In altered rocks, magnesian mineral species such as chlorite, montmorillonite, and serpentine occur. Sedimentary forms of magnesium include carbonates such as magnesite and hydromagnesite, the hydroxide brucite, and mixtures of magnesium with calcium carbonate. Dolomite has a definite crystal structure in which calcium and magnesium are present in equal amounts.

The alteration of magnesian olivine (forsterite) to serpentine can be written

$$5Mg_2SiO_4 + 8H^+ + 2H_2O = Mg_6(OH)_8Si_4O_{10} + 4Mg^{+2} + H_4SiO_4.$$

This is somewhat analogous to the reactions shown previously for weathering of feldspar and produces a solid alteration product, serpentine. The reaction, like that for the alteration of other silicates, is not reversible and cannot be treated as a chemical equilibrium. Released products, however, can be expected to participate in additional reactions.

FORM OF DISSOLVED MAGNESIUM

The magnesium ion, Mg^{+2}, will normally be the predominant form of magnesium in solution in natural water. This ion is hydrated and could be represented as $Mg(H_2O)_6^{+2}$, but it is customary to omit the water of hydration in writing the formula. Data given by Sillén and Martell (1964, p. 41–42) show that the complex $MgOH^+$ will not be significant below about pH 10.00. The ion pair $MgSO_4(aq)$ has about the same stability as the species $CaSO_4(aq)$, and magnesium complexes with carbonate or bicarbonate have approximately the same stability as the similar species of calcium. The sulfate ion pair and the bicarbonate ion pair will not be significant unless the solution contains more than 1,000 mg/l of sulfate or bicarbonate.

SOLUBILITY CONTROLS

Magnesium carbonate solubility relationships cannot be represented as simply as those of calcium carbonate, because there are many different forms of magnesium carbonates and hydroxy-carbonates and they may not dissolve reversibly. Magnesite, $MgCO_3$, from solubility products given by Sillén and Martell (1964, p. 136, 137), seems to be about twice as soluble as calcite. However, the hydrated species nesquehonite, $MgCO_3 \cdot 3H_2O$, and lansfordite, $MgCO_3 \cdot 5H_2O$, are considerably more soluble than magnesite. The basic carbonate hydromagnesite, $Mg_4(CO_3)_3(OH)_2 \cdot 3H_2O$, may be the least soluble species under some conditions. Magnesite apparently is not usually precipitated directly from solution, and as noted by Hostetler (1964), a considerable degree of supersaturation with respect to all magnesium carbonate species can exist for long periods of time.

Most limestones contain a moderate amount of magnesium. The activity of magnesium which should be assigned to such impure solids so that they may be evaluated as controls over the solubility of their components by equilibrium calculations is uncertain. It was suggested by Garrels and Christ (1964, p. 47–49) that the activities of the cations in the solid are not unity, but could be approximated from their mole fractions. More work, however, is needed to evaluate more closely the range of compositions for which this generalization can be used. Berner (1967) found numerous complications in studies of solubility of calcium carbonate in the presence of magnesium.

The process by which the large volumes of dolomite rock that are now in existence were formed is not completely evident, but conditions under which dolomite will precipitate directly from solution seem not to be common in nature at present. Dolomite precipitates have been reported from saline lakes in various places. Hostetler (1964) cited reports from South Australia, Austria, Utah, and the U.S.S.R., but noted that the amounts of magnesium, calcium, and carbonate ions in water associated with dolomite was often far above saturation, as calculated from values given in the literature.

The solubility product for dolomite given by Garrels, Thompson, and Siever (1960) is $10^{-19.33}$. This value was obtained in a laboratory study in which equilibrium was approached only from unsaturation. Several investigators (Hsu, 1963; Holland and others, 1964; Barnes and Back, 1964a) have used analyses of ground water in association with dolomite to calculate an ion-activity product near $3 \times 10^{-17.0}$. The general characteristics of ground water from dolomitic terrane suggest that saturation with respect to calcite also occurs in many such waters. Yanat'eva (1954) reported dolomite was somewhat more soluble than calcite under a partial pressure of CO_2 approaching that of ordinary air.

Magnesium hydroxide (brucite) occurs naturally, and the solubility product at 25°C s between 10^{-10} and 10^{-11} (Sillén and Martell, 1964, p. 41–42).

Eaton, McLean, Bredell, and Doner (1968) proposed that the solubility of magnesium in irrigation drainage water might be limited by precipitation of a magnesium silicate of unspecified composition.

OCCURRENCE OF MAGNESIUM IN WATER

Some indications of the properties of water associated with dolomite can be obtained from the analyses in table 16. Water from dolomite which is at or below saturation should contain nearly equal concentrations of calcium and magnesium in terms of milliequivalents per liter, because in the solution process, equal amounts of the two ions will be dissolved. Water that is near or above saturation, however, may have lost some calcium by calcite precipitation, so the water attains a concentration of magnesium greater than that of calcium. Analysis 1, table 16, is somewhat below saturation with respect to calcite and dolomite. Analysis 10, on the other hand, shows a substantial supersaturation with respect to both. Calcium and magnesium equivalents are nearly equal in analysis 1, but magnesium substantially exceeds calcium in analysis 10. Both analysis 2 and 7 represent water showing some influence of dolomitic minerals, but both waters are well below saturation. The calculations for all these are subject to some uncertainty because the pH values were obtained in the laboratory some time after the samples were collected.

Waters in which magnesium is the predominant cation are somewhat unusual. Large concentrations, such as shown by analysis 6, may occur in brines associated with evaporites. Analysis 4 represents water than has participated in reactions with magnesium silicate minerals. Analysis 3 represents a more dilute solution affected by weathering of olivine.

Some control over the relative amounts of magnesium and calcium present in solution may be exerted by the coprecipitation of magnesium in calcite. The extent to which this may occur is not clear from published literature. In discussing the chemistry of irrigation water, Eaton (1950) postulated the nearly complete precipitation of calcium and magnesium as carbonates from irrigation water in the soil when the soil moisture was depleted. Although general behavior of magnesium in irrigation water seems to be in accord with Eaton's postulate, there are very few data on the composition of carbonates deposited in the soil or subsoil.

Ion-exchange minerals in rocks and in soil may adsorb magnesium a little more strongly than calcium, but this effect evidently is not very important as a control over magnesium concentrations.

TABLE 16.—*Analyses of waters in*

Analyses by U S. Geological Survey except as indicated. Date under sample number is date of collection.
and Barker (1962,

Constituent	(1) Feb. 27, 1952		(2) April 11, 1952		(3) Sept. 29, 1948		(4) Oct 18, 1957	
	mg/l	meq/l	mg/l	meq/l	mg/l	meq/l	mg/l	meq/l
Silica (SiO$_2$)	8.4		18		31		175	
Aluminum (Al)	1.4							
Iron (Fe)	24		1.4					
Calcium (Ca)	40	1.996	94	4.69	20	1.00	34	1.70
Magnesium (Mg)	22	1.809	40	3.29	42	3.45	242	19.90
Sodium (Na)	.4	.017	17	.74	19	.83	{184	8.00
Potassium (K)	1.2	.031	2.2	.06			{ 18	.46
Carbonate (CO$_3$)	0		0		0		0	
Bircarbonate (HCO$_3$)	213	3.491	471	7.72	279	4.57	1,300	21.31
Sulfate (SO$_4$)	4.9	.102	49	1.02	22	.46	6.6	.14
Chloride (Cl)	2.0	.056	9.0	.25	7	.20	265	7.47
Fluoride (F)	.0	.000	.8	.04	.2	.01	1.0	.05
Nitrate (NO$_3$)	4.8	.077	2.4	.04	2.5	.04	.2	.00
Dissolved solids:								
Calculated	190		466		281		[1]1,580	
Residue on evaporation	180		527					
Hardness as CaCO$_3$	190		400		222		1,080	
Noncarbonate	16		13		0		14	
Specific conductance (micromhos at 25° C)	326		764		458		2,500	
pH	7.4		6.7		8.2		6.5	
Color	5		5					

[1] Includes 18 mg/l of boron (B).
[2] Contains 0.01 mg/l of manganese (Mn) and 0.9 mg/l of lithium (Li).
[3] Density, 1.345 g/ml.

1. Spring 2½ miles northwest of Jefferson City, Tenn. Flow, 5,000 gpm. From Knox Dolomite. Temp, 14 4°C.
2. Well number 5, City of Sidney, Ohio. Depth, 231 ft. Water-bearing formation, Niagara Group. Temp, 11 7°C.
3. Spring in Buell Park, Navajo Indian Reservation, Ariz. Water-bearing formation, olivine tuff-breccia; 18 gpm. Temp, 12 2 °C
4. Main spring at Siegler Hot Springs, NE¼ sec. 24, T. 12 N., R. 8 W., Lake County, Calif. Water issues from contact with serpentine and sediments. Temp, 52.5°C.

Analysis 5, table 16, represents a water from Carlsbad Caverns, N. Mex., which has lost calcium by precipitation of calcite on the cave formations. It is considerably enriched in magnesium as a result of this process. A similar apparent enrichment may occur in water samples that precipitate calcium carbonate on the inside of the container during storage (analysis 2, table 11). The calcium carbonate deposited in this way is usually almost pure.

In most natural fresh water, the magnesium concentration is much lower than the calcium concentration, and Livingstone's (1963) mean composition of river water (table 3) also suggests that magnesium concentration (4.1 mg./l) is normally less than the sodium concentration (6.3 mg/l). Table 1, showing abundances of elements in rocks, indicates that magnesium is substantially less abundant than calcium in all rock types and, therefore, should be less available for solution in water.

Table 7, showing residence times of the elements in the ocean, points up another factor of some importance: Magnesium has a very long residence time compared to calcium. This means the total supply of

SIGNIFICANCE OF WATER PROPERTIES AND CONSTITUENTS

which magnesium is a major constituent

Source of data: numbers 1, 3–7, and 9, unpub. data, U.S. Geol. Survey files; numbers 2, 8, and 10, Scott p. 19, 87, 113)]

(5) Apr. 20, 1945		(6) Feb. 7, 1939		(7) May 21, 1952		(8) May 27, 1952		(9) Mar. 11, 1952		(10) Oct. 29, 1954	
mg/l	meq/l	mg/l	meq/l	mg/l	meq/l	mg/l	meq/l	mg/l	meq/l	mg/l	meq/l
11	....	....	....	4.2	....	12	....	13	....	18	....
....	....	....	....	....	....	1.0	....	....	....	.2	....
.04	....	....	....	.13	....	.03	....	.01	....	.39	....
16	0.80	130	6.49	18	0.898	23	1.148	140	6.99	35	1.747
71	5.84	51,500	4,236.39	8.5	.699	12	.987	43	3.54	33	2.714
7.8	.34	{59,000 / 2,810	2,566.60 / 71.85	2.5 / 2.4	.109 / .061	1.7 / 1.6	.074 / .041}	21	.91	{28 / 1.3	1.218 / .033
21	.70	....	....	0	....	0	....	0	....	0	.000
320	5.25	1,860	30.49	90	1.475	127	2.081	241	3.95	241	3.950
21	.44	299,000	6,225.18	9.5	.198	1.6	.033	303	6.31	88	1.832
8	.23	22,600	637.55	1.5	.042	2.0	.056	38	1.07	1.0	.028
1.0	.05	....	....	.0	.000	.0	.000	.8	.04	.9	.047
19	.31	....	....	.8	.013	.9	.014	4.1	.07	1.2	.019
333	....	436,000	....	92	....	119	....	682	....	²326	....
321	....	510,000	....	106	....	108	....	701	....	329	....
332	....	....	....	80	....	107	....	526	....	224	....
34	....	....	....	6	....	3	....	329	....	27	....
570	....	(³)	....	165	....	197	....	997	....	511	....
8.3	....	....	....	7.0	....	7.6	....	7.4	....	8.2	....
....	....	....	....	35	....	5	....	....	....	10	....

5. Green Lake in Carlsbad Caverns, N. Mex. Pool of ground-water seepage.
6. Test well number 1, SW¼ sec. 24, T. 25 S., R. 26 E., Eddy County, N. Mex. Water from 142–195 ft.
7. Wisconsin River at Muscoda, Grant County, Wis. Flows through area of magnesian limestone.
8. Spring, SW¼ sec. 26, T. 16 S., R. 7 E., Calhoun County, Ala. Supplies city of Anniston. Flow, 46 cfs. Water-bearing formation, quartzite. Temp, 17.8°C.
9. Oasis flowing well, SW¼ sec. 15, T. 11 S., R. 25 E. Chaves County, N. Mex. Depth, 843 ft. Flow over 9,000 gpm when drilled. Water-bearing formation, San Andres Limestone (limestone and dolomitic limestone with minor amounts sandstone, gypsum, and anhydrite).
10. Drilled well, NW¼ sec. 6, T. 6 N., R. 21 E., Milwaukee County, Wis. Depth, 500 ft. Water-bearing formation, Niagara Dolomite. Temp, 10.0°C.

calcium is in more rapid circulation than magnesium. The long residence time for magnesium shows that it precipitates less readily than calcium.

SODIUM

Sodium is the most abundant member of the alkali-metal group. The other naturally occurring members of this group are lithium, potassium, rubidium, and cesium. In igneous rocks, sodium is only slightly more abundant than potassium, but in sediments, sodium is much less abundant. The amounts of sodium held in evaporite sediments and in solution in the ocean are an important part of the total.

When sodium has been brought into solution, it tends to remain in that status. There are no important precipitation reactions that can maintain low sodium concentrations in water, in the way that carbonate precipitation controls calcium concentrations. Sodium is retained by adsorption on mineral surfaces, especially by minerals with high cation exchange capacities such as clays. The effect of ion-exchange reactions, however, is only to replace one kind of solute ions with another, and, therefore, these reactions do not directly control the solubility of ions.

SOURCES OF SODIUM

According to Clarke's (1924b, p. 6) estimate, about 60 percent of the body of igneous rock in the earth's outer crust consists of feldspars. The feldspars are tectosilicates, with some aluminum substituted for silica and with other cations making up for the positive-charge deficiency that results. The common feldspars are orthoclase, $KAlSi_3O_8$, and plagioclase, which has a composition ranging from the pure sodium form albite to the pure calcium form anorthite.

Potassium feldspar is very resistant to chemical attack; however, species containing sodium and calcium are more susceptible and upon attack yield the metal cation and silica to solution and form a clay mineral with the aluminum and part of the original silica. Besides the kaolinite shown as a product in earlier discussions of this reaction in this paper, other clays such as illite or montmorillonite may be formed. Feth, Roberson, and Polzer (1964), in studies of water in the Sierra Nevada of California and Nevada, found that in general the calcium and sodium ions were the most important in stream and spring water and that these tended to reflect abundance of the ions in the type of rock and the rate at which the minerals were attacked. Analyses 1--3, table 12, represent water from igneous terrains in which moderate sodium concentrations are reached. Sodium is the principal cation in each of these waters.

In resistate sediments, sodium may be present in unaltered mineral grains as an impurity in the cementing material or as crystals of readily soluble sodium salts deposited with the sediments or left in them by saline water that entered them at some later time. The soluble salts go into solution readily and usually are rather quickly removed from coarse-grain sediments after environmental changes, such as an uplift of land surface or decline of sea level, impose a freshwater leaching regime. During the early stages of the leaching process, the water leaving the formation may have high concentrations of sodium in solution. The last traces of marine salt or connate water may persist for long periods where circulation of water is impaired.

In hydrolyzate sediments, the particles normally are very small, and the circulation of water through the material is greatly impaired. Thus, the water trapped in the sediment when it was laid down may be retained with its solute load for long periods. The hydrolyzates include a large proportion of clay minerals with large cation-exchange capacities. Several kinds of effects on water composition are attributable to sediments of this type. Analysis 4, table 17, represents low flow in the Moreau River in western South Dakota, where the country rock is fine grained and contains soluble material. In the process of weathering, the exposed surfaces of these rocks may develop noticeable efflorescences of salts by evaporation of water between rainy periods. Precipitation heavy enough to wash away these deposits occurs occasionally, and at the same time the layer of leached sediment at the surface also may be stripped away. Colby, Hembree, and Jochens (1953) have discussed these effects in greater detail.

Wells that penetrate clay and shale to reach lower, more permeable beds may at times obtain water high in dissolved solids directly from the shale layers, as seems indicated in analysis 9, table 17. Such wells may also obtain mixtures of water from sands and clays with a deterioration in water quality sometimes being noted as the more permeable beds are dewatered and the proportion contributed from fine-grained material increases. Kister and Hardt (1961) observed this effect in irrigation wells of the Santa Cruz basin in Arizona.

The Dakota Sandstone is a well-known artesian aquifer, which yields water of a rather high sodium and dissolved-solids content over much of the area of discharge. The geology and hydrology of the system are complex (Swenson, 1968) and at least some of the solutes in water from the sandstone must originate from rocks other than the Dakota. Analysis 8, table 14, shows the composition of water obtained from the Dakota Sandstone in southeastern North Dakota. The water is high in sodium and in sulfate.

Ion exchange and membrane effects associated with clays could influence ground- and surface-water composition in other ways as well. Some of these effects are not yet well understood. Hanshaw (1964), for example, showed that when compacted, clays may preferentially adsorb sodium, but when dispersed in water, they may preferentially adsorb calcium. The role of clay membranes in producing electrical potentials that can be observed between strata penetrated by wells and the probability that clay membranes may allow water to pass through while holding back ions have already been suggested. Some aspects of this osmotic effect were discussed by Hanshaw and Zen (1965).

SOLUBILITY AND FORM OF DISSOLVED SPECIES

Sodium bicarbonate is one of the less soluble of the common sodium salts. At ordinary room temperature, however, a pure solution of this salt could contain up to about 15,000 mg/l of sodium. In natural water, the conditions required for precipitation of pure sodium bicarbonate are unlikely to be attained, although water in some closed basins may attain high concentrations of carbonate and bicarbonate and leave a residue of solid forms of sodium carbonate. These are somewhat more soluble than the bicarbonate Sodium carbonate evaporite deposits occurring in southern Wyoming are mined as a source of soda ash, and

TABLE 17.—Analyses of water in

[Analyses by U.S. Geological Survey. Date below sample number is date of collection Source of data: 1, Water-Supply Paper 1162 (p. 457); 5 and 6, U.S. Geol. Survey, Water-

Constituent	(1) June 3, 1952		(2) Oct. 3, 1949		(3) May 13, 1952		(4) Jul. 2-3, 1949	
	mg/l	meq/l	mg/l	meq/l	mg/l	meq/l	mg/l	meq/l
Silica (SiO_2)	22		16		22		8.2	
Iron (Fe)	.20		.15		.00		.04	
Calcium (Ca)	2.5	0.12	3.0	0.15	49	2.45	40	2.00
Magnesium (Mg)	2.1	.17	7.4	.61	18	1.48	50	4.11
Sodium (Na)	1182	7.90	857	37.28	168	7.29	699	30.40
Potassium (K)			2.4	.06			16	.41
Carbonate (CO_3)	30	1.00	57	1.90	0		26	.87
Bicarbonate (HCO_3)	412	6.75	2,080	34.09	202	3.31	456	7.47
Sulfate (SO_4)	3.5	.07	1.6	.03	44	.92	1,320	27.48
Chloride (Cl)	9.5	.27	71	2.00	246	6.94	17	.48
Fluoride (F)	1.7	.09	2.0	.11	.1	.01	1.0	.05
Nitrate (NO_3)	.6	.01	.2	.00	2.2	.04	1.9	.03
Boron (B)			.40				.60	
Dissolved solids:								
Calculated	457		2,060		649		2,400	
Residue on evaporation	452				651		2,410	
Hardness as $CaCO_3$	15		38		196		306	
Noncarbonate	0		0		31		0	
Specific conductance (micromhos at 25°C)	718		2,960		1,200		3,140	
pH	8.7		8.3		7.7		8.2	
Color	1				2			

¹ Density, 1.019 g/ml at 20°C.
² Density, 1.21 g/ml.

1. Well at Raleigh-Durham airport, Wake County, N.C. Depth, 184 ft. Water-bearing formation, Coastal Plain sediments.
2. Well, SE¼NE¼ sec. 2, T. 22 N., R. 59 E., Richland County, Mont. Depth, 500 ft. Water-bearing formation, Fort Union Formation (sandstone and shale).
3. Irrigation well, SE¼ sec. 3, T. 1 N., R. 5 E., Maricopa County, Ariz. Depth, 500 ft. Water-bearing formation, valley fill. Temp, 20.5°C.

a brine associated with the deposits is represented by analysis 2, table 18. This water contains 22,700 mg/l of sodium.

A higher solubility limit on sodium concentration is exerted by the separation of solid sodium chloride, or halite. When saturated with respect to halite, a solution could have as much as 150,000 mg/l of sodium and about 230,000 mg/l of chloride, but concentrations this high are seldom reached in natural environments. Analysis 8, table 17, is a natural brine associated with a halite deposit. In this solution, the sodium concentration is 121,000 mg/l.

Because of the high sodium concentrations that can be reached before any precipitate is formed, the sodium contents of natural water have a very wide range, from less than 1 mg/l in rainwater and dilute stream runoff in areas of high rainfall to the very high levels found in brines of closed basins where more than 100,000 mg/l may be present.

The sodium of dilute waters where total solids concentrations are below 1,000 mg/l is generally in the form of the Na^+ ion. In more concentrated solutions, however, a variety of complex ions and ion pairs is possible. Species for which stability constants are given in Sillén and Martell (1964, p. 136, 235) include $NaCO_3^-$, $NaHCO_3(aq)$, and $NaSO_4^-$.

SIGNIFICANCE OF WATER PROPERTIES AND CONSTITUENTS 149

which sodium is a major constituent

3, 7, and 8, unpub data U S. Geol Survey files; 2, Torrey and Kohout (1956, p. 44); 4, U S. Geol. Survey Supply Paper 1253 (p 205, 214), 9, Griggs and Hendrickson (1951, p. 111)]

(5)		(6)		(7)		(8)		(9)	
1951-1952		1951-1952		Dec. 8, 1951		Jan 31, 1938		Oct 15, 1946	
mg/l	meq/l	mg/l	meq/l	mg/l	meq/l	mg/l	meq/l	mg/l	meq/l
18		31		46				13	
.01		.01							
48	2.40	353	17.61	505	25.20	722	36.03	30	1.50
14	1.15	149	12.25	291	23 94	2,490	204.83	31	2.55
150	6.52	1,220	53.05	10,100	439 35	121,000	5,263 50}	279	12.13
5 8	.15	9 8	.25	170	4.35	3,700	94.61}		
0		0		0		63	2.10	0	
153	2 51	355	5.82	1,520	24 91	40	.66	445	7.29
50	1.04	1,000	20.82	899	18 72	11,700	243.59	303	6.31
233	6.57	1,980	55.84	16,200	457.00	189,000	5,331.69	80	2.26
.4	.02	1 9	.10					1.2	.06
2.4	.04	24	.39					17	.27
.15		2.4		17					
597		4,940		¹29,000		²329,000		973	
611									
178		1,490		2,460				202	
52		1,200		1,210				0	
1,090		7,620		41,500		225,000		1,510	
				7.1					

4. Moreau River at Bixby, S. Dak ; composite of two daily samples, mean discharge 1.7 cfs. Drains Pierre Shale, Fox Hills Sandstone, and Hell Creek Formation.
5. Salt River below Stewart Mountain Dam, Ariz Weighted average 1952 water year; mean discharge 362 cfs.
6. Gila River at Gillespie Dam, Ariz Weighted average 1952 water year, mean discharge 71.1 cfs
7. Spring entering Salt River at Salt Banks near Chrysotile, Ariz Water-bearing formation, quartzite and diabase Temp, 21 1°C
8. Test well 3, sec 8, T. 24 S , R. 29, E., Eddy County, N. Mex. Depth, 292 ft. Brine from Salado Formation and Rustler Formation.
9. Well in SW¼ sec. 7, T. 17 N., R. 26 E., San Miguel County, N. Mex. Depth, 50 ft. Water-bearing formation, shale of the Chinle Formation.

In sea water, as noted by Garrels and Christ (1964, p. 105–106), the proportions of the total anion contents which may be in the form of sodium ion pairs can be substantial.

OCCURRENCE OF SODIUM IN WATER

Water exposed to evaporite rocks may attain high sodium concentrations. Analysis 8, table 17, as noted previously, contains 121,000 mg/l of sodium as a result of contact with halite. Analysis 7 in this table represents saline spring water containing 10,100 mg/l of sodium. Water like this discharges to the Salt River in the area northeast of Globe, Ariz. The water of the river shows the effect of these saline inflows where it is withdrawn from reservoirs for irrigation in the Phoenix area (analysis 5, table 17). The use and reuse of the water in the irrigated area causes the final effluent below the irrigated area to be very high in dissolved sodium (analysis 6, table 17).

Analyses 1 and 2, table 17, represent water that has been naturally softened by cation exchange, probably within the saturated zone. A water that has lost most of its calcium and magnesium by cation exchange may later participate in chemical reactions that raise the pH.

The pH of such a solution can rise to rather high levels because buffering by calcium carbonate precipitation becomes relatively ineffective. The pH reported in analysis 1 is 8.7.

Many water analyses in the literature report a computed value for sodium rather than an actually determined one. The computed value represents the difference between the sum of the determined anions in milliequivalents per liter and the determined cations expressed in the same units. Obviously, the computation cannot be made unless all the other major ions have been determined. Potassium is commonly lumped with sodium in the computation, and the value reported as "sodium and potassium, as sodium." The principal reason for omitting sodium from the analysis in former years was that the determination was difficult, tedious, and expensive by the procedures then available. In more recent times, the flame photometer and related flame spectrochemical methods have made the sodium determination one of the quickest and easiest in the analytical chemists' repertoire, and thus analyses with computed sodium values are no longer common. The principal objection to the computation of sodium is that it prevents any really effective check of the accuracy of the determinations of the major dissolved components; therefore analyses reporting computed sodium values are more likely to contain major errors.

POTASSIUM

As shown in table 1, potassium is slightly less common than sodium in igneous rock, but potassium is the more abundant in all the sedimentary rocks. In the ocean, the concentration of potassium, though substantial, is far less than that of sodium. These figures point up the very different behavior of these two alkali metals in natural systems. Sodium tends to remain in solution rather persistently once it has been liberated from silicate-mineral structures. Potassium is liberated with greater difficulty from silicate minerals and exhibits a strong tendency to be reincorporated into solid weathering products, especially certain clay minerals. In most natural water, the concentration of potasssium is much lower than the concentration of sodium.

SOURCES OF POTASSIUM

The principal potassium minerals of silicate rocks are the feldspars orthoclase and microcline ($KAlSi_3O_8$), the micas, and the feldspathoid leucite ($KAlSi_2O_6$). The potassium feldspars are very resistant to attack by water, but presumably they are altered to silica, clay, and potassium ions by the same process as other feldspars, only more slowly.

In sediments, the potassium commonly is present in unaltered feldspar or mica particles or in illite or other clay minerals. Evaporite rocks locally include beds of potssium salts and constitute a source for high potassium concentration in brines.

Potassium is an essential plant nutrient and is present in the ash of plants, sometimes in rather high percentages. A few data are given by Lovering (1959) in comments on the role of plants in the weathering of silicates.

OCCURRENCE OF POTASSIUM IN WATER

Although geologic terranes in which the potassium content of rocks is greatly in excess of the sodium content are not unusual, there are few fresh waters in which the potassium concentration nearly equals or even exceeds the sodium concentration. Several of the analyses in the tables of this report, notably analysis 9, table 14, and anlyses 1, 7, and 8, table 16, do show this relationship. Two represent ground water from sedimentary rock, and one represents a ground water from quartzite. The other is a dilute runoff water in a sedimentary area.

In the more dilute waters, where sodium contents are below 10 mg/l, the potassium concentration may commonly be half or a tenth that of sodium. Concentrations of potassium more than a few tens of milligrams per liter, however, are decidedly unusual except in water with very high dissolved-solids concentrations or in water from hot springs. The concentration of potassium in the two most saline waters represented in table 17 are much above the usual levels for more dilute solutions.

The rather narrow range of concentration of potassium observed in natural water suggests a significant chemical control mechanism may be involved. The nature of such a mechanism, however, is not well enough known so that any calculations can be made. The two general principles already referred to are the resistance to solution exhibited by potassium feldspar and the apparent preferential incorporation of potassium into clay or mica mineral structures.

Regarding the rate of solution of potassium feldspar, Garrels (written commun., 1967) has commented that potassium feldspar is nearly inert and tends to remain in the form of solid particles during weathering of igneous rock.

The potassium ion is substantially larger than the sodium ion, and it would normally be expected to be adsorbed less strongly than sodium in ion-exchange reactions. Actually, however, potassium is incorporated in a special way into some clay-mineral structures. In illite, potassium ions are incorporated in spaces between crystal layers where they are not removable by further ion-exchange reactions (Buckman and Brady, 1960, p. 81). A considerable amount of work on potassium behavior has been done by soil scientists, but water chemists have not studied potassium behavior extensively. More research on this subject is certainly desirable.

Hicks (1921) suggested that leaching of the ashes of grass destroyed in prairie fires played a part in the concentration and transportation of potassium and its accumulation in closed-basin lakes in the sandhill region of Nebraska. The element is removed from the soil by plants, but under natural conditions is returned to the soil when the plant dies. The effect of plants and addition of fertilizers to the soil in the circulation of potassium in water may be significant in some intensively cultivated areas. It should be pointed out, however, that most others who have tried to explain the accumulation of potassium in the Nebraska lakes have not accepted Hicks' interpretation. It may be of interest to note that streams of the north-central prairie region of the United States carry consistently higher potassium concentrations than streams of comparable dissolved-solids content in other parts of the United States. U.S. Geological Survey water-quality records also show a tendency for the potassium concentration in these streams to be nearly the same at low and high flows. Steele (1968a, p. 72) observed a tendency for rather high potassium concentrations in direct storm runoff in Pescadero Creek in San Mateo County, Calif.

There are deficiencies in basic information regarding chemistry of potassium in water. Many of the available chemical analyses of water and some of the analyses included in this report do not include potassium determinations. Some of the old analytical methods for potassium require considerable analytical skill, and some of the earlier published values for potassium concentrations appear to be of questionable accuracy. In more recent years, better methods have become available, but potassium has remained one of the most difficult ions to determine accurately.

ALKALINITY

In water chemistry, alkalinity is defined as the capacity of the solution to neutralize acid. To complete this definition, some end-point pH needs to be specified. Normally, the pH designated is a selected value between 5.1 and 4.5, or that of the methyl-orange end point, (about pH 4.0–4.6). Sometimes, however, an alkalinity above the phenolphthalein end point (about pH 8.3) is also specified. Thus one may find terms such as "methyl-orange alkalinity," or its equivalent "total alkalinity," and "phenolphthalein alkalinity."

Several different solute species contribute to the alkalinity of water as defined above. A noncommittal statement is often made by reporting alkalinity in terms of an equivalent quantity of calcium carbonate. In most natural water, the alkalinity is practically all produced by dissolved carbonate and bicarbonate ions. A more meaningful and useful statement of the alkalinity in such water is obtained by expressing the results of the determination as concentrations of bicarbonate and carbonate. The analyses in this book are expressed in this way, as are all water analyses made and published by the U.S. Geological Survey. Conversion factors for changing alkalinity concentrations from one form of expression to another are given in table 8.

The alkalinity defined here is a capacity function which therefore has a different chemical basis than the intensity function, pH. A solution whose pH is neutral may have considerable titratable alkalinity.

SOURCES OF ALKALINITY

Any ion that enters into a chemical reaction with strong acid can contribute to titrated alkalinity, provided the reaction takes place significantly above the pH of the specified end point. For the most part, alkalinity is produced by anions or molecular species of weak acids which are not fully dissociated above a pH of 4.5.

The most common of the weak acids in natural water is carbonic acid, formed when carbon dioxide is dissolved. The relevant reactions were listed in the section "Hydrogen-ion activity (pH)." The addition of acid (represented by H^+) to solutions containing carbonate or bicarbonate reverses these reactions. The hydrogen ions produced by dissociation of the dissolved carbon dioxide species are the principal source of the H^+, which is often written as a reactant in equations representing processes of rock-mineral dissolution. Where the dissolved carbon dioxide species are relatively plentiful, they are instrumental in maintaining the pH of the solution at a fairly constant value, although the pH also is influenced by many other reactions involving solids, other solutes, and H^+.

Other weak acids which may contribute species to titratable alkalinity include silicic, phosphoric, and boric acids. In some waters there also may be a contribution from organic acids or a wide variety of dissolved or suspended materials normally present only in trace quantities. The direct contribution of hydroxide is determinable from the pH.

The user of water analyses should remember that under any system of reporting titrated alkalinity now in use the effects of all the anions which may react when strong acid is added are lumped together and reported as an equivalent amount of a single substance or in terms of postulated ions. Thus, although alkalinity values are generally interpreted in terms of specific concentrations of bicarbonate and carbonate ions, the determination is indirect, and the actual ionic species present may not be exactly what the analysis states. The most common deviations probably are in highly alkaline ground waters like analysis 1, table 18. At the pH of 9.4 reported in this analysis, as much as half the reported silica may be present as ionic silicate. In such cases alkalinity titration is strongly influenced by silicate, and the actual

TABLE 18.—*Analyses of waters having various alkalinity-acidity-pH relationships*

[Analyses by U.S. Geological Survey. Date under sample number is date of collection Source of data. 1 and 4, unpub. data, U.S. Geol. Survey; 2, Lindeman (1954); 3, White, Hem, and Waring (1963, p. F46)]

Constituent	(1) Sept. 9, 1954		(2) December 1935		(3) August 31, 1949		(4) August 23, 1963	
	mg/l	meq/l	mg/l	meq/l	mg/l	meq/l	mg/l	meq/l
Silica (SiO_2)	75				213		14	
Aluminum (Al)					56	6.23		
Iron (Fe)	.05				33	1.18	.03	
Manganese (Mn)	.08				3.3	.12		
Calcium (Ca)	1.3	0.065	0	0	185	9.23	12	0.599
Magnesium (Mg)	.3	.025	20	1.65	52	4.28	2.9	.239
Sodium (Na)	72	3.131	22,700	987.45	6.7	.29	8.3	.361
Potassium (K)	2.4	.061	160	4.09	24	.61	5.2	.133
Hydrogen (H)					13	12.6		
Carbonate (CO_3)	38	[1] 1.266	17,800	593.27	0			
Bicarbonate (HCO_3)	20	.328	5,090	83,43	0		10	.164
Sulfate (SO_4)	32	.666	780	16.24	1,570	32.69	13	.271
Chloride (Cl)	6.5	.183	10,600	299.03	3.5	.10	12	.339
Fluoride (F)	16	.842			1.1	.06	.1	.005
Nitrate (NO_3)	0	.0			.0	.00	36	.581
Dissolved solids:								
Calculated	254		57,100				109	
Residue on evaporation	239							
Hardness as $CaCO_3$	4		82				42	
Noncarbonate	0		0				34	
Specific conductance (micromhos at 25°C)	328		([2])		4,570		164	
pH	9.4				1.9		5.2	
Color								
Acidity as H_2SO_4 (total)					913			

[1] Probably about 0 6 meq/l of the total alkalinity is actually present in the form of $H_3SiO_4^-$.
[2] Density, 1.046 g/ml.

1. Spring NW¼ sec. 36, T. 11 N., R. 13 E., Custer County, Idaho. Water-bearing formation, quartz monzonite. Temp. 57 2°C.
2. Brine well MFS 1, NW¼ sec. 26, T. 18 N., R. 107 W., Sweetwater County, Wyo. Depth, 439 ft. Water-bearing formation, evaporates.
3. Lemonade Spring, Sulfur Springs, Sandoval County, N. Mex. Water-bearing formation, volcanic rocks. Temp. 65.6°C. Fumaroles emit H_2S and SO_2 in vicinity.
4. Spring at Winnsboro city well field, Winnsboro, Franklin County, Tex.; 25 gpm. Temp. 18.3°C.

concentrations of carbonate and bicarbonate are substantially less than the analysis reports.

OCCURRENCE AND DETERMINATION OF ALKALINITY

The pH of a water indicates directly the ratios of certain ionic species to one another if equilibrium is established. The major equilibria among dissolved carbon dioxide species, hydrogen and hydroxide ions, have been given in the section on pH. The equilibria are represented graphically by a species-distribution diagram, figure 19. This diagram shows the percentages of activities of undissociated carbonic acid and of bicarbonate and carbonate ions in the total of dissolved species of carbon dioxide as a function of pH. Activities of the dissolved species are in moles per liter and standard conditions of 25°C and 1 atmosphere of pressure are specified. The intercepts of the boundaries with the pH grid shows the percentages of the species present. For example, at a pH of 7.0, about 18 percent of the total is H_2CO_3 and 82 percent is HCO_3^-.

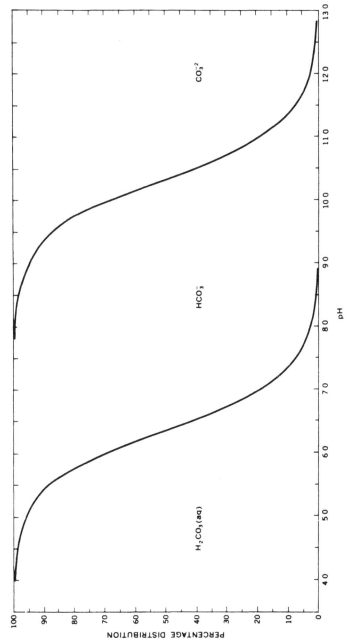

FIGURE 19.—Percentages of total dissolved carbon dioxide species in solution as a function of pH, 25°C; pressure 1 atmosphere.

The species-distribution diagram aids in the interpretation of the chemistry of the alkalinity determination. For example, when bicarbonate activity and pH are known, the activities of undissociated carbonic acid and of carbonate ions can be readily estimated. The diagram shows why the concentration of carbonate cannot be determined very accurately by titration. The pH at which carbonate constitutes 1 percent of the total dissolved carbon dioxide species, about 8.3, is where the titration end point for carbonate would generally be placed. This is a low enough pH that about 1 percent of the total now also is in the form of H_2CO_3. If a water contains much bicarbonate and only a little carbonate, the overlapping of the two steps in the vicinity of pH 8.3 may make it impossible to determine the carbonate even to the nearest milliequivalent per liter. Because of the overlap, the change in pH as acid is added may be very gradual rather than abrupt at this end point. Usually, if the carbonate concentration is small compared to the bicarbonate concentration, a value for carbonate can be calculated from the equilibrium equations more accurately than it can be measured by titration.

Figure 19 uses activities rather than ionic concentrations and, therefore, can be strictly applied only to very dilute solutions. The effects of ionic strength can be evaluated by means of the Debye-Hückel equation and appropriate calculations.

Species-distribution diagrams can be prepared for other ions of interest in natural-water chemistry. Techniques for preparing such diagrams were described by Butler (1964, p. 120). Other ionic species may contribute significantly to alkalinity, and for some purposes their contribution may need to be calculated. Orthophosphate may be significant in some solutions. From data given by Sillén and Martell (1964, p. 180-182) the predominant form of orthophosphate between pH 12.3 and 7.2 is HPO_4^{-2} and between pH 7.2 and 2.1 is $H_2PO_4^-$. Thus, during the alkalinity titration, any orthophosphate present would be converted to the dihydrogen form. Boric acid, with a first dissociation constant near $10^{-9.2}$ (Sillén and Martell, 1964, p. 105), will seldom be present in quantities sufficient to influence titrated alkalinity. The same comment applies to the numerous other possible weak acids. However, if the total concentration of the weak-acid species has been determined by other analytical procedures, the contribution to alkalinity can generally be calculated from the pH. The contribution of free hydroxide ions to alkalinity can be computed directly from the pH, the ion-activity product for water, and the ionic strength of the solution. At a pH of 9.0, the concentration of free hydroxide is not more than a few tenths of a milligram per liter. Hydroxide complexes such as $Al(OH)_4^-$ can contribute to alkalinity in some solutions.

SIGNIFICANCE OF WATER PROPERTIES AND CONSTITUENTS 157

The general subject of bicarbonate equilibria is discussed extensively in the literature. A good introduction to the topic is given by Garrels and Christ (1964, p. 74–92).

Although alkalinity titrations commonly are made to a fixed end point, a more accurate determination can be made by means of potentiometric titration where the end point is taken as the point at which the pH change is most abrupt for a given addition of titrating acid. This procedure is required for the accurate determination of alkalinity because of the effects of ionic strength and of reagent used in establishing the free H^+ content of the solution (Barnes, 1964).

OCCURRENCE OF BICARBONATE AND CARBONATE

The carbon dioxide which is dissolved by naturally circulating water appears in chemical analyses principally as bicarbonate and carbonate ions. Carbon which follows this path represents a linkage between the carbon cycle and the hydrologic cycle. The large supply of atmospheric carbon dioxide is partly intercepted by photosynthesizing vegetation and converted to cellulose, starch, and related carbohydrates. These products are later returned to carbon dioxide and water with a release of stored energy. The carbon dioxide released within the soil by respiration and decay is capable of producing low pH in circulating water if minerals that act as proton acceptors are scarce. At 1 atmosphere of pressure of pure CO_2 gas over distilled water, the solution would have a pH near 3.4. In the presence of an excess of calcite, however, figures 17 and 18 indicate that the solution would contain some 350 mg/l of dissolved calcium, and its pH would be near 6.0.

Soils of humid, temperate regions may become depleted in calcium carbonate by leaching, and the pH of ground water at shallow depths may be rather low. Analysis 4, table 18, shows this effect in ground water in northeastern Texas. The soil minerals in such areas may absorb H^+, which could be released from time to time by addition of soil amendments or by other changes in chemical environment, to reinforce the hydrogen-ion content of ground-water recharge.

In more calcareous environments, the circulation of water rich in carbon dioxide may produce solutions that are highly supersaturated when exposed to the air. Such solutions may deposit large quantities of calcium carbonate as travertine near their points of discharge. Blue Springs, represented by analysis 3, table 11, deposit travertine in the bottom of the lower section of the Little Colorado Canyon in Arizona. The springs issue from deeply buried cavernous limestone.

Because of the obvious need for some potent source of carbon dioxide to explain travertine-depositing springs, some geologic literature suggests the metamorphism of limestone might be occurring in some places to liberate CO_2. This suggestion seems to be most common in literature of the U.S.S.R. A similar mechanism is sometimes proposed to explain the discharge of gas under high pressure with a high carbon dioxide content in certain areas. Sodium carbonate can accumulate as an evaporite in closed basins, and a water very high in carbonate as a result of contact with this kind of deposit is represented by analysis 2, table 18.

The quantity of carbon dioxide produced by photosynthesis in various environments and some other quantities and rates in the carbon cycle have been estimated by Lieth (1963). His figures include maximum respiration and decay rates for different vegetation types at a steady-state condition. For a temperate, humid area with abundant vegetation, such as a European forest, he estimated that 15 metric tons of carbon dioxide per hectare (or 6.7 tons per acre) per year would be produced by respiration and decay. In a tropical rain forest, the production might be twice as great, but the amount produced by grassland is much less—only 1 or 2 metric tons per hectare per year. These figures give some idea of the upper limits for carbon dioxide availability in weathering solutions. If all the carbon dioxide produced in the temperate forest could be trapped in soil moisture and used to attack limestone, and the products were all removed in runoff, the rate of outflow of calcium and bicarbonate ions would be about 55 metric tons per hectare (or 25 tons per acre) per year. This would be equivalent to 5,500 metric tons per square kilometer, or 16,000 tons per square mile.

A considerable part of the released carbon dioxide must return directly to the atmosphere, and obviously the efficiency of utilization of the dissolved portion will be far below the theoretical maximum. The loads of bicarbonate and calcium ions carried by streams are nowhere known to approach the maximum values. Water-quality records for streams in the United States indicate that the maximum rate for calcium and bicarbonate removal is near 400 tons per square mile per year, but most streams carry far less than half this much. The figures do suggest that under the most favorable circumstances, limestone may be rather rapidly eroded. It may also be of interest to note that a considerable part of the anionic load of many streams is a contribution from the carbon dioxide of the atmosphere rather than from the rocks of the drainage basin.

The bicarbonate concentration of natural water generally is held within a moderate range by the effects of carbonate equilibria. The concentration in rainwater commonly is below 10 mg/l and sometimes less than 1.0 mg/l. Most surface streams contain less than 200 mg/l, but in ground waters somewhat higher concentrations are not uncommon. Concentrations over 1,000 mg/l sometimes occur in waters which are low in calcium and magnesium and especially where processes releasing carbon dioxide (such as sulfate reduction) are occurring in the ground-water reservoir.

ACIDITY

The term "acidity," as applied to water, is defined by the American Society for Testing and Materials (1964, p. 364) as "the quantitative capacity of aqueous media to react with hydroxyl ions." The definition of alkalinity, or "basicity," given by that reference is the same with the substitution of the word "hydrogen" for "hydroxyl." As noted in the discussion of alkalinity, a statement of the end-point pH or the indicator used is required to interpret the results of an alkalinity titration. The same requirement applies to determinations of acidity. The acidity titration, however, measures a property that is somewhat difficult to describe. Some solutes react with hydroxide ions at slow rates, and some of the products are unstable; often interpretation of the results is thus very uncertain. The usual acidity titration cannot be interpreted in terms of any single ion, and in any event, the solutes contributing to acidity are usually separately determined by other procedures. In contrast, the alkalinity determination can be defined for almost all waters as a determination of the carbonate- and bicarbonate-ion concentrations.

The titrated acidity of a water may be expressed in terms of milliequivalents per liter of hydrogen ions or as equivalent concentrations of $CaCO_3$ or of sulfuric acid. The determination has no particular geochemical usefulness. It may, however, provide some indication of the behavior of a water in certain treatment processes.

SOURCES OF ACIDITY

Some of the major constituents of natural water may react with both hydrogen ions and hydroxide ions and could thus be defined as either acidity or alkalinity. For example, bicarbonate may react with hydroxide to form carbonate,

$$HCO_3^- + OH^- = CO_3^{-2} + H_2O,$$

or with hydrogen ions to form undissociated carbonic acid,

$$HCO_3^- + H^+ = H_2CO_3(aq).$$

The $H_2CO_3(aq)$ form also is a contributor to acidity. If, for the acidity titration, and end-point pH of 4.5 is selected or the methyl-orange end point is used the effect of carbon dioxide species will generally be avoided.

The direct contribution of free H^+ to titratable acidity can be calculated directly from the pH and the ionic strength of the solution and is only a few hundredths of a milliequivalent per liter at a pH of 4.5. As pH decreases, however, the milliequivalents per liter of H^+ increase. The contribution of H^+ to the ionic balance must be taken into account in the analysis of strongly acid water. Free H^+ is sometimes determined by titration to the methyl-orange end point and reported as "free mineral acid." The determined and calculated values may not, however, coincide. As in the titration of alkalinity, the titrated species include all the contributors to buffering capacity.

Among dissolved species which may contribute to acidity on titration are undissociated or partly dissociated acids other than carbonic acid. One of the most common species of this type is the bisulfate ion, HSO_4^-, which may constitute about half of the total dissolved-sulfate species at a pH of 2.0. Undissociated hydrofluoric acid could be present in significant proportions at this pH. Metal ions which combine with hydroxide and precipitate affect the acidity titration. Metal-ion species also may undergo oxidation during the titration and produce H^+ that combines with the titrating base. As noted earlier, a water containing iron standing in contact with air will tend to decrease in pH as the ferrous ions are oxidized and precipitated as ferric hydroxide.

In the titration of acidity in a water containing iron, the reactions observed may proceed very slowly at the methyl-orange end point. Determinations of total acidity are sometimes made by heating the sample aliquot and titrating to a pH of 8.3 or the phenolphthalein end point. The reactions of iron should be faster under these conditions. End points other than 4.5 or 8.3 also are used by some analysts.

Some other ions that may enter into reactions with hydroxide include

$$NH_4^+ + OH^- = NH_4OH(aq)$$

and

$$H_2S(aq) + OH^- = HS^- + H_2O.$$

Neither of these two species, however, is conveniently determinable by titration with base.

Calculation of the cation-anion balance of an acid water requires the correct formula, concentration, and ionic charge for all dissolved species, and it may be difficult to attain a balance in some of the more complicated acidic solutions.

ORIGIN OF ACID WATERS

The importance of hydrogen ions in the attack of natural waters on rock minerals has been indicated previously by the presence of hydrogen ions in most of the reaction equations which have been shown here. The three sources of most of the H^+ that participates in such reactions are hydrolysis, as in

$$H_2O = H^+ + OH^-,$$

dissociation of acidic solutes,

$$H_2CO_3 = H^+ + HCO_3,$$

oxidation reactions,

$$H_2S + 4H_2O = SO_4^{-2} + 10H^+ + 8e^-.$$

Once formed, the H^+ may be stored and redistributed by cation exchange, but this is not a process by which H^+ ions are produced. Acid also may be present in water as a result of pollution.

In natural systems where the production of H^+ is more rapid than the rate at which it can react with the minerals available, the pH may remain more acid than the range which can be reached at equilibrium in carbonate systems. Most naturally occurring water that is strongly acid is in thermal areas where a low pH is maintained by solution of acidic gases and by oxidation of sulfur species. An example of this type of water is given in analysis 3, table 18. The concentration of hydrogen ions present in this water as calculated from the pH is 13 mg/l. The spring area yields only small amounts of water, but there are fumaroles nearby from the H_2S and SO_2 gases are emitted. Studies by Ehrlich and Schoen (1967) suggested that sulfur-oxidizing bacteria in the soil in hot-spring areas may produce sulfate through oxidation of sulfur and H_2S and that bacterially mediated oxidation may produce much of the acidity in some hot-spring areas.

Other analyses of acidic water in this report include a shallow well in Arkansas (No. 1, table 13) where the high sulfate may indicate oxidation of sulfur species and a dilute spring water from Franklin County in northeastern Texas (No. 4, table 18). In the latter analysis, the observed pH is high enough, so it may be controlled by carbon dioxide equilibria. Water of his type can occur only where soil and underlying rocks are impoverished in carbonate minerals. Rainwater in industrialized areas may have a pH in the vicinity of 4, going to solution of sulfur dioxide or other impurities from smoke and stack discharges. Water draining from mines often has a low pH owing to oxidation of sulfides.

SULFATE AND OTHER SULFUR SPECIES

The element sulfur when dissolved in water generally occurs in the fully oxidized (S^{+6}) state complexed with oxygen as the anion sulfate SO_4^{-2}. The reduced form S^{-2} is sometimes found as the HS^- ion or as dissolved undissociated H_2S. Conversion of oxidized to reduced species or vice versa often is associated with biochemical processes, and in fact some reactions of sulfur such as the reduction of sulfate hardly can be made to take place in natural systems unless certain species of bacteria, along with a suitable food or energy supply, are present. Suspensions of free sulfur are sometimes produced in the oxidation of sulfides, and various polysulfide species occur naturally as solids, the most common being pyrite, FeS_2, where the sulfur is present as S^-.

Because of the slowness of sulfur oxidation or reduction reactions and because some require special conditions to take place, nonequilibrium forms of sulfur can persist for long periods. It is not possible to predict from equilibrium assumptions what the oxidation state of sulfur will be in all systems. Therefore, the Eh-pH diagram does not present a completely reliable evaluation of sulfur behavior. Nevertheless, figure 20 does have value for indicating boundary conditions and the general features of the chemistry of sulfur species.

As figure 20 shows and as Garrels and Naeser (1958) stated, the only thermodynamically stable forms of sulfur in water at 25°C and 1 atmosphere of pressure are SO_4^{-2}, HSO_4^-, free sulfur, HS^-, $H_2S(aq)$, and S^{-2}. The latter occurs only at very high pH. The activity of sulfur species assumed for figure 20 is 10^{-3} molar, equivalent to about 96 mg/l as SO_4^{-2}, the same value used in preparing figures 14 and 15.

Occasional analyses in the literature report sulfur ionic species of intermediate oxidation state, including sulfite (SO_3^{-2}) or various thionates such as $S_4O_6^{-2}$. These analyses generally represent thermal springs where the Eh-pH relationships for 25°C may not be applicable. Cloke (1963) expressed the opinion that polysulfide species may be important in transport of metals in hydrothermal solutions.

The only boundaries in figure 20 which would be changed by changing the specified total concentration would be those around the field of stability of free sulfur. This field becomes larger if sulfur activities in solution increase and becomes smaller if the sulfur activities are decreased.

SOURCES OF SULFATE IN NATURAL WATER

Sulfur is not a major constituent of the earth's outer crust, but is widely distributed in reduced form both in igneous and sedimentary rocks as metallic sulfides. Concentrations of these sulfides often constitute ores of economic importance. In weathering in contact with aerated water, the sulfides are oxidized to yield sulfate ions which are carried off in the water. Hydrogen ions are produced in considerable quantity in this oxidation process. Pyrite crystals often occur in sedimentary rock and constitute a source of both ferrous iron and sulfate in ground water. Pyrite is particularly commonly associated with biogenic deposits such as coal, which were deposited under strongly reducing conditions.

SIGNIFICANCE OF WATER PROPERTIES AND CONSTITUENTS 163

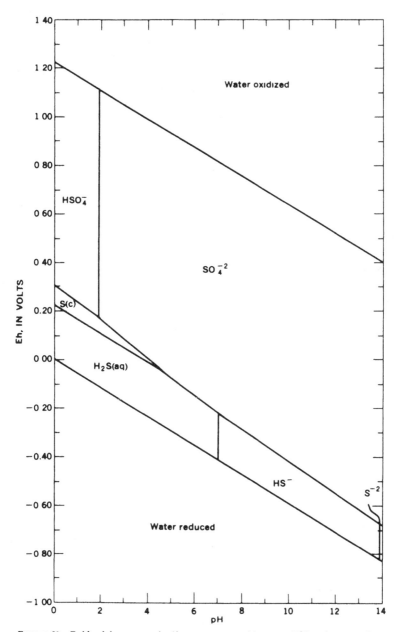

FIGURE 20.—Fields of dominance of sulfur species at equilibrium at 25°C and 1 atmosphere pressure. Total dissolved sulfur activity 96 mg/l as SO_4^{-2}.

Sulfate occurs in certain igneous-rock minerals of the fieldspathoid group, but the most extensive and important occurrences are in evaporite sediments. Calcium sulfate as gypsum, $CaSO_4 \cdot 2H_2O$, or as anhydrite, which contains no water of crystallization, makes up a considerable part of many evaporite-rock sequences. Barium and strontium sulfates are less soluble than calcium sulfate, but are relatively rare. Sodium sulfate occurs in some closed basins.

CIRCULATION OF SULFATE IN NATURAL WATER

The sulfate ion is chemically stable in aerated water and forms salts of low solubility with only a few metals, most of which are not common. The principal complications in sulfate chemistry of concern in natural water are related to the tendency for sulfate to form ion pairs and complexes with metal ions and to the involvement of sulfur in biological processes. Enough is known about sulfate ionic species that the most important ones can be evaluated. The involvement of sulfur in the biosphere, however, causes a general interconnection between the sulfur cycle and the hydrologic cycle that is not well understood.

One of the most interesting aspects of sulfate circulation is the occurrence of the sulfate ion in rainfall and other forms of precipitation. Concentrations frequently exceed 1 mg/l and are generally considerably greater than chloride concentrations except in rain falling on or near the ocean. Concentrations of sulfate in excess of 10 mg/l in rainfall frequently have been reported.

The sulfate in rainfall has been attributed to a number of factors by different writers. Conway (1943) thought it reached the atmosphere through emission of H_2S from shallow ocean water near the continental margins. The H_2S that reaches the atmosphere is ultimately oxidized to sulfur dioxide and thence to sulfate. The additional knowledge of rainfall composition obtained since 1943, however, seems conclusively to indicate that other factors are involved, because the sulfate content of rainfall inland is consistently higher than that near the ocean.

The effect of air pollution, especially the contribution from combustion of fuel, is noticeable in many places, and Junge (1960) attributed about 30 percent of the sulfate in rainfall to this source. Rain falling through unpolluted inland rural air, however, still contains considerably larger amounts of sulfate than rain near the ocean. Junge (1960) suggested this might be explained by assuming a more rapid rate of sulfur oxidation in the atmosphere over land, owing to catalytic effects of dust particles in cloud droplets. Terrestrial sources of sulfur, however, would seem to offer a simpler explanation.

Among the land-based factors which may be significant are the sulfur emitted by volcanoes, fumaroles and springs, and the solution of dust particles. Also, a major possible source whose magnitude seems to be poorly understood is the influence of land biota which extract sulfur from the soil and return some to the atmosphere. Until more information is available on the magnitude of such effects, speculation about the sulfur cycle and attempts to assign quantities to incompletely measured parts of it will remain highly uncertain.

Several writers have expressed the belief that the amount of sulfate brought to the land surface in precipitation is a major part of the total sulfate content of river runoff. To imply that this means sulfate is largely cyclic and little is currently being leached from rocks by weathering, however, is an oversimplification. In some areas, no doubt, a considerable part of the sulfate load of streams does come from rainfall. In other areas only a minor fraction of the sulfate load can be attributed to such a source.

In regions where the country rock was initially well supplied with sulfides, as most shales and fine-grained sediments are when freshly raised above sea level, the natural processes of weathering bring about oxidation from the surface down to or below the water table, and the sulfate produced is available for transport away from the area. The rate at which the sulfate is removed is a function of the runoff rate, however, and may lag behind the rate at which sulfate is produced. In humid regions, the upper layers of soil and rock are kept thoroughly leached, and as fast as they are formed the soluble products are removed from the area in a solution diluted because the amount of water available is large in proportion to the supply of solutes.

In semiarid and arid regions on the other hand, the soils are usually not fully leached, and surplus solutes may accumulate near the surface. The amount of drainage water that leaves such an area is a small fraction of the total received in precipitation. Because of these factors, the supply of solutes is relatively large in proportion to the water volume in which it can be carried away. As a result, surface and underground waters in semiarid regions tend to be comparatively high in dissolved solids. Sulfate is a predominant anion in many places.

From the time of the earliest explorations in the western half of the United States, aridity has been cited as a cause of the high dissolved-solids content of many of the streams, and comments about "alkali" occur in all the early reports on such explorations. In this context alkali meant any white efflorescence; it is mostly sodium sulfate.

Where rocks do not contain unstable minerals or other major sources of readily soluble matter, the solutes may not accumulate in soil or ground water. Except for basins with interior drainage, where the solutes cannot escape, the water occurring in regions where the rocks are of igneous or metamorphic origin may be of very good quality where annual precititation is no more than 5 inches. In some of the desert regions of southern Arizona, the ground water has less than 300 mg/l of dissolved solids.

When an area of low rainfall and accumulated solutes is reclaimed by irrigation, the increased water supply tends to leach away the solutes, and they appear in drainage water or return flow. The process is an acceleration of natural leaching and will increase the dissolved-solids concentrations and loads in the residual water of the affected area for a considerable period. Even where the soil is relatively free from soluble salts, the sulfate concentrations and chloride concentrations of the residual water draining from an irrigated area are generally much higher than they were in the original water supply.

Gypsum and anhydrite are sufficiently soluble to cause water in contact with them to be rather high in dissolved solids. Analysis 3, table 15, represents water from a spring issuing from a gypsum bed. Gypsum is abundant in the Pecos River basin, and the quality of the river water shows the effect plainly (analysis 8, table 15).

Extensively eroded terrains of fine-grained sediments, such as the badlands of certain areas in the western United States, provide a source of soluble matter that is constantly renewed as the insoluble detritus is mechanically removed by water running off the exposed rock surfaces, and fresh surfaces containing soluble material mixed with the insoluble components are exposed for future solution. Such a source is at least partly responsible for the sulfate in the water of the Moreau River represented by analysis 4, table 17, and for water in the Rio Grande during summer flood periods (fig. 4). Salts also are brought to the surface of such rocks through capillary action and surface evaporation (Colby, Hembree, and Jochens, 1953, p. 12).

Pyrite oxidation can be a potent source of sulfate as well as of hydrogen ions and iron. Analyses showing this effect strongly are No. 7, table 14, and No. 4, table 13.

FORMS OF DISSOLVED SULFATE

The dissociation of sulfuric acid is not complete in the lower pH range of natural water, and in some acid waters the bisulfate ion HSO_4^- constitutes a considerable part of the total sulfate concentration. As noted in figure 20, the HSO_4^- ion predominates below about pH 1.90. At a pH one unit higher (2.90) about 10 percent of the total sulfate would be in that form, and at a pH of 3.90 only 1 percent. Thus, above a pH of 3.9, the contribution of HSO_4^- will be insignificant. An exact calculation of the concentration of HSO_4^- can be made if pH, total sulfate, and ionic strength of the solution are known (Hem, 1960).

The usual analytical procedures for sulfate do not discriminate between the SO_4^{-2} and HSO_4^- forms, but the amount present as HSO_4^- may need to be computed to attain a satisfactory anion-cation balance in the analysis of an acid water. If no other sulfate complexes of importance are present, the two equations required are

$$[H^+][SO_4^{-2}] = [HSO_4^-] \times 1.26 \times 10^{-2} *$$

and

$$C_{SO_4} = \frac{[SO_4^{-2}]}{\gamma_{SO_4^{-2}}} + \frac{[HSO_4^-]}{\gamma_{HSO_4^-}}.$$

Square brackets indicate molar activities, or thermodynamic concentrations, and C_{SO_4} is the analytical concentration of sulfate reported. The value of $[H^+]$ can be directly obtained from pH, and the ion activity coefficients, the γ terms, can be calculated from the ionic strength of the solution by using the Debye-Hückel equation.

Analysis 3, table 18, represents a water in which a considerable part of the sulfate probably is present as HSO_4^-. The exact amounts of the various sulfate species present, however, may also depend on the concentrations of other sulfate-metal complexes.

Sulfate is itself a complex ion, but displays a strong tendency to form further complex species. The most important of these in natural-water chemistry are associations of the type $NaSO_4^-$ and $CaSO_4^0$. These generally are referred to as "ion pairs," and as sulfate concentrations increase, an increasing proportion of the sulfate in solution becomes tied up in this way. Thermodynamic data on sulfate ion pairs given by Sillén and Martell (1964, p. 232-251) show that the strongest ones are formed with divalent or trivalent cations. For calcium, the relationship

$$\frac{[CaSO_4^0]}{[Ca^{+2}][SO_4^{-2}]} = 10^{2\ 31}$$

implies that solutions containing 10^{-2}–10^{-3} moles per liter of sulfate (~1,000–100 mg/l) will contain significant amounts of the ion pair. The ionic balance of the analysis is not affected if species of this type are present, and they are usually ignored in chemical analyses. The ion pairs do influence solubility of solids such as gypsum, however, and because the ion pairs have lower charges than the free ions (actually zero for the $CaSO_4$ form), their presence complicates calculation of dissolved solids from conductivity determinations and influences the behavior of ions in the chemical analysis of the solution.

---

*The equilibrium constant given here is based on free-energy values quoted by Latimer (1952, p 74) The values for the constant quoted by Sillén and Martell (1964, p. 232-235) seem to cluster near $10^{-1.99}$, or $1\ 02 \times 10^{-2}$.

Figure 21 shows the calculated solubility of gypsum in solutions of sodium chloride, based on a report by Tanji and Doneen (1966). The calculations used a solubility product for gypsum of 2.4×10^{-5} and the ion-pair stability given above and considered the effects of ionic strength from the four ionic species Na^+, Ca^{+2}, Cl^-, and SO_4^{-2}. The data apply at 25°C. Natural waters contain other ions which may influence gypsum solubility.

The concentrations of calcium and sulfate are equivalent in the simple system represented by figure 21; under this condition the sulfate concentration would be about 1,480 mg/l in the absence of sodium and chloride, and in the presence of 2,500 mg/l of sodium plus chloride the sulfate would be 1,800 mg/l. Analysis 3, table 15, coincides rather closely with the predicted calcium and sulfate concentrations of figure 21. This water is presumably saturated with respect to gypsum, and constituents other than calcium and sulfate are relatively unimportant.

As noted earlier, strontium sulfate is sparingly soluble, and barium sulfate is relatively insoluble in water. The solubility products for these solids listed by Sillén and Martell (1964, p. 236) generally are near $10^{-6.5}$ for $SrSO_4$ and $10^{-10.0}$ for $BaSO_4$. Thus, a water containing $\sim$10 mg/l of Sr^{+2} should have no more than a few hundred milligrams per liter of sulfate, and a water containing 1 mg/l of barium should have only a few milligrams per liter of SO_4^{-2}. These are rough approximations given only to indicate the general effects of barium and strontium on sulfate solubility. More exact solubilities can be calculated from thermodynamic data in the literature. The influence of barium and strontium on the sulfate concentration of natural waters is seldom very important. More commonly, low sulfate concentrations result from bacterial reduction of sulfate which can occur in anaerobic sediments or ground-water aquifers. A water which shows this effect is represented by analysis 2, table 17.

Magnesium sulfate and sodium sulfate are highly soluble, and waters containing large amounts of these components with little calcium can attain sulfate concentrations in excess of 100,000 mg/l. An extreme instance is shown in analysis 6, table 16, a brine high in magnesium.

REDUCED SULFUR SPECIES

The areas of stability for sulfide solute species shown in figure 20 indicate that below a pH of 7.0 in reducing environments, the most likely form is undissociated H_2S. In alkaline solutions, the ion HS^- is predominant. The ion S^{-2} requires a very high pH, above the range normally to be expected in natural water.

Most sulfide-bearing ground water contains H_2S or HS^-. The rotten-egg odor of H_2S gas is distinctive and can be detected in water

SIGNIFICANCE OF WATER PROPERTIES AND CONSTITUENTS 169

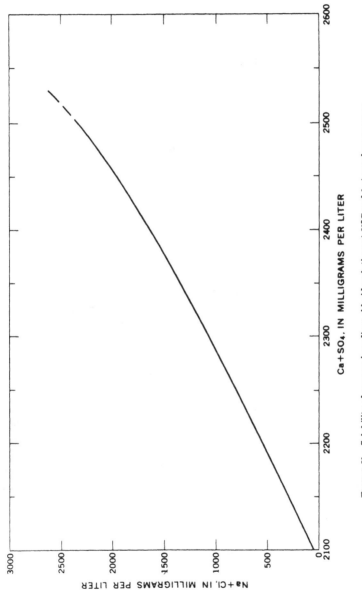

FIGURE 21.—Solubility of gypsum in sodium chloride solutions at 25°C and 1 atmosphere pressure.

containing only a few tenths of a milligram per liter of sulfide. The sulfides of most metals are low in solubility, and sulfide often seems to be lost from solution in ground water, probably by precipitation of sulfides, before the water has moved any great distance. In any event, ground water that is low in sulfate, presumably because of sulfate reduction, commonly contains little or no sulfide in solution either.

Some brines, especially some of the ones associated with petroleum, may contain several hundred milligrams per liter of dissolved H_2S. In general, the process of sulfate reduction requires the presence of bacteria and organic matter. The reaction may be represented as

$$SO_4^{-2} + CH_4 = HS^- + HCO_3^- + H_2O.$$

The thermodynamic data in the literature for these species show that this reaction should proceed spontaneously to the right, releasing energy. It, therefore, can be a source of energy for anaerobic bacterial species. If a boundary for the reduction of carbon from carbon dioxide species to more reduced forms, such as methane, were drawn in figure 20, it would fall within the stability field for reduced sulfur species.

CHLORIDE

The element chlorine is a member of the halogen group of the elements. Other important elements in this group are fluorine, bromine, and iodine. Chlorine, however, is by a wide margin the most important and most widely distributed of the halogens in natural water. Generally, it is present as the chloride ion, Cl^-. The chloride ion can form complexes with a few of the positively charged ionic species of natural water. Such complexes are weak, however, and generally are insignificant unless chloride concentrations in solution are very high.

Chlorine gas dissolves readily in water, forming hypochlorite (ClO^-). Species with higher oxidation states of chlorine, extending to perchlorate, ClO_4^-, in which the oxidation number of chlorine is $+7$, are well known. Chlorine is widely used in water treatment as a disinfectant or biocide. Amounts used in public water supplies generally are such that material reacting with chlorine is destroyed, and a few tenths milligram per liter of unreacted, or residual, chlorine are left throughout the water-distribution system.

Although some of the oxidized chlorine species, when once formed, are relatively stable, they are not found in significant amounts in any natural water. The presence of chlorine residuals may influence laboratory tests for some other ions, however.

Chlorine in aqueous solution is a stronger and usually faster acting oxidizing agent than dissolved oxygen. According to Latimer (1952, p. 53), the couple

$$Cl_2 + 2e^- = 2Cl^-$$

has a standard potential of 1.3595 volts in acid solutions. This means that the stability region for dissolved chlorine is outside the water-stability field even at pH=0, and chlorine does slowly decompose water. Most of the other species of chlorine, except for chloride ions, also are unstable within the water-stability region. For this reason no Eh-pH diagram is included here for chloride, as it would be dominated by the large field of stability for Cl^-.

SOURCES OF CHLORIDE

According to data in table 1, chloride is present in the various rock types in lower concentrations than any of the other major constituents of natural water. Among the chloride-bearing minerals occurring in igneous rock are the feldspathoid sodalite, $Na_8[Cl_2(AlSiO_4)_6]$, and the phosphate mineral, apatite. Johns and Huang (1967) recently summarized the available data on chlorine content of rocks. Shand (1952, p. 34) quoted data indicating that residual water contained in pores in granite or included within crystals in the rock may contain chloride. Available data on igneous-rock composition may not always dependably indicate the importance of chloride held in this way. Kuroda and Sandell (1953) suggested that chloride may replace hydroxide in biotite and hornblende and may be in solution in glassy rocks, such as obsidian. Noble, Smith, and Peck (1967) showed that most of the halogens present in glassy rocks were lost on crystallization.

On the whole, it must be concluded that igneous rocks, at least those available to the geochemist for sampling and analysis, cannot yield very high concentrations of chloride to normally circulating natural water. Considerably more important sources are associated with sedimentary rocks, particularly the evaporites. Chloride may be present in resistates as the result of inclusion of connate brine and is to be expected in any incompletely leached deposit laid down in the sea or in a closed drainage basin. When porous rocks are submerged by the sea at any time after their formation, they become impregnated with soluble salts, in which chloride plays a major role. Fine-grained marine shale might retain some of this chloride for very long periods. In all these rock types, the chloride is mostly present either as sodium chloride crystals or as a solution of sodium and chloride ions.

Billings and Williams (1967), from analyses of deeply buried rock, computed an average chloride content of 1,466 parts per million for shale. This contrasts with the figure of 170 parts per million given in table 1 and similar values in other compilations of this type. Billings and Williams believed that most analyses of shale represent specimens that have lost chloride by leaching from near-surface exposure and that their higher value for chloride is more realistic for the bulk of this type of rock.

More than three-fourths of the total quantity of chlorine known to be present in the earth's crust to a depth of 16 kilometers and in the hydrosphere is in solution in the oceans. This rather striking fact is first of all a result of the characteristic chemical behavior of chloride ions. Chlorine has been used as a key to certain features of the geochemical history of the earth. The literature contains numerous papers speculating on the manner in which the chloride reached the ocean, and the general consensus of these papers is that, owing to its volatility, the elemental chlorine of the earth became separated from the crustal rock at an early stage. The various aspects of this subject form an interesting area for speculation, but will not be taken up here. The amounts of chlorine which might still be present at depths greater than 16 kilometers are uncertain owing to absence of actual analyses of deep material.

OCCURRENCE AND CHEMISTRY OF CHLORIDE IN WATER

The chemical behavior of chloride in natural water is very tame and subdued compared to the other major ions. Chloride ions do not significantly enter into oxidation or reduction reactions, form no important solute complexes with other ions, do not form salts of low solubility, are not significantly adsorbed on mineral surfaces, and play few vital biochemical roles. The circulation of chloride ions in the hydrologic cycle is largely through physical processes. The lack of complications is illustrated by experiments with tracers in ground water described by Kaufman and Orlob (1956). These investigators found chloride ions moved with the water through most soils tested with less retardation or loss than any of the other tracers tested—including tritium—that had actually been incorporated into the water molecules. This behavior, however, should not be expected where movement is through compacted clay or shale.

Chloride is present in all natural waters, but mostly the concentrations are low. In most surface streams, chloride concentrations are lower than those of sulfate or bicarbonate. Exceptions occur where streams receive inflows of high-chloride ground water or industrial waste or are affected by oceanic tides. It would seem, from the lack of complications in chemical behavior, that chloride circulation in natural water could be rather fully and simply explained, and some literature suggests that there are no difficulties in making this kind of explanation. Unfortunately, some serious problems lie in wait for the investigator who makes this assumption.

Rainwater close to the ocean commonly contains from 1.0 mg/l to several tens of milligrams per liter of chloride, but the concentrations observed generally decrease rapidly in a landward direction; the average over the United States, as indicated by Junge and Werby

(1958), is only a few tenths of a milligram per liter. Whether these figures give a completely reliable basis for computing the total landward flux of chloride during a given year could be questioned. Air masses containing sodium chloride particles may move inland and drop out salt by mechanisms other than by washing out in rainfall. Eriksson (1960) and other investigators have mentioned the effects of direct fallout and interception of airborne sodium chloride by vegetation. Most geochemists who have discussed this topic in recent years have assumed that all the chloride in river water reaching the ocean comes from rainfall or other forms of precipitation and is therefore cyclic— that is, it represents recycled chloride from the ocean. Relatively few investigations, however, have included enough actual measurements of quantities of chloride brought into a drainage basin in rainfall as well as quantities carried out in streamflow to ascertain how closely this assumption can be verified. In all studies to date in the United States, where enough data have been obtained to permit reliable computations to be made, the amount of chloride brought in by rainfall seems inadequate to explain the amounts appearing in runoff.

Gambell and Fisher (1966) made a comparison of loads of ions brought in by precipitation to the drainage basins of four North Carolina streams and the loads carried out of the basins in streamflow by the four streams. These results covered a period of 1 year. During this period the chloride outflow was more than four times the influx from precipitation. Gambell and Fisher suggested this indicates a substantial contribution of chloride from erosion of the crystalline rock that underlies most of the area studied. The degree to which the chloride loads into and out of the basins studied varied from year to year was not known, however, and final conclusions cannot be drawn from this study alone. It seems obvious the chloride discharged in any 1 year would include contributions from precipitation that occurred before that year. Fisher (1968), in reporting on a continuation of this investigation for a second year, however, found again that the chloride outflow was about four times as great as the amount brought in by rainfall. Fisher pointed out that the total sales of salt in the area studied accounted for a considerable part of the discrepancy and suggested human activities might be a major factor in chloride circulation. An important source of chloride in some areas is the use of salt for deicing highways.

Van Denburgh and Feth (1965) used the data of Junge and Werby (1958) on composition of rainfall and water-quality records for 5 years for river basins covering 545,000 square miles in the western United States to determine solute loads removed from the area and the percentage of the chloride load that could be attributed to rainfall and snowfall. They concluded that most of the chloride load in all basins came from sources other than rain and snow. The percentage of the chloride load derived from precipitation ranged from 17 percent for the Rogue River basin of Oregon to 1.6 percent for the Pecos River basin in New Mexico.

The influence of chloride from hot springs and volcanic gases is very evident in the chloride loads of some streams. For example, the Gila River system in Arizona receives inflows from deep-seated springs, especially those which occur along the Salt River in the central highlands of the State. Feth and Hem (1963) pointed out that several hundred tons per day of sodium chloride enter the stream from such springs and strongly influence the composition of the river water.

The widely published assumption that chloride loads of rivers represent recycled oceanic chloride is not entirely vitiated by individual examples such as those mentioned, but these examples certainly imply that the route by which the oceanic chloride reaches the streams is not always by way of rainfall. The chloride of marine evaporite sediments and connate water also is of marine origin and is recycled when it goes into solution in streamflow. Even some of the chloride emitted by volcanoes may be from the ocean. Opportunity for penetration of sea water at the bases of volcanic islands might well exist. The assumption, however, that the bulk of the chloride now in the ocean was driven out of near-surface rocks in past geologic time by some natural process is certainly open to the interpretation that the process may not have entirely stopped, and careful studies might reveal places where, on a small scale, the process is still active. The quantity of new chloride gained by the ocean each year could now be so small in comparison to the total already stored there that its effect is not measurable and could be overshadowed by processes in oceanic sediments which convey oceanic chloride to rock minerals, from which their re-solution will be very long delayed.

Some investigators have ascribed the chloride accumulations of well-known hydrologically closed basins, such as the Dead Sea, to oceanic chloride transported inland in the atmosphere. Geologists have not agreed on the importance of this effect in most areas, however. Most inland closed basins have short lives in a geologic sense, and those with important salt lakes or saline accumulations contain exposures of evaporite sediments and saline springs that contribute solutes to runoff. Airborne sea salt seems to be agreed upon as the most likely source of salinity in parts of the arid interior of Australia. Anderson (1945) noted that surface waters in the arid parts of Australia bore a resemblance to sea water, and most Australian literature accepts the atmospheric circulation mechanism as the most likely one. Yaalon (1961) attributed the salinity of soil in arid regions of Israel to accumulations of airborne oceanic salt.

Chloride ions may be concluded characteristically to be retained in solution through most of the processes which tend to separate out other ions. Residual brines, therefore, would be high in chloride concentration for this reason alone. Because the chloride ion is physically large compared to many of the other major ions in water, it could be expected to be held back in interstitial or pore water in clay and shale while water itself was transmitted.

The differential permeability of clay and shale may be a major factor in the behavior and composition of saline ground water associated with fine-grained sediments. For example, chloride held back while water molecules passed through a clay layer might accumulate until high concentrations were reached. The selective behavior of such a layer also influences the residual concentration of cations. The more strongly retained ions in such a solution would be the ones most strongly attracted to cation-exchange sites. Calcium is usually the ion preferentially held. Thus, a mechanism is suggested for the origin of calcium chloride brines, an example of which is represented by analysis 4, table 15. Brines of this type have been discussed by many geochemists. Valyashko and Vlasova (1965) described calcium chloride brines occurring in the U.S.S.R. and cited mechanisms for their formation.

A less extreme instance of an altered water in which calcium and chloride are the principal components is represented by analysis 6, table 15. This water is from an irrigation and drainage well in the Salt River Valley of Arizona. The water applied in irrigating this area came originally from the Salt River and had a composition like that represented by analysis 5, table 17. The river water, with its ions concentrated by evapotranspiration, is the source of recharge for the ground water. However, ion-exchange reactions in the soil and subsoil zones, and probably other processes, have changed the relative amounts of cations in solution.

The most common type of water in which chloride is the dominant anion is one in which sodium is the predominant cation. Waters of this type range from dilute solutions influenced by rainfall near the ocean (analysis 4, table 6) to brines near saturation with respect to sodium chloride. Analysis 8, table 17, is a brine of this type which has 189,000 mg/l of chloride and 121,000 mg/l of sodium.

Sea water has a chloride concentration near 19,000 mg/l, but where mixing is impaired, higher or lower values may be observed.

Although a few acid waters in which chloride is the dominant anion are known, none of the analyses in this paper have such characteristics. White, Hem, and Waring (1963, p. F44) gave analyses for some waters of this type.

ACCURACY OF DETERMINATION

The determination of chloride is commonly assumed to be one of the simplest and most dependable procedures in water analysis. However, the most commonly used procedure, the Mohr titration, which uses a standard silver nitrate solution with chromate to indicate the end point, has fundamental limitations which have not always been adequately recognized. The optimum range of concentration for this procedure is from about 20 to about 5,000 mg/l. Accuracy and precision of the Mohr procedure are inadequate for determining concentrations of chloride below about 10 mg/l unless the sample is concentrated by evaporation of a large aliquot. Methods more satisfactory for very low or for very high concentrations are readily available, however.

FLUORIDE

The element fluorine is the lightest member of the halogen group of elements. In a number of respects, its chemical behavior is different from that of other halogens, and some of these differences are evident in its behavior in natural water. Fluorine is the most electronegative of all the elements. In solutions, it forms F^- ions. Other oxidation states are not found in natural aqueous systems. Fluoride ions have the same charge and nearly the same radius as hydroxide ions, and thus the ions may tend to replace each other. Fluoride forms strong solute complexes with many cations, and some fairly common mineral species of low solubility contain fluoride.

A significant fact noted earlier in summarizing the geochemistry of chlorine is that more than 75 percent of the total amount of that element known to be present in the outer part of the earth is contained as chloride in solution in the ocean. Fluorine is, on the other hand, almost all tied up in rock minerals, and a very small percentage of the total is contained in sea water. Table 1 shows that the amount of fluorine in rocks exceeds the amount of chlorine. When all forms of the two elements are considered, however, chlorine is by far the more abundant.

The element fluorine is utilized by higher life forms in the structure of bones and teeth. The importance of fluoride in forming human teeth and the role of fluoride intake from drinking water in controlling the characteristics of tooth structure has been realized only within the past 30 or 40 years. During this period, in research on tooth structure and decay, the fluoride content of natural water has been very extensively studied. Therefore, although the element usually is present only in small concentrations, a vast amount of information is available on fluoride concentrations in natural water.

SOURCES OF FLUORIDE IN WATER

The calcium fluoride fluorite (CaF_2) is a common fluoride mineral. This mineral has a rather low solubility and occurs both in igneous and sedimentary rock. Apatite commonly contains fluoride. Amphiboles, such as hornblende and some of the micas, may contain fluoride replacing part of the hydroxide. According to Rankama and Sahama (1950, p. 758) rocks rich in alkali metals, and also obsidian, are as a class higher in fluoride content than most other igneous rocks. Fluoride often is associated with volcanic or fumarolic gases, and in some areas these may be important sources of fluoride for natural water.

Mineral particles that contain fluoride are widespread constituents of resistate sediments. Ground water containing fluoride concentrations exceeding 1.0 mg/l is found in many places in the United States and is obtained from a wide variety of geologic terranes.

CHEMISTRY OF FLUORIDE IN WATER

The analyst reports fluoride concentrations in terms of the free F^- ion. This form probably is the predominant one in most water, but other varieties of dissolved fluorine are certainly possible. At low pH the form HF^0 could occur. From dissociation constants given by Sillén and Martell (1964, p. 256–257), it would appear that below a pH of 3.5 the HF^0 form probably would be predominant. Data in the same reference suggest that strong fluoride complexes would be formed with aluminum, beryllium, and ferric iron, and a series of mixed fluoride-hydroxide complexes is possible in the presence of boron. In acid solutions, the fluoride could well be associated with silica in six-coordinated or four-coordinated structures such as SiF_6^{-2} or SiF_4^0; however, the conditions required for stability of these two species probably are seldom reached in natural water. Calculations by the writer (Hem, 1968) show that aluminum fluoride complexes are likely to be found in waters whose pH is somewhat below neutrality. Although most of the dissolved aluminum in such solutions would be complexed, the concentration of fluoride commonly would exceed that of aluminum, and the proportion of the total fluoride present as complexes might not be great.

The solubility product for fluorite given by Smyshlyaev and Edeleva (1962) is $10^{-10.57}$ at 25°C. This means that in the presence of an activity of 40 mg/l of Ca^{+2}, the equilibrium activity of F^- would be

$$[F^-]^2 = 10^{-10.57} \times 10^{3.0} = 10^{-7.57}$$

$$[F^-] = 10^{-3.78} = 1.66 \times 10^{-4} \text{ moles per liter}$$
$$= 3.2 \text{ mg/l}.$$

The effect of ionic strength and of other fluoride complexes in solution might increase the solubility somewhat. No study of the extent to which fluoride concentrations in natural water might be controlled by the fluorite solubility limit has been made. The higher concentrations of fluoride observed in natural water generally are in waters that are relatively low in calcium.

Experiments by Roberson and Hem (1969) showed that cryolite ($Na_3 AlF_6$) may precipitate from solutions containing Na^+, Al^{+3}, and F^- in moderate concentrations, and under some conditions cryolite solubility might control the activity of fluoride in solution. Most natural waters do not meet these conditions, however, as they are most unlikely to contain enough dissolved aluminum to have a major influence on fluoride solubility.

RANGE OF CONCENTRATION

The concentration of fluoride in most natural water, which has a total dissolved-solids content less than 1,000 mg/l, is less than 1 mg/l. Concentrations as high as 50 mg/l have been reported, however, in solutions that might otherwise be considered potable. The highest concentration in the analyses in this report is 32 mg/l in the water represented by analysis 1, table 19. The sample was obtained from an abandoned flowing well near San Simon in southeastern Arizona. Ground waters containing more than 10 mg/l of fluoride are found in a few other places in the United States. Bond (1946, p. 43) reported 67 mg/l in a water from the Union of South Africa. Higher values are reported in water affected by volcanism by White, Hem, and Waring (1963, p. F44). Fluoride concentrations in river water seldom are greater than a few tenths of a milligram per liter, but there are some exceptions. The Gila River above the Safford Valley in southeastern Arizona is characteristically high in fluoride and carried an average concentration greater than 1 mg/l for the whole 5-year period of daily sampling record from 1943 to 1948.

The ionic radius of fluoride is nearly the same as that of hydroxide, and the two ions sometimes substitute for one another in crystal structures. Some mineral surfaces may be capable of absorbing anions, and if such a surface carried fluoride ions, they could perhaps be available for release by substitution of hydroxide ions for them from water of high pH. In some places, the high fluoride concentrations of ground water may result from this effect. Studies relating to fluoride sorption on soils were made by Bower and Hatcher (1967). Some ground waters that contain rather large amounts of fluoride do have a high pH. Although analysis 1, table 19, did not have a pH determination, the ratio of carbonate to bicarbonate in the water suggests that perhaps the pH was greater than 9. Analysis 1, table 12, represents a

SIGNIFICANCE OF WATER PROPERTIES AND CONSTITUENTS 179

TABLE 19.—*Analyses of waters containing fluoride, nitrogen, phosphorus, or boron in unusual amounts*

[Analyses by U.S. Geological Survey. Date below sample number is date of collection. Source of data: 1 Hem (1950, p.87); 2, 4, 5, 8, and 9, unpub. data, U.S. Geol. Survey files; 3, Berry (1952); 6, U.S. Geol. Survey Water-Supply Paper 1198 (p. 24); 7, U.S. Geol. Survey Water-Supply Paper 1945 (p. 591)]

Constituent	(1) Apr. 29, 1941		(2) Aug. 14, 1952		(3) Sept. 16, 1947		(4) Jan. 31, 1946		(5) May 19, 1952		(6) Oct. 1950– Sept. 1951		(7) Oct. 3, 1961		(8) Sept. 26, 1960		(9) Mar. 26, 1957	
	mg/l	meq/l	mg/l	meq/l	mg/l	meq/l	mg/l	meq/l	mg/l	meq/l	mg/l	meq/l	mg/l	meq/l	mg/l	meq/l	mg/l	meq/l
Silica (SiO_2)			23		27				18		15		17		314		72	
Aluminum (Al)															.22			
Boron (B)					.28						.05				48		660	
Iron (Fe)															.52			
Manganese (Mn)															.00			
Arsenic (As)									.00						4.0		.02	
Strontium (Sr)															.67			
Calcium (Ca)	5.5	0.27	92	4.59	64	3.19	36	1.80	42	2.10	49	2.45	24	1.20	3.6	0.18	7.0	0.35
Magnesium (Mg)	4.4	.36	38	3.12	19	1.56	18	1.48	19	1.56	14	1.15	11	.90	.0	.00	22	1.81
Sodium (Na)	157	6.81	110	4.80	{114	4.96	102	4.44	29	1.26	5.4	.23	34	1.48	660	28.69	1,100	47.85
Potassium (K)					9.5	.24			.7	.02	3.1	.08	6.2	.16	65	1.66	33	.84
Lithium (Li)									.3	.04					7.0	1.01	4.8	.69
Carbonate (CO_3)	58	1.93	0	0	0	0	0	0	0									
Bicarbonate (HCO_3)	163	2.67	153	2.51	402	6.59	303	4.97	65	1.07	168	2.75	129	2.11	312	5.11	2,960	48.51
Sulfate (SO_4)	42	.87	137	2.85	74	1.54	34	.71	114	2.37	40	.83	32	.67	108	2.25	454	9.45
Chloride (Cl)	10	.28	205	5.78	30	.85	32	.90	13	.37	5.4	.12	18	.51	874	24.65	690	19.46
Fluoride (F)	32	1.68	.6	.03	.1	.01	.4	.02	5.0	.26	4.3	.23	.4	.02	2.6	.14	1.0	.05
Nitrate (NO_3)			83	1.34	60	.97	68	1.10	.3	.00	.2	.01	13	.21	2.7	.04	.0	.00
Ammonium (NH_4^+)									30	.95	14	.23	14	.1	.0	.24		
Phosphate (PO_4)																		
Dissolved solids:																		
Calculated	389		764		596		440		303		251		236		2,240		4,990	
Residue on evaporation					578		164		318		180				2,360				476	26.38
Hardness as $CaCO_3$	32		386		238				183		42		106		9.0		108	
Noncarbonate	0		260		0		0		130				0		0		0	
Specific conductance (micromhos at 25°C)	660		1,320		875		724		413		365		361		3,430		7,060	
pH					7.4				7.2				7.5		8.9		7.5	
Color									20				20					

1. Flowing well NE¼ sec. 24, T. 13 S., R. 30 E., Chochise County, Ariz. Depth, 850 ft. Water-bearing formation, valley fill. Temp. 18.3°C.
2. Irrigation well, SE¼ sec. 25, T. 2 N., R. 2 W. Maricopa County, Ariz. Depth, 275 ft. Water-bearing formation, valley fill. Temp. 29.4°C.
3. Well, SE¼ sec. 21, T. 12 S., R. 10 W., Lincoln County, Kans. Depth, 32 ft. Water-bearing formation, alluvium. Temp. 14.4°C.
4. Well, NW¼ sec. 2, T. 8 S., R. 5 W., Maricopa County, Ariz. Depth, 495 ft. Water-bearing formation, valley fill.
5. Peace Creek at State Highway 17 bridge, Salfa Springs, Fla.; 140 cfs.
6. Iowa River at Iowa City, Iowa. Discharge-weighted average of composites of daily samples, Oct. 1, 1950 to Sept. 30, 1951; Mean discharge, 2,543 sec-ft.
7. Powder River, 4.5 miles north of Baker, Baker County, Oreg.
8. Nevada Thermal well 4, Steamboat Springs, Washoe County, Nev. Depth, 746 ft. Bottom temperature, 186°C. Also contained bromide (Br) 1.5 mg/l and iodide (I) 0.6 mg/l.
9. Spring at Sulphur Bank, sec. 5, T. 13 N., R. 7 W., Lake County, Calif. Temp., 77°C. Also contained H_2S, 3.6 mg/l; Br, 1.4 mg/l; I, 3.6 mg/l.

water with a pH of 9.2 and with a content of 22 mg/l of fluoride; however, not all fluoride-bearing water has a high pH.

NITROGEN

Small amounts of nitrogen are present in rocks, but the element is concentrated to a greater extent in soil or biological material. A considerable part of the total nitrogen of the earth is present as nitrogen gas in the atmosphere. Nitrogen can occur at all oxidation states ranging from -3 to $+5$, although not all these are represented by species that are important in natural water. In ammonia, NH_3, the oxidation state is N^{-3}, and aqueous forms include NH_4^+ and $NH_4OH(aq)$. Amino, or organic, nitrogen in forms such as protein or amino acid may occur in water that contains organic waste. Urea, NH_2CONH_2, is a common constituent of nitrogenous waste containing the NH_2 or amino group. Hydroxylamine, NH_2OH, contains nitrogen in the -1 oxidation state. On oxidation, the reduced nitrogen species may be converted to N_2 gas, or to nitrite NO_2^- in which nitrogen is in the N^{+3} state, and finally to nitrate (NO_3^-) where nitrogen is in the N^{+5} form.

In addition to these species, nitrogen forms certain complex inorganic ions that are probably not significant in natural systems, but they may enter water supplies through industrial waste disposal. An important example is the cyanide ion CN^-. The oxidation state of nitrogen in this ion is uncertain, because it depends on the arbitrary assignment of an oxidation number to carbon, but in any event, the ion is stable over a considerable range of chemical conditions and forms strong complexes with many metal cations.

The oxidation and reduction of aqueous nitrogen species are closely tied to biological activity, and both the paths followed and the end products of such reactions depend very strongly on kinds and numbers of biota present. Although pH-Eh diagrams can be prepared for nitrogen (Morris and Stumm, 1967), the influence of biota on the reactions of the element and the extensive departure from equilibrium conditions that may be observed, severely limit the value of this approach in predicting species to be expected in natural water. In general, the oxidation of organic nitrogen in air can be expected to produce nitrite and finally nitrate. Some mediating organisms reduce nitrate nitrogen and can produce nitrogen gas instead of, or in addition to, ammonia. In ground water, nitrate appears usually to be the only form of nitrogen of significance, although nitrate-bearing water may occur in reducing environments. In part, this observation may be the result of incomplete information, as in routine analysis of groundwater samples nitrate nitrogen is usually the only form of nitrogen that is looked for. Weart and Margrave (1957) found ammonia nitrogen in a number of ground waters used for public supplies in Illinois, and small quantities may not be uncommon in ground waters whose other chemical properties (low sulfate or high iron concentrations) might suggest they had been exposed to a reducing environment.

Nitrogen concentrations are determined and reported in different ways in published analyses. Most laboratories engaegd in studies of organic pollution report ammonia, amino and organic nitrogen, and nitrite either separately or as a combined figure and in terms of equivalent concentration of elemental nitrogen. Other laboratories that have been more interested in the inorganic contents of water have determined and reported only nitrate, usually in terms of concentration of the nitrate ion, NO_3^-. The result of this selectivity has been that the total nitrogen content of many natural waters is not determined. There also is a lack of information on the stability of the various species when more than one form is present. Analyses in table 19 report nitrate or ammonium concentrations in terms of milligrams per liter of the ionic species indicated. None of the analyses report total dissolved nitrogen, and none report nitrite concentrations because the ion was not determined. Nitrite is seldom present in amounts large enough to influence the ionic balance to a noticeable degree.

SOURCES OF NITRATE IN WATER

The literature on sources of nitrogen has been reviewed by Feth (1966), whose paper summarizes the subject in more detail than is possible here. The relationships of nitrogen to water pollution are considered by Ingram, Mackenthun, and Bartsch (1966).

Nitrogen in the form of dissolved nitrate is a major nutrient for vegetation, and the element is essential to all life. Certain species of bacteria in soil, especially those living on roots of legumes, and the blue-green algae, and other microbiota occurring in water, can extract nitrogen from air and convert it into nitrate. Some nitrate occurs in rainwater, although the former belief that a significant fraction of this was produced by lightning discharges now seems incorrect (Junge, 1958). Nitrate in the soil that is utilized by plants is partly returned to the soil when the plants die, although some nitrate is lost from the cycle in drainage and runoff and apperas in river water. The nitrate in the soil is artificially increased by man when nitrate fertilizers are used. In spite of the great increase in the use of such fertilizers in recent years, however, the records of river-water composition that were available in 1967 do not show any well-defined recent upward trend in the nitrate concentration in runoff (McCarty and others, 1967).

Evidence of the importance of soil leaching in producing the nitrate concentrations observed in river water can be gathered from records of river-water quality for streams in the more highly productive agricultural regions. The Iowa River at Iowa City, Iowa, for example, had average concentrations of nitrate near or above 10 mg/l during most years of a 7-year period from 1944–1951, and highest concentrations commonly occurred during periods of above-average runoff, when much of the flow must have from runoff from the surfaces of cultivated fields. Analysis 6, table 19, represents the discharge-weighted average for the 1951 water year.

Farm animals produce considerable amounts of nitrogenous organic waste that tends to concentrate in places where large numbers of animals are confined. The occurrence of high nitrate concentrations in shallow ground water in certain areas in Kansas has been attributed to leachings from livestock corrals by rainfall (Durum in Berry, 1952). The recent general trend toward confining many animals in small areas, such as in feeding pens for beef cattle, probably will bring about more occurrences of this type. In past years, most investigators, however, seem to have thought this effect had only local singificance and have stated that the high nitrate content of ground water in extensive areas cannot be explained by barnyard pollution. Stewart, Viets, Hutchinson, and Kemper (1967) made quantitative studies of soil moisture and shallow ground water in the South Platte Valley of Colorado and found substantial contributions of both nitrogen and phosphorus reached the ground water beneath irrigated fields, and particularly large contributions were associated with feed lots.

Analysis 2, table 19, is for water containing 83 mg/l of NO_3^- from a well in the Salt River Valley area of Arizona, which is intensively irrigated and fertilized. It seems probable that the occurrence of nitrate in many wells in that area is explained by leaching of irrigated soil. Analysis 4, table 19, is for water from a well about 12 miles south of Gila Bend, Ariz., in a desert basin, which was undeveloped at the time of sampling. The high nitrate content of this water (68mg/l) is much more difficult to explain. Some species of desert vegetation are legumes, and possibly nitrate could accumulate in successive soil zones as a basin is filled with rock debris if precipitation were not sufficient to keep the soils leached of soluble salts. The extensive nitrate deposits of northern Chile occur in an arid environment, but the manner in which the deposits were formed is not fully known.

Nitrate from certain other sources is worth mentioning. Limestone caves that are used for shelter by large numbers of bats may accumulate guano that serves as a nitrogen source to ground water in the vicinity. Analysis 5, table 16, represents water from a pool in Carlsbad Caverns, N. Mex., containing 15 mg/l of NO_3. Pools nearer the section of the Caverns frequented by bats have been known to contain water with over 1,000 mg/l of NO_3^- (U.S. Geol. Survey, unpub. data).

An instance of industrial pollution which greatly increased the nitrate content of a stream was cited by McCarty and others (1967). The Dolores River near Cisco, Utah, had average nitrate concentrations near 25 mg/l for the whole year of 1963, owing to release of nitrate by a uranium-processing plant upstream.

REDUCED NITROGEN SPECIES

The sources of reduced forms of nitrogen in natural water presumably are similar to the sources of nitrate, and the state of oxidation of nitrogen probably is controlled by biochemical processes. Although the reduced forms normally would be transformed to nitrate in most natural-water environments, there is considerable evidence that a significant amount of reduced nitrogen often is present in many waters.

The pH at which the transformation of aqueous ammonia to ammonium ion is half completed is about 9.24 (Sillén and Martell, 1964, p. 150). This is above the usual pH of natural water and suggests that in most environments any ammonia nitrogen in solution would have the form NH_4^+. Analysis 9, table 19, represents water from a thermal spring in California that is high in ammonia. Oil-field brines sometimes also display this property.

As noted by Feth (1966), most of the nitrogen dissolved in rainwater appears to occur in the form of ammonium ions. The circulation in the atmosphere of this form is facilitated by its volatility.

More studies are needed to evaluate the rates at which species such as ammonium and amino or nitrite nitrogen are converted to nitrate in surface water. The organic forms are partly removed from water by filtration in some instances. Data that are available suggest that the total nitrogen content of polluted surface streams may include a significant proportion of reduced species. Some nitrogen may be associated with the organic coloring material present in some unpolluted natural water.

A study of the stability of ammonia nitrogen in water from Sulphur Bank, Calif., (analysis 9, table 19) and another spring area where high ammonia contents were observed was made by Roberson and Whitehead (1961). Oxidation of ammonia to nitrite and nitrate occurred both in spring flow as it moved out of the discharge area and in stored samples exposed to air.

PHOSPHORUS

The form in which phosphorus is likely to be present in natural water is somewhat uncertain, but the most probable species would appear to be phosphate anions, complexes with metal ions, and colloidal particulate material. Phosphorus is most often reported in terms of an equivalent amount of orthophosphate (PO_4^{-3}). Like nitrogen, a wide range of oxidation states is possible for phosphorus, but no strong resemblance in aqueous chemical behavior between the two elements is apparent. The most common oxidation state in natural systems is probably P^{+5}. Both phosphorus and nitrogen are essential nutrients for plant growth. Concentrations of phosphate normally present in natural water are far less than those of nitrate; however, the published data on water quality contain relatively few analyses for phosphate, and specific information is therefore incomplete.

Phosphorus has a strong tendency toward formation of polymeric complexes and forms ionic complexes and compounds of low solubility with many metals.

SOURCES OF PHOSPHATE

The most common rock mineral in which phosphorus is a major component is apatite, a general name for mineral species that are principally calcium orthophosphate and which also contain fluoride, hydroxide, or chloride ions. These minerals are widespread, both in igneous rock and in marine sediments. When apatite is attacked by water, the phosphorus species released probably recombine rather rapidly to form other minerals or are adsorbed by hydrolyzate sediments, especially clay minerals, in soil. Phosphate is made available for solution in water from several kinds of cultural applications of phosphate in the activities of man, and these pollution sources probably are the most important causes of high concentrations of phosphate in surface water.

Phosphate fertilizer is manufactured at many localities, commonly close by the areas where phosphate-bearing rock is mined and concentrated. The wastes from all these processes are important sources of phosphate in river water (analysis 5, table 19). Phosphate fertilizer applications, especially where irrigation is practiced, are a potential source of phosphate in drainage water, but no specific data showing the effect were available at the time this paper was written.

Phosphorus is a component of sewage, as the element is essetinal in metabolism, and it is always present in animal metabolic waste. In recent years, the use of sodium phosphate as a builder in household detergents probably has greatly increased the output of phosphate by sewage-disposal plants. Again, few data are available that clearly show this effect. Analysis 7, table 19, represents water from a small stream that appears to contain organic pollution. Both phosphate and nitrate concentrations are high, and some ammonium was reported.

Sodium polyphosphates are readily prepared by heat treatment of sodium orthophosphate and often are used in small amounts in water treatment to inhibit calcium carbonate or iron hydroxide precipitation. Polyphosphate anions are unstable in solution and in time revert to orthophosphate. The amounts that may be present in finished water as a result of this treatment are usually less than 1 mg/l. Reduced forms of phosphorus are used in synthesizing some insecticides and other chemicals. These species may be stable in water and may not be detected by the orthophosphate determination.

CHEMISTRY OF PHOSPHATE IN WATER

The orthophosphate ion (PO_4^{-3}) is the final dissociation product of phosphoric acid, H_3PO_4. The dissociation of the acid occurs in steps, and four solute species are possible: $H_3PO_4(aq)$, $H_2PO_4^-$, HPO_4^{-2}, and PO_4^{-3}. Figure 22 is a species-distribution diagram showing the proportions of total activity of phosphorus present in each form from $pH=0$ to $pH=14$. For solutions of low ionic strength, in general those having less than 1,000 mg/l of dissolved solids, the diagram can be used to give a reasonable estimate of the proportion of total phosphate to assign to the various species when pH is known. In natural water whose pH is 7.21, the phosphate activities would be evenly divided between $H_2PO_4^-$ and HPO_4^{-2}. In the process of titrating alkalinity in such a water, all the HPO_4^{-2} will be converted to $H_2PO_4^-$, and that fraction would appear in the alkalinity value as an equivalent quantity of bicarbonate.

Phosphate ions form complexes with many of the other soluets present in natural water. A calculation of solubility controls over phosphate activity is difficult when so many solute species containing phosphorus would be possible and the form of solids that might be produced is uncertain. The meager data available, however, suggest that phosphate contents of natural water are not generally restricted by precipitation of a sparingly soluble inorganic compound. More likely, the utilization of phosphorus by aquatic vegetation and perhaps the adsorption of phosphate ions by metal oxides, especially ferric hydroxide, are the major factors that prevent concentrations greater than a few tenths or hundredths of a milligram per liter from being present in solution in most waters.

NITROGEN AND PHOSPHORUS AS NUTRIENTS

As noted in the preceding sections, nitrogen and phosphorus are essential nutrients for plant growth. Aquatic vegetation of the free-floating types, such as algae, depends on dissolved nitrogen and phosphorus compounds for its nutrient supply. Growth of these species may also be influenced by the availability of other required elements. Dense, rapidly multiplying algal growths or blooms sometimes occur in water bodies that periodically receive increased concentrations of nitrogen or phosphorus These dense growths are generally undesirable to water users and may interfere with other

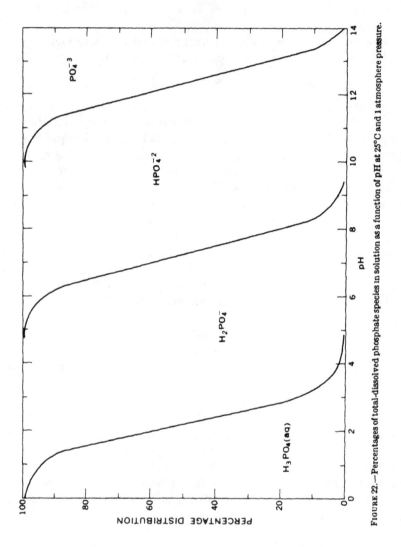

FIGURE 22.—Percentages of total-dissolved phosphate species in solution as a function of pH at 25°C and 1 atmosphere pressure.

forms of aquatic life, especially if the water body becomes overloaded with oxidizable debris as a result of the sudden dieback of an algal bloom.

The enrichment of a water body with nutrients is termed "eutrophication" by limnologists and is accompanied by a high rate of production of plant material in the water. Troublesome production rates of vegetation presumably can only occur when optimum supplies of all nutrients are present and available, but it seems obvious that the rates at which nutrients can become available also need to be considered.

BORON

Although it is a minor constituent of most water, a considerable amount of information exists on the concentration of boron in natural water. Most of the determinations of this element were made because boron is important in agriculture. Small amounts are essential to plant growth. Excessive concentrations of boron in soil and in irrigation water, however, are harmful, and for some plants the toxic concentration is very low.

SOURCES OF BORON

The most widely distributed mineral of igneous rocks in which boron is an essential constituent is tourmaline. This mineral is highly resistant to chemical attack and may appear in resistate sediments, as well as in original igneous rock. The mineral is typically a constituent of grantitic rocks and pegmatites. Boron, however, also may be present as an accessory constituent of biotite and the amphiboles.

Boron is a light element and its ionic and solute species tend to be somewhat volatile. Boron may be liberated in volcanic gases in the form of orthoboric acid, H_3BO_3, or as halogenides such as BF_3. Water in volcanic areas and the water of many thermal springs, therefore, may contain considerable concentrations of boron. Ocean water has 4.6 mg/l of boron, probably mostly in the form of undissociated boric acid. This compound has an appreciable vapor pressure at ordinary earth-surface temperatures, and some is vaporized into the air over the ocean as a result (Gast and Thompson, 1959). Boron is a contributor to the buffer capacity of the ocean and some natural waters.

Certain evaporite deposits of closed basins, especially in southeastern California, contain considerable amounts of boron and constitute ores from which boron salts are extracted. Sodium tetraborate is widely used as a cleaning aid and detergent; hence boron may be present in sewage and in industrial wastes.

CHEMISTRY OF BORON IN NATURAL WATER

In the usual detailed chemical analysis of water, boron is reported in terms of concentration of the element B, and no effort is made to describe the solute species which could be present. The aqueous chemistry of boron is intricate and not very well known. The simplest assumption, however, in the absence of better information is that orthoboric acid species are the most likely. Dissociation of orthoboric acid (H_3BO_3) has been extensively studied, as indicated by the large number of references in Sillén and Martell (1964, p. 104–106). The results generally indicate the acid's first hydrogen is not readily released. Most values for the dissociation constant at 25°C are near $10^{-9.2}$, a value which implies the acid would be half dissociated at a pH of 9.2. At a pH of 8.2, the ratio of $H_3BO_3(aq)$ to $H_2BO_3^-$ would be about 10:1. It is evident that the undissociated form will be predominant in the pH range of ordinary natural water.

Complex solute species containing boron probably occur in some natural waters, but they do not appear to have been extensively studied. Colemanite, a calcium borate with the formula $Ca_2B_6O_{11} \cdot 5H_2O$, and kernite, $Na_2B_4O_7\ 4H_2O$, a partly dehydrated form of borax, are common in the evaporite boron deposits of southeastern California and Nevada. Solubility data for colemanite seem to be lacking, but the boron content of most natural water does not seem to conform to any simple solubility limitation.

The high boron content of water from thermal springs has been noted. Schofield (1960) attributed the high boron content of water in the Waikato district of New Zealand to thermal effects. Koga (1957) observed high boron contents in the hot springs of Beppu, Japan.

Analyses 8 and 9, table 19, represent water from thermal springs which are relatively high in boron. Analyses in earlier tables in this report suggest that boron in concentrations up to a few tenths of a milligram per liter can be expected in a good many types of surface and ground water.

MINOR AND TRACE CONSTITUENTS

There is no precise definition of the terms "minor" or "trace" with reference to constituents of natural water, but substances that typically occur in concentrations of less than 1.0 mg/l are commonly thought of as being in this category. Some minor constituents, for example fluoride, are commonly determined, and their occurrence and chemical behavior are well known. Constituents discussed in this section are those less commonly determined and, therefore, whose occurrence in natural water cannot yet be fully evaluated. Although it is reasonable to suppose that all the naturally occurring elements will be found in natural water if a sufficiently sensitive method of analysis is available, many elements are rare and do not have readily soluble forms. Hence the concentrations of many elements certainly are extremely small and difficult to detect. The elements that will be discussed here include only those which have been detected or looked for in a considerable number of waters. A small, but growing, body of information also exists on the content of organic compounds present in trace quantities, but information on this subject is not sufficient to merit much discussion as yet.

The history of research on minor constituents of natural water contains discoveries that some supposedly insignificant trace constituent was actually of vital importance to human health, plant nutrition, or other area of general interest. Some of these discoveries have already been cited—for example, the relation of fluoride content of drinking water to the occurrence of tooth decay was discovered in the 1930's, and prior to that time fluoride concentrations in water were rarely determined. Other discoveries relating to minor constituents may be made in the future, as more continues to be learned about their occurrence and importance.

AVAILABILITY OF ANALYTICAL DATA

Concentrations of minor elements in water have, for the most part, been determined in four different kinds of studies:

1. Efforts to learn more about circulation and distribution of minerals in the rocks and waters of the earth.
2. Studies seeking to establish relationships between water composition and public health, either related to water pollution or to natural conditions.
3. Studies of plant nutrition and pathology.
4. Prospecting for ore deposits.

Data collected for one purpose may not satisfy the diversified interests of all investigators, and the amount of information on some elements is much greater than that on others. A significant influence on progress of research in this field has been the improving ability of the analytical instruments to detect low concentrations. Recent major advances in instrumentation will undoubtedly be reflected in great increases in volume of data on minor constituents in water.

One of the techniques of geochemical prospecting for mineral deposits involves determining concentrations of minor elements in river or underground water. A valuable indication of the occurrence of some metals in natural water can be gained from the large amount of data, published and unpublished, that has been obtained in this work. The U.S. Geological Survey has abstracted the literature on geochemical prospecting published through June 1957 (Harbough, 1953; Erickson, 1957; Markward, 1961). Some of this kind of data is pertinent to general levels of metal contents to be expected in natural water and will be referred to in describing the occurrence of these metals.

ANALYSIS TECHNIQUES

The emission spectrograph has been used extensively for determining minor constituents in rocks and water. The procedure involves vaporization of the sample at the high temperature of the electric arc and measuring the intensity of spectral lines produced. The method can determine very low concentrations, especially for some metals, but is not suitable for all elements. Pretreatment of the sample is required. The water sample may be evaporated to dryness and mixed with a suitable matrix material (Haffty, 1960), or the elements of interest may be precipitated and separated from solution before arcing (Silvey, 1961).

The atomic-absorption flame spectrophotometer recently has come into wide use in this field. Although it is generally less sensitive than the arc spectrograph, the atomic-absorption technique requires less sample manipulation and is influenced less by the mutual enhancement or quenching effects than emission spectrography. Atomic-absorption spectroscopy, however, is not free from interference effects. For some elements, notably zinc and cadmium, the atomic-absorption technique has a higher sensitivity than the arc spectrograph.

Other procedures, including ultraviolet fluorescence, infrared spectrophotometry, chromatography, and various wet methods, hold some potential in the field of minor-element determination and are the only methods suitable for some solute species.

SAMPLE COLLECTION AND TREATMENT

Sample-collection and treatment techniques are very important in analyses for minor constituents. If proper care is not taken to avoid sample contamination or loss of the minor constituents during manipulation misleading results will be obtained. Any analyst familiar with minor-element analysis is aware of these problems. Water samples for minor-element analysis are usually acidified after collection to minimize loss of metals by adsorption on container walls. If the container is not carefully cleaned or the acid used is impure, contamination results.

Surface waters pose special sampling problems. The usual technique is to filter a sample and then acidify it; the filter used most widely is a membrane type with pores 0.45 micron in average diameter (Skougstad and Scarboro, 1968). There is good evidence that particulate metal hydroxides and probably aluminosilicates exist in most surface streams in the form of particles smaller than 0.45 micron. These and the larger particles may carry a considerable part of the minor-metal content of the water, perhaps as adsorbed metal ions. If a solution containing such particulate material is acidified, the adsorbed metal ions will be brought into solution, but metal hydroxide will tend to dissolve also. The chemical significance of the results of analyzing this sample may be difficult to establish. Coarser suspended material also carries adsorbed ions. The best procedure for determining minor element contents of surface streams remains to be decided, as noted by Eichholz, Galli, and Elston (1966), who have commented on some of the problems. Many published data on minor-element contents of water contain artifacts resulting from sample treatment, but in most instances it is impossible to evaluate the importance of such influences.

A paper by Glagoleva (1959) suggested that iron, manganese, vanadium, chromium, nickel, gallium, and zinc are largely transported by rivers as suspended colloidal oxide particles either independently or adsorbed on other sediment particles but that copper, strontium, and barium are transported as solutes. The elements iron, manganese, phosphorus, and possibly nickel, copper, lead, and tin may be carried as organic complexes in water containing organic matter. Turekian and Scott (1967) published data on the content of chromium, silver, molybdenum, nickel, cobalt, and manganese in suspended material retained by the 0.45 micron filter for 18 rivers in the eastern United States.

SOURCES OF GENERAL INFORMATION

Analyses for minor constituents are scattered through the literature and until recent years usually have been included in water analyses only when the water under study appeared to have unusual properties. Much of the earlier information, therefore, referred only to unusual conditions. Minor-element contents of major rivers or of ground water of low to moderate dissolved-solids content were for the most part unknown.

An international program for determining the minor- and major-element composition of water of large rivers of the world was launched in 1957, and analyses made for that program included about 24 minor elements, determined spectrographically. Reports summarizing the data include a progress report by Durum, Heidel, and Tison (1960), a tabulation of data by Durum and Haffty (1961), and some interpretation of the results by Durum and Haffty (1963). Somewhat similar information for major rivers of the U.S.S.R. was being obtained at that time, and some of those results were published by Konovalov (1959). Kroner and Kopp (1965) reported contents of 17 minor elements in major United States streams as determined in the National Water Pollution Surveillance System. Their report is a compilation of material from annual summaries of the data from this system formerly published by the U.S. Public Health Service and, more recently, by the U.S. Federal Water Pollution Control Administration.

Chemical analyses for the water supplies of the 100 largest cities of the United States were compiled by Durfor and Becker (1964). Their report contains data for 26 minor constituents detected in these waters by means of the emission spectrograph. Silvey (1967) studied the occurrence of 17 minor elements in surface- and ground-water sources in California. Analyses of water made in geochemical-prospecting studies have been published in many reports.

The discussion of occurrence of radioactive substances in water is widely scattered in the literature. A large number of analyses for the purpose of determining natural radioactivity in ground water were compiled by Scott and Barker (1962).

Interest in minor elements in water is widespread, and scientists from many countries have published reports containing data on the subject. The work of Penchev (1960), in Bulgaria, might be cited as an example. Studies have been made by many investigators in the U.S.S.R. and in Japan. In this preliminary survey of the field, individual studies cannot be completely listed. Data from all the sources cited here have been utilized as a basis for the generalized statements on individual minor constituents.

The identification of individual organic compounds remains in the early stages of development. Very sensitive procedures have been developed for some herbicides and pesticides, but little is yet known of specific amounts of any of the possible organic compounds in natural water that has not been affected by pollution.

MINOR ELEMENTS IN SEA WATER

A large amount of work has been done on minor-element occurrence and chemistry in sea water. Table 2 contains concentration values for many elements about whose occurrence in fresh water almost no information is available. The uncertainties faced in determining solubility controls in sea water are perhaps greater than the ones that must be dealt with in natural water of lower dissolved-solids content.

The conclusion reached by Krauskopf (1956b), which seems reasonable for sea water, was that for some 13 minor metals the ocean is considerably below chemical saturation, and other mechanisms need to be invoked to explain the observed values. Whether similar conclusions are justified for fresh water remains uncertain.

OCCURRENCE OF MINOR CONSTITUENTS

The minor constituents considered here will be taken up by groups, as follows:
Alkali metals
Alkaline earths
Other metals
Nonmetallic elements
Radioactive elements
Organic constituents

ALKALI METALS

The alkali metals not already considered are lithium, rubidium, and cesium. Lithium is the lightest and smallest ion of the three. It tends to be concentrated in pegmatites and evaporites, and its ionic radius permits some substitution for magnesium (Rankama and Sahama, 1950, p. 427). The common ion-exchange minerals in soil apparently absorb lithium less strongly than they do other common elements (Kelley, 1948, p. 61). Therefore, when lithium is brought into solution by weathering reactions, it should tend to remain in the dissolved state. Rubidium is a much more abundant element than lithium. Its geochemical behavior probably is somewhat like that of potassium, according to Rankama and Sahama (1950, p. 439-440), who also stated that rubidium is generally more strongly held at cation-exchange sites than potassium. This fact might tend to prevent the appearance of much rubidium in solution in natural water. Cesium is a rarer element and is more strongly absorbed by clays than either rubidium or lithium.

Lithium is readily determined either by flame photometry or by atomic-absorption photometry, and a considerable number of water analyses containing lithium values exist. The determination is included in some of the analyses in tables 12, 13, 16, and 19. Concentrations of a few milligrams per liter occur in some waters having high dissolved-solids concentrations. Data on rubidium concentrations are much less plentiful.

Lithium is potentially toxic to plants. Bingham, Page, and Bradford (1964) studied this subject. A limit of 5 mg/l for lithium in irrigation water was suggested by the U.S. Federal Water Pollution Control Administration (1968). It seems unlikely water that high in lithium and not otherwise unsuitable for irrigation will be found.

The data for major rivers of North America (Durum and Haffty, 1963) show a slightly higher median concentration of rubidium than lithium (1.5 µg/l versus 1.1 µg/l). Durfor and Becker (1964), however, reported a median value of 2.0 µg/l for lithium and 1.05 µg/l for rubidium. On the basis of these rather fragmentary data, it seems that lithium and rubidium occur in about the same general concentration range in dilute natural water. Fewer data are available for cesium, and the detection limit is somewhat higher than that for rubidium. Cesium was not detected in any of the analyses of Durfor and Becker. Probably it is logical to conclude that cesium is normally present in natural water at lower concentrations than lithium or rubidium.

ALKALINE EARTHS

The elements in this group discussed here are beryllium, strontium, and barium. Radium is a member of the group also, but it is of major significance as a contributor of radioactivity and will be considered under that topic. Calcium and magnesium, the common members of the alkaline-earth group, are major constituents of nearly all water.

BERYLLIUM

Although nominally included in the alkaline-earth group, the element beryllium has little in its chemistry that is in common with the typical alkaline-earth metals. Beryllium ions are small enough to replace silicon in igneous-rock minerals. One of the more important of the minerals in which beryllium is an essential constituent is beryl, a silicate of aluminum and beryllium that is found most commonly in pegmatites. Other silicates or hydroxy-silicates may also be important sources of beryllium.

Beryllium oxide and hydroxide species have very low solubilities, and although several oxide forms are possible, the concentrations of uncomplexed Be^{+2} at a pH of 7.0 can be calculated from data in Sillén and Martell (1964, p. 41) to be from 10^{-7} to 10^{-10} molar at equilibrium, or from 1 $\mu g/l$ to about 0.001 $\mu g/l$. Complex hydroxide ions may increase the solubility somewhat. Beryllium ions in solutions at low pH probably tend to be strongly adsorbed by clay surfaces and by other mineral species present in surface water, and at higher pH the ions would have a strong tendency to form polynuclear complex species with hydroxide. Beryllium should be present in particulate rather than in dissolved form in most natural environments.

Durum and Haffty (1963) found no beryllium in the water they examined. Durfor and Becker (1964) reported finding only one sample with a detectable concentration (0.75 $\mu g/l$). Beryllium concentrations were below the detection limit (0.3 $\mu g/l$) in all samples analyzed by Silvey (1967). Although beryllium is highly toxic, its chemical properties seem highly unfavorable for the occurrence of any but extremely low concentrations in natural water.

STRONTIUM

The chemistry of strontium is much like that of calcium. Strontium is a fairly common element in minor amounts, replacing calcium or potassium in igneous-rock minerals although apparently favoring those species that are typical of granitic and syenitic rocks rather than ultrabasic rocks (Rankama and Sahama, 1950, p. 476). The carbonate strontianite and the sulfate celestite are common in sediments. According to Kulp, Turekian, and Boyd (1952), the strontium to calcium ratio in most limestone is less than 1:1,000, although fossils in the limestone tend to be enriched in strontium.

The usual wet-chemical analysis methods for calcium also are sensitive to strontium and include any strontium present as an equivalent amount of calcium. As a result, the presence of small to moderate amounts of strontium does not cause analytical problems and usually goes unreported unless special attention is given to determining strontium separately.

Mostly because the radioactive isotope strontium-90, which is a product of nuclear fission, is a very undesirable impurity in drinking water, attention began to be focused on the chemistry and occurrence of strontium in water in the 1940's and 1950's. As a result, a good deal of information on its occurrence is now available.

From free-energy data quoted by Garrels and Christ (1964, p. 423) for strontianite, the value of K in the relationship

$$\frac{[HCO_3^-][Sr^{+2}]}{[H^+]} = K$$

is calculated to be 15.1. Thus at a pH of 8.0 when the activity of bicarbonate is 10^{-3} molal (about 61 mg/l), the strontium activity would be about 28 mg/l. Figure 17 shows the solubility of calcium in relation to calcite under these conditions would be about 40 mg/l. Thus strontianite is a little less soluble than calcite.

Solubility products for strontium sulfate quoted by Sillén and Martell (1964, p. 236) are mostly near $10^{-6.50}$ at 25° C. This value shows that a solution containing an activity of 280 mg/l of strontium could contain an activity of only about 10 mg/l of sulfate at equilibrium. If a higher sulfate concentration occurred, the maximum strontium concentration would decrease. For example, a sulfate activity of 100 mg/l would permit only 28 mg/l for the strontium activity. The analytical value for sulfate concentration in such a solution could be considerably above 100 mg/l owing to effects of ionic strength and formation of sulfate ion-pairs.

The strontium concentration in most natural water does not approach the solubility limit of either strontianite or celestite. Some ground water, however, contains rather large amounts of strontium. Analysis 1, table 20, represents water from a well at Waukesha, Wis. The strontium content of this water (52 mg/l) is one of the highest reported for potable water, although some brines (analysis 4, table 15) contain much more. Nichols and McNall (1957) found that many wells in the eastern part of Wisconsin yielded water containing more than 1 mg/l of strontium. Rather high concentrations also are known to be present in ground water at certain localities in Ohio (Feulner and and Hubble, 1960) and in Florida (Odum, 1951). Calculations taking into account the effects of ionic strength show the water from the Waukesha well is very close to saturation with respect to strontium sulfate. Durum and Haffty (1963) found the median value for strontium in major North American rivers to be 0.06 mg/l. Skougstad and Horr (1963) published the most intensive study to date of the occurrence of strontium in natural water, both surface and underground, in

TABLE 20.—*Analyses of waters containing unusual concentrations of metals and other constituents*

[Analyses by U.S. Geological Survey. Date below sample number is date of collection. Source of data: 1, Lohr and Love (1954a), 2 and 3, Emmons (1917), 4, 5, unpub data, U.S. Geol. Survey files]

Constituent	(1) May 2, 1952		(2) About 1911		(3) About 1917		(4) Mar. 29, 1963		(5) Dec 8, 1934	
	mg/l	meq/l	mg/l	meq/l	mg/l	meq/l	mg/l	meq/l	mg/l	meq/l
Silica (SiO_2)	8.7		56		23		10		21	
Aluminum (Al)			433	48.65	12	1.32				
Iron (Fe)	.37		2,178	77.99	143		10		.04	
Manganese (Mn)	.05		.2	.01						
Copper (Cu)			312	9.82						
Zinc (Zn)			200	6.12	345	10.55				
Calcium (Ca)	60	2.99	68	3.39	260	12.97	5.0	0.25	452	22.55
Magnesium (Mg)	31	2.55	41	3.37	49	4.03	1.0	.08	555	45.64
Strontium (Sr)	52	1.19								
Sodium (Na)	12	.52	23	1.00	13	.57	136	5.92	1,900	82.65
Potassium (K)	4.0	.10	20	.51	3.2	.08	1.2	.03	17	.43
Carbonate (CO_3)							10	.33		
Bicarbonate (HCO_3)	285	4.67	0		0		296	4.85	476	7.80
Sulfate (SO_4)	111	2.31	6,600	138.65	1,650	34.35	10	.21	6,330	131.79
Chloride (Cl)	12	.34	1	.00	3.7	.10	34	.96	288	8.12
Fluoride (F)	5	.03					.8	.04		
Nitrate (NO_3)	8	.01					.7	.01	244	
Orthophosphate (PO_4)							2.6			
Arsenic (As)							1.3			
Selenium (Se)									1.98	
Boron (B)							4.6			
Dissolved solids										
Calculated	440		9,990		2,500		363		10,900	
Residue on evaporation										
Hardness as $CaCO_3$	337						17		3,410	
Noncarbonate	104						0			
Specific conductance (micromhos at 25°C)	658						594			
pH	7.6						8.5			
Acidity as H_2SO_4			130		252					

1. North Street well, Waukesha, Wis Public supply Depth, 1,907 ft Water-bearing formation, sandstones of Cambrian and Ordovician age
2. Water from first level below black copper workings, Burra-Burra mine, Ducktown, Tenn
3. Water from Victor mine, Joplin district, Missouri
4. Well, NW¼SW¼ sec 11, T 19 S, R 3 W, Lane County, Oreg. Depth, 150 ft Water-bearing formation, Fisher Formation
5. New drainage ditch about 4 miles north of Mack, Mesa County, Colo.

the United States. Their paper contains more than 200 anlayses. The median content of strontium for the larger United States public water supplies is 0.11 mg/l.

BARIUM

Table 1 indicates that barium is somewhat more abundant in igneous rocks than strontium but that the carbonate rocks contain considerably less barium than strontium. Barium ions have a larger radius than strontium ions and probably cannot readily enter the crystal lattices of the carbonates of lighter alkaline-earth metals. The solubility of barium carbonate is about the same as that of calcite (Sillén and Martell, 1964, p. 138).

A likely control over the concentration of barium in natural water would be the solutility of barite ($BaSO_4$), which is a fairly common mineral. The solubility product for this material is near 10^{-10} (Sillén and Martell, 1964, p. 236), and at sulfate molar activities near 10^{-4} ($\sim$10 mg/l) or 10^{-3} ($\sim$100 mg/l) the corresponding equilibrium molar activities of Ba^{+2} would be 10^{-6} ($\sim$0.14 mg/l) or 10^{-7} ($\sim$0.014 mg/l), respectively.

Another factor which seems likely to influence the concentration of barium in natural water is adsorption by metal oxides or hydroxides. Barium seems commonly to be found in deep-sea manganese nodules and also in fresh-water manganese oxide deposits (Ljunggren, 1955).

The median concentration for barium in public water supplies reported by Durfor and Becker (1964, p. 78) is 0.043 mg/l, a value which is in the range that would be expected were the concentration under the control of barium sulfate solubility equilibria. Durum and Haffty (1963) reported a median concentration of 0.045 mg/l in the larger rivers of North America.

If the concentration of barium in natural water actually tends to be controlled by $BaSO_4$ solubility, one would expect a narrower range between the upper and lower extreme values than for elements whose concentration is more likely to be a function of the availability of the element in the environment. Strontium is more likely to fit in the latter category. It is interesting to note that in the two compilations cited barium does indeed have a narrower range of concentration from maximum to minimum observed than the other elements determined.

The suggestion that barium concentrations may tend to be controlled by solubility equilibria is by no means a firm conclusion. It should be tested by additional studies.

OTHER METALLIC ELEMENTS

The elements grouped here for purposes of this discussion include more than half the periodic table. Very little information, however, is available on the concentrations of most of the elements in the group in natural water. The three whose chemical behavior in natural water has been most closely studied, aluminum, manganese, and iron, already have been discussed individually. Others considered here include only those for which some data are given in published studies of natural-water composition.

TITANIUM

Although titanium is one of the 10 most abundant elements in the earth's crust, the titanium content of natural water, from what information is available, seems to be extremely small. From thermodynamic data published by Garrels and Christ (1964, p. 426) and solubility data in Sillén and Martell (1964, p. 45), it is evident that the chemistry of titanium is unfavorable for solution in natural water. The element has the oxidation state of +4 over practically all the Eh range to be expected in natural water. Solubility of the oxide TiO_2 (rutile) is very low except in a strongly acid solution and should not exceed 1 µg/l as Ti^{+4}, unless the pH is below 5. The most likely form of dissolved titanium is the cation $TiOOH^+$ (Sillén and Martell, 1964, p. 45).

The median concentration of titanium reported in major rivers of North America by Durum and Haffty (1963) is 8.6 µg/l. The median for public water supplies of the United States is reported by Durfor and Becker (1964, p. 78) as "less than 1.5 µg/l." Somewhat similar concentrations were reported by Silvey (1967). The higher values reported in both sets of data may represent colloidal material not separated from the samples by the filtration techniques used. In general, the concentrations observed support the tentative conclusion that titanium concentrations in natural water are controlled by the solubility of oxide and hydroxide species of Ti^{+4}. The concentration of titanium in sea water reported in table 2 is 1 µg/l.

VANADIUM

The chemistry of vanadium is relatively complicated. Aqueous species that might be stable in natural water include both anionic and cationic forms of V^{+3}, V^{+4}, and V^{+5} (Evans and Garrels, 1958). Although many of these possible forms are readily soluble, the V^{+3} and V^{+4} species have a minimum solubility near or below 1 µg/l near neutral pH and under mildly reducing conditions. These conditions could occur in ground water having an Eh low enough to maintain a concentration of ferrous iron amounting to 0.5 mg/l or more. The V^{+5} species are anionic and generally of higher solubility.

It would appear that vanadium should be most soluble in surface water, or ground water containing a measurable concentration of oxygen. Concentrations as high as 70 µg/l were reported by Durfor and Becker (1964) in some ground water used for public supplies. The concentrations generally observed in surface water seem to be considerably lower than this. Most likely the concentrations of vanadium observed in water are related to the rarity of the element and probably seldom reach saturation. Several investigators have reported tendencies for vanadium to be concentrated in vegetation.

CHROMIUM

Information published on the occurrence of chromium in water mainly relates to industrial pollution. Numerous incidents of chromium wastes entering and contaminating ground water have been described. The inclusion of a maximum limit for chromium in the U.S. Public Health Service standards for drinking water has resulted in a considerable number of determinations of concentration for various water supplies.

Chromium has several possible oxidation states. Thermodynamic data given by Latimer (1952, p. 246) show that Cr^{+3} and Cr^{+6} species are stable in water and that solubility of chromium is less than 10^{-8} moles per liter (equivalent to 0.5 µg/l of Cr) only between pH 8.0 and 9.5 and in solutions having a little lower Eh than is normally imparted by saturation with atmospheric oxygen. The anionic species dichromate $Cr_2O_7^{-2}$ and chromate CrO_4^{-2} are apparently relatively stable in ground-water environments. In rather strongly reducing environments below a pH of 8.0, the predominant form is the complex cation $CrOH^{+2}$.

Chromium is a relatively common element, but the chromate and dichromate forms probably are not often found except in polluted water. The median value for the public water supplies studied by Durfor and Becker (1964, p. 78) was 0.43 µg/l as Cr, and for North American rivers the median reported by Durum and Haffty (1963) was 5.8 µg/l. Information is insufficient to permit any conclusions regarding the possible solubility controls over the observed concentrations.

MOLYBDENUM

Multiple valence states and tendency for polymeric hydroxide anion species are complicating factors in molybednum chemistry. The anionic (molybdate) forms are probably predominant in natural water. Molybdenum is a rather rare element, and thus one might expect no very effective solubility control over molybdenum concentrations, but rather that concentrations observed in water would be somewhat directly related to the abundance of molybdenum in solid mineral species in the environment. Some writers have felt that for this reason, and related reasons, molybdenum was a good element to use for tracing mineral anomalies in geochemical prospecting (Hawkes and Webb 1962, p. 227).

The median concentrations of molybdenum reported in the two compilations quoted here are 0.35 µg/l in the North American rivers and 1.4 µg/l in public water supplies. Vogeli and King (1969) found concentrations greater than 1 µg/l in water from a few streams in Colorado.

Molybdenum is accumulated by vegetation, and forage crops, especially legumes, raised on land irrigated with water that contains small amounts of this element may become toxic to grazing animals. Incidents of molybdenum poisoning of cattle have been documented, and the proposed U.S. Federal Water Pollution Control Administration (1968, p. 152) standards for irrigation water fix an upper limit of 0.01 mg/l. This is equivalent to 10 µg/l and is probably exceeded by many ground waters. Present drinking-water standards (U.S. Public Health Service, 1962) contain no limit for molybdenum. The normal concentrations of this element in potable water are very poorly known and need closer investigation than they have yet received. A study by Barnett, Skougstad, and Miller (1969) showed molybdenum concentrations over 100 µg/l in water from Dillon Reservoir, one of the sources of the Denver, Colo., public supply.

Vinogradov (1957) described an area in the U.S.S.R. where background, or normal, concentrations of molybdenum in ground water were near 3 µg/l and where anomalous concentrations useful for geochemical prospecting ranged from 10 µg/l to 10 mg/l.

COBALT

The element cobalt is known to be essential in plant and in animal nutrition, but is required only in very small quantities. Cobalt has some chemical properties similar to those of iron, for example, in having both $^{+2}$ and $^{+3}$ valence states. Co^{+3} ions generally, however, are not compatible with redox potentials likely to occur in natural water; hence the most probable oxidation state is Co^{2+}.

The solubility of $Co(OH)_2$ is similar to that of ferrous hydroxide and probably is not a control of cobalt concentrations in natural water. The solubility of $CoCO_3$, however, is much lower than that of siderite and could be an important factor. Thus, at a pH of 8.0 in the presence of about 100 mg/l of HCO^-_3, the solubility of cobalt would be 10^{-7} molal, or about 6 µg/l. Cobalt forms many complex ions which tend to increase the solubility above calculated values that do not allow for complexing effects.

According to Rankama and Sahama (1950, p. 684), cobalt generally is more common in ultrabasic rocks than in other types of igneous rock. Although cobalt carbonate has a very low solubility, the data of Graf (1962) show that it is a very minor constituent of carbonate rocks.

Few data on cobalt concentrations in natural water are available, partly because in many of the waters studied by Durum and Haffty (1961) and Durfor and Becker (1964), the amounts present were below the limit of detection. In any event, the concentrations appear to be limited by factors other than simple solubility relationships. Maliuga (1950) reported average concentrations of 20 µg/l in water of mineralized zones in the southern Ural area of the U.S.S.R.

One mechanism that probably is involved in controlling observed concentrations of cobalt is adsorption by colloidal particles of oxides and hydroxides of manganese or iron. An excellent discussion of this effect for cobalt and other minor metals in sea water has been published by Goldberg (1954). The nature of this process is probably different in fresh water, but ferric hydroxide and manganese oxides are present in stream sediment and soils. Jenne (1968) has cited studies showing that both cobalt and nickel tend to be concentrated on particles of manganese oxide in soils. The low concentrations of cobalt in natural water may be brought about by adsorption reactions of this type.

NICKEL

The general geochemical behavior of nickel is similar to that of cobalt. The oxidation state in natural water can be expected to be Ni^{+2}. The nickel ion is probably strongly adsorbed by iron and manganese oxides, and concentrations may thereby be held below the solubility levels of the hydroxide $Ni(OH)_2$. This hydroxide evidently is less soluble than $Co(OH)_2$, although values for the solubility product given by Sillen and Martell (1964, p. 56) have a rather wide range. Some of the disagreement over the solubility values can probably be ascribed to changes in the precipitated hydroxide during aging. The solubility of nickel carbonate appears to be much greater than that of $CoCO_3$.

The median concentration of nickel in North American rivers reported by Durum and Haffty (1963) was 10 µg/l. A value of "less than 2.7 µg/l" was reported for the larger public water supplies by Durfor and Becker (1964). Both studies gave higher values for nickel than for cobalt. In part, this could be a reflection of the greater abundance of nickel in rocks, as indicated by data in table 1. Other compilations such as that of Silvey (1967) show a similar tendency for nickel to occur in water in higher concentrations than cobalt. Maligua (1950) found an average of 40 µ/l of nickel in water from a mineralized region in the southern Urals of the U.S.S.R., a value twice as great as the average concentration for cobalt.

COPPER

The Eh-pH diagrams for copper given by Garrels and Christ (1964, p. 239-240) suggest that cupric oxide, or hydroxy-carbonate minerals, would tend to limit the solubility of copper in aerated water at a pH of 7.0 to about 10^{-6} moles per liter, equivalent to 64 µg/l. At a pH of 8.0, the solubility would be about one tenth as great, or about 6.4 µg/l. Copper solubility is generally lower in reducing systems than in oxidizing systems, especially if reduced sulfur species are present.

The amounts of copper reported in river water by Durum and Haffty (1963) are within the general range predicted from the solubility calculated by Garrels and Christ (1964, p. 239-240). Copper is a widely distributed element, and besides the natural occurrences, the metal is used extensively in water treatment and in fabrication of pipe, valves, and pumping equipment. Small amounts of copper are commonly added to water stored in reservoirs to inhibit algal growth. Most of the higher concentrations reported by Durfor and Becker (1964) for public supplies were in water which had been impounded in storage reservoirs and treated in this way. Many waters that are used for public supplies probably have obtained some copper from cultural sources.

High concentrations of copper may occur in water from mines or in water that has leached ore dumps or mill tailings. Analysis 2, table 20, is an example of water of this kind. Copper contents of ground and surface water in areas where ore bodies occur may be rather large. Maliuga (1950) found an average of 200 µg/l of copper in water from a mineralized region in the southern Urals of the U.S.S.R. Geochemical-prospecting literature contains many data on copper concentrations in water from such areas.

Copper is essential in the nutrition of plants and animals, and the circulation of the element in the hydrosphere is obviously influenced to some extent by biota. No mechanism for copper removal besides inorganic precipitation, however, is needed to explain the general levels of concentration in fresh water.

SILVER

The U.S. Public Health Service (1962) standards for drinking-water quality include a recommended maximum concentration of 0.05 mg/l (50 µg/l) for silver. The redox potential for the couple

$$Ag^+ + e^- = Ag(c)$$

given by free-energy data of Latimer (1952, p. 190) is +0.797 volts. This means that metallic silver is stable over most of the Eh-pH field for water, unless very low activity of dissolved silver is assumed. An activity of Ag^+ of 1.0 µg/l ($\sim 10^{-8}$ molar) will be present in equilibrium with metallic silver at an Eh of 0.323 volts. This general level of Eh is often observed in aerated water at a pH of ~ 8. Lower Eh would decrease silver solubility.

The solubility of silver oxide is low enough to prevent high concentrations of silver at high pH, and silver chloride has a low enough solubility to exert a major control where chloride concentration exceeds 10^{-3} molar (35 mg/l).

These solubility relationships suggest that in dilute aerated water the equilibrium concentration of silver ought to be between 0.1 and 10 µg/l. The data on public supplies by Durfor and Becker (1964) show a median concentration of 0.23 µg/l, and the river-water analyses of Durum and Haffty (1963), a median of 0.09 µg/l. Silver is a rather rare element, and occurrence of amounts below saturation would not be unexpected.

Silver has been used as a disinfectant for water, and Woodward (1963) quoted data showing that as little as 10 µg/l in alkaline water was effective in removing *Escherichia coli,* although at this low concentration the reaction was slow. The possibility of residual silver in water from biocidal treatment, therefore, exists, although the element is not widely used for this purpose.

GOLD

The solubility indicated for gold by Garrels and Christ (1964, p. 257–258) is extremely low within the Eh-pH region of water stability, except where reduced sulfur species are present which can form complex ions with gold. Colloidal suspensions of metallic gold are readily prepared in the laboratory and may exist in nature.

There are virtually no data on gold content of natural water, and it seems likely the amounts which might be present in solution are below detection by most of the currently used methods.

ZINC

The most common zinc mineral probably is the sulfide sphalerite, which is also the most important zinc ore. In some of the igneous-rock minerals, zinc may replace iron or magnesium, and it is commonly present in carbonate rocks. The general geochemistry of zinc has been described by Wedepohl (1953). The weathering of reduced zinc minerals in contact with air releases zinc ions. Analysis 3, table 20, represents a mine water that contained 345 mg/l of zinc. The recommended upper limit for zinc in drinking water is 5.0 mg/l (U.S. Public Health Service, 1962). Zinc is a widely used metal and may be dissolved from galvanized pipe. It may also be present in industrial waste.

The solubility products for zinc hydroxide ($Zn(OH)_2$) quoted by Sillén and Martell (1964, p. 61–63) are generally in the vicinity of 10^{-17} at 25°C. The equilibrium activity of Zn^{+2} at a pH of 8.00 in the presence of this solid would be 10^{-5} moles per liter, or about 650 µg/l. As pH increases, the solubility of $Zn(OH)_2$ passes through a minimum and then increases as the anion $Zn(OH)_3^-$ is formed. The solubility of zinc carbonate is in the same general range as the hydroxide. Concentrations of zinc ranging considerably above 1 mg/l can be chemically stable at neutral pH.

Natural water seldom seems to contain the concentrations of zinc which would be present in equilibrium with hydroxide or carbonate. The concentrations that are observed thus should tend to reflect the availability of zinc in the rock and soil through which the water has passed and be useful as an indication of possibly economically valuable ores. Zinc can be determined conveniently at very low concentrations by wet-chemical procedures and by means of atomic-absorption spectrophotometry to at least 2 µg/l. The detection limit for zinc in the arc spectrograph, however, is too high for this technique to be useful for natural water.

Among many reports of studies of prospecting procedure for ores using analyses of zinc in natural water is a paper by Kennedy (1956) on the mineralized area of southwestern Wisconsin. A background of zinc and lead (determined together) of about 10 µg/l applies to the ground water of that area. Some of the springs near known ore deposits had a total of more than 1,500 µg/l of these metals. Concentrations of zinc in water from nonmineralized areas probably are generally considerably below 10 µg/l.

CADMIUM

Amounts of cadmium present in rocks are much less than the amounts reported for zinc (table 1). The concentrations of cadmium that have been found in natural water are very small. Durfor and Becker (1964) found none above their detection limit in the public water supplies they studied. Kroner and Kopp (1965) found detectable quantities, generally below 10 µg/l in a few samples from the Colorado, Mississippi, and Columbia Rivers. The element is considered toxic, and a recommended upper concentration limit of 10 µg/l for drinking water has been given (U.S. Public Health Service, 1962). The lower limit of detection for cadmium by the atomic-absorption spectrophotometer is 10 µg/l, but treatment of the sample can increase the sensitivity.

Information on the occurrence of cadmium in water is too incomplete to justify detailed consideration of possible solubility controls. The element is relatively rare, its abundance in rocks being comparable to mercury. It is used industrially, however, mostly as a plating metal and possibly could be found in polluted water. Krauskopf (1956b) noted that cadmium apparently is present in sea water in concentrations farther below saturation than any other element of the group he considered.

MERCURY

Very few natural waters contain detectable concentrations of mercury. The element may be introduced into water through disposal of mining and metallurgical or other industrial waste. Because of its toxicity, mercury is an undesirable impurity in water, but dangerous concentrations are not likely to be reached in water that is potable in other respects.

Because of its volatility, mercury may be transported more readily in hot than in cold water, and cinnabar (HgS) is deposited by some hot springs (White, Hem, and Waring, 1963, p. F50). The concentration in water associated with such deposits, however, seems to be uniformly below readily detectable limits, probably less than 10 μg/l.

GERMANIUM

Although germanium is a rather rare element, it has been reported in detectable quantities in some of the California waters studied by Silvey (1967). It was present in amounts greater than 50 μg/l in water from most of 30 springs and was believed by Silvey to indicate a deep-seated origin for part of the water.

Germanium seems likely to occur as an anionic (germanate) species—at least this form is suggested for sea water (Goldberg, 1963). The element has been reported to occur in relatively large quantity in coal ash (Rankama and Sahama, 1950, p. 736), and probably its behavior in the hydrosphere is influenced by biochemical factors. The data on the occurrence of germanium in natural water are still very incomplete.

LEAD

The chemistry of lead in natural water generally may be indicated by a series of Eh-pH diagrams by Garrels and Christ (1964, p. 234–238). Over most of the water-stability region, the oxidation state of lead at equilibrium is Pb^{+2}, and the most likely control of solubility in oxidizing systems is the carbonate, cerussite. $PbSO_4$ also is relatively insoluble and could control the solubility of lead in oxidizing systems. In the presence of reduced sulfur species, the lead sulfide, galena, has a very low solubility. In general, there would seem to be a high probability that lead concentrations in natural water would be limited mainly by these solubility restrictions, which easily could lower equilibrium activity of Pb^{+2} to 10^{-8} moles per liter (about 2 µg/l).

Lead has been widely used in plumbing. The word itself is derived from the Latin word for lead, plumbum. Although modern practice minimizes the opportunity for drinking water to contact lead, older systems still in use may provide opportunities for objectionable amounts to be dissolved, especially by water with a low pH. The element can replace calcium in aragonite and potassium in feldspar (Rankama and Sahama, 1950, p. 733).

A median value of 3.7 µg/l of lead for public water supplies was reported by Durfor and Becker (1964), although lead was not detected at all in many samples. The median value for North American rivers reported by Durum and Haffty (1963) was 4.0 µg/l. The wide use of lead as a gasoline additive and its dispersal in engine exhaust into the atmosphere has considerably increased the availability of the element for solution in rainfall and fresh water. The importance of this effect, however, is not known at present.

NONMETALS

Elements considered in this group include arsenic, selenium, bromine, and iodine. All these occur as anions in natural water.

ARSENIC

In rocks, arsenic may occur in reduced form in mineral veins and as oxide. Arsenate (AsO_4^{-3}) occurs in some metallic ores and can replace phosphate in apatite. From thermodynamic data it can be shown that the arsenate species $H_2AsO_4^-$ and $HAsO_4^{-2}$ are the equilibrium forms over the pH range of most natural water. $HAsO_4^{-2}$ is predominant above a pH of 7.2, and $H_2AsO_4^-$ below. In reducing environments, however, the form $HAsO_2(aq)$ may be present. At a pH of 7.0 the Eh at which the change occurs is about 0.05 volt. Arsenic may be added to water supplies through waste disposal and is present in certain insecticides and herbicides.

The concentration of arsenate ions at equilibrium in water depends on the cation concentrations and the solubility of the various arsenates that could be precipitated. Solubility products for metal arsenates quoted by Sillén and Martell (1964, p. 205) indicate that many of these compounds have low solubilities in terms of the anion AsO_4^{-3}. This form predominates only above pH 11.5, however, and in the usual range of pH of natural water, the solubility of calcium or magnesium arsenate is sufficient to permit several tens of milligrams of arsenic to be retained in solution. Minor metals can afford more effective controls. An activity of Cu^{+2} of 65 µg/l, for example, would limit the equilibirum solubility for arsenic to a few tenths of a milligram of As per liter. Probably the sorption of arsenate on precipitated ferric hydroxide or other active surfaces is also an important factor limiting arsenic solubility in natural water systems. The reaction of arsenate with ferric hydroxide may represent the formation of a ferric hydroxy-arsenate. Such compounds are known for ferric iron and orthophosphate and have a very low solubility (Sillén and Martell, 1964, p. 185).

The U.S. Public Health Service (1962) drinking-water standards recommend an upper limit of 0.05 mg/l of As. Concentrations up to 1.0 mg/l have reportedly been present in water used for drinking and, at least for short periods of time, have produced no apparent ill effects (McKee and Wolf, 1963, p. 140), but long-term use of a concentration of 0.21 mg/l was reported to be poisonous. Deaths among cattle as a result of drinking natural water containing arsenic have been reported from New Zealand. Local reports of similar deaths of cattle in the western United States are heard, but generally the presence of arsenic in the suspected "arsenic spring" has not been demonstrated. About 20 milligrams of arsenic per pound of animal is considered a lethal dose.

Arsenic is commonly present in water of thermal springs. Analysis 8, table 19, for example, reports 4.0 mg/l. Significant concentrations have been found in ground water in Lane County, Oreg. (unpub. data, U.S. Geol. Survey files). Analysis 4, table 20, represents a water from the latter area which has 1.3 mg/l of arsenic and, a dissolved-solids concentration of 363 mg/l.

SELENIUM

The chemistry of selenium is similar in some respects to that of sulfur, but selenium is a much less common element. In the 1930's it was discovered that a disease of livestock in certain areas of the western United States was actually caused by an excessive intake of selenium, and an extensive research program began. The agricultural significance of selenium has been summarized by Anderson, Lakin, Beëson, Smith, and Thacker (1961). The discussion here is based largely on this summary.

The intake of selenium by animals is mainly from vegetation. Some species of the genus *Astragalus* are particularly notable for taking up selenium from the soil, and some plants have been found to contain several thousand milligrams of selenium per kilogram of dry plant parts. Drainage water from seleniferous irrigated soil has been reported to contain as much as 1 mg/l of dissolved selenium. This type of water (analysis 5, table 20) is so high in dissolved solids, however, that it would not be likely to be used as drinking water.

Thermodynamic data show that above a pH of 6.6 in aerated water, with the Eh to be expected in such a solution, the stable form of selenium is the anion selenite, SeO_3^{-2}. Under mildly reducing conditions, however, the equilibrium species is elemental selenium, which presumably has a low solubility. It is to be expected that selenium concentrations in water in a reducing environment would be very low. It seems likely, in any event, that the occurrence of selenium in water is influenced by controls other than selenite solubilities, although little information is avaliable concerning this aspect of the subject. An Eh-pH diagram for selenium species is given by Anderson, Lakin, Beeson, Smith, and Thacker (1961, p. 11).

Many analyses of water from streams draining seleniferous areas have been made, and Anderson, Lakin, Beeson, Smith, and Thacker (1961) reported 80 $\mu g/l$ in one sample from the Gunnison River in the vicinity of Grand Junction, Colo. In most samples from the Colorado River and its tributaries, however, the concentrations were much lower. The maximum concentration of selenium for drinking water recommended by the U.S. Public Health Service (1962) standards is 0.01 mg/l.

BROMINE

Bromine is similar in chemical behavior to chlorine, but much less abundant. Bromine in natural water is always present as the bromide ion Br^-. The concentration present in sea water is 65 mg/l, an amount sufficient to make bromide an important constituent. The total quantity in the ocean probably is a major fraction of the quantity present in the outer 16 kilometers of the earth's crust, but the available data do not form a satisfactory basis for deciding whether the extraction of bromine into the ocean has been as complete as for chlorine. The literature on geochemistry of bromine was reviewed by Correns (1956).

Bromide is present in major concentrations in some brines. Analysis 4, table 15, which contains 3,480 mg/l of bromide, represents such a brine. The concentration of bromide in a chloride brine can increase as a result of evaporation, even after the solution reaches saturation with respect to sodium chloride. Bromide may possibly be selectively concentrated by the clay-membrane effects noted previously for enrichment of brines in other ions, as the Br^- ion is larger than the Cl^- ion.

Bromide, in trace concentrations, is present in rainwater, and apparently the Br·Cl ratio is generally higher than in sea water. According to Winchester and Duce (1966), bromide and iodide occur in smaller aerosol particles in the atmosphere than chloride. Bromide must occur in traces in river water, but little information is available on the bromide concentrations of dilute natural water.

IODINE

Although iodine is not a particularly abundant element, it is essential in nutrition of the higher animals, including man, and concentrations in natural water have received considerable attention. The circulation of the element appears to be strongly influenced by biochemical processes. Nearshore marine vegetation, especially kelp, concentrates iodide from sea water, which itself has only 0.06 mg/l of iodide. The volatility of the element and some of its solid salts probably adds to the general tendency for iodide to be circulated in the atmosphere. Contents of iodide in rainwater reported by Rankama and Sahama (1950, p. 767) were 1–3 μg/l. Konovalov (1959) reported concentrations ranging from 42.4–3.3 μg/l for principal rivers of the U.S.S.R. Many data on iodide in natural fresh water have been obtained by Soviet investigators.

Iodide is concentrated in some brines; analysis 4, table 15, reports 48 mg/l.

A silver iodide aerosol has been used in experiments to increase rainfall or snowfall in many places in recent years. The amount of iodide which might be introduced into natural water by this process generally is too small to be detected by ordinary means, but Warburton and Young (1968), using neutron activation, were able to detect silver in rain from areas seeded with silver iodide. The silver-detection limit for this technique was about 0.02 μg/l.

RADIOACTIVE ELEMENTS

To discuss the subject of radioactivity in any detail is beyond the scope of this paper. The treatment here is necessarily brief and emphasizes naturally occurring radioactive nuclides. A very large number of radioactive nuclides is produced in the process of nuclear fission of certain elements. These fission products constitute problems of containment, storage, utilization, and disposal that have received extensive scientific study. For more comprehensive discussions of radioactivity, the reader is referred to standard texts such as Friedlander and Kennedy (1955).

Radioactivity is the release of energy by changes occurring within atomic or nuclear structures. Certain arrangements within these structures are inherently unstable and spontaneously break down to form more stable arrangements. The most unstable configurations disintegrate rapidly and do not exist in measurable amounts in the present earth crust. Other unstable nuclides, however, have a very slow rate of decay and still exist in significant quantity.

Radioactive energy is released in various ways. The three types of radiation of principal interest in natural-water chemistry are (1) alpha radiation, consisting of positively charged helium nuclei, (2) beta radiation, consisting of electrons or positrons, and (3) gamma radiation, consisting of electromagnetic wave-type energy similar to X-rays.

The principal sources of radioactivity in water are dissolved ions, but tritium (H^3), which is present in place of normal hydrogen in occasional water molecules, may give rise to small amounts of activity.

Three isotopes of high atomic weight, uranium-238, thorium-232, and uranium-235, which exist naturally, are spontaneously radioactive and give rise to most of the naturally occurring radioactivity in water. They disintegrate in steps forming a series of radioactive nuclide products, mostly short lived, until a stable lead isotope is produced. The uranium-238 series probably produces the greatest part of the radioactivity observed in natural water, although the thorium-232 series also may be significant in some places. The uranium-235 or actinium series is less important than the others because only a very small fraction of natural uranium is composed of this isotope.

A few other naturally occurring nuclides are radioactive. The principal ones are potassium-40 and rubidium-87. Both constitute small fractions of the natural element and are very long lived.

Alpha-emitting substances in natural water are mainly isotopes of radium and radon, which are products of the uranium and thorium series. Beta and gamma activity is evidenced by some members of these series and also is characteristic of potassium-40 and rubidium-87. Many of the fission products are strong beta emitters. Among the ones of special interest in water are strontium-89, strontium-90, iodine-131, cesium-45, phosphorus-32, and cobalt-60.

It has been known since the early years of the 20th century that some natural waters are radioactive, especially those of certain thermal springs The first measurements of radioactivity in natural water were made with the electroscope, which is sensitive to small ionizing effects and is still used by some modern investigators. The Geiger-Müller tube and various scintillation devices, with counters and scalers, and other ionization measuring techniques are now more extensively employed

Where possible, radioactivity data are expressed in terms of concentration of specific nuclides. General measurements of total or gross alpha or beta and gamma activity also are often reported. The element uranium is considered a radioactive constituent, but is most conveniently measured by chemical means. For some elements, the radiochemical techniques permit detection of concentrations that are far smaller than any chemical method can attain.

SIGNIFICANCE OF WATER PROPERTIES AND CONSTITUENTS 211

Various terms and units are used in reporting radioactivity. Some studies have merely reported counts per mniute or disintegrations per minute for specific volumes of sample. These data cannot be considered quantitative unless amounts of sample used, counter efficiency, and various details of the conditions of measurement are given. To standardize results and make comparisons possible, the radioactivity of water is usually expressed in terms of an equivalent quantity of radium, or in terms of the rate of radioactive disintegration (curies). One curie is defined as 3.7×10^{10} disintegrations per second, the approximate specific activity of 1 gram of radium in equilibrium with with its disintegration products. This unit is very large for the purpose of expressing natural radioactivity levels, and such data are often expressed in picocuries (curies $\times 10^{-12}$) for this reason. Other units are occasionally seen in the older literature such as the Rutherford (2.7 $\times 10^{-5}$ curies), the Eman (10^{-10} curies per liter), and the Mache unit (3.6×10^{-10} curies per liter).

Radioactivity expressed in picocuries or concentrations of specific nuclides determined by radiochemical procedures presents some specific problems of detection limits. Counting procedures are limited in sensitivity by background effects and the manner in which the counting is done. Many samples do not give a reading that is significantly above background, and tabulated data contain many entries of "less than" some specific activity level which represented the background. The user of such data should remember that the number so quoted is not indicative of a specific activity in the sample. It is essentially a statement that the activity present was below the detection limit for the determination in that particular sample.

The U.S. Geological Survey has conducted studies of radioactivity in water for a considerable time. Results of analyses of natural ground water from a wide variety of aquifers over the entire conterminous United States have been published (Scott and Barker, 1962), and methods used for determining uranium (Barker and others, 1965), radium (Barker and Johnxon, 1964), and beta-gamma activity (Barker and Robinson, 1963) have been described. The methods used for tritium were described by Hoffman and Stewart (1966), and methods for strontium-90, by Johnson and Edwards (1967). Results of monitoring stream waters in the United States have been published in the reports by the U.S. Federal Water Pollution Control Administration for the National Water Pollution Surveillance Network.

URANIUM

Natural uranium is composed of several isotopes, of which uranium-238 is predominant. This nuclide is the starting point in a radioactive decay series that ends with the lead isotope lead-206. The half life of uranium-238 is 4.5×10^9 years, a value which makes the nuclide rather weakly radioactive. In the amounts in which uranium occurs in natural water, chemical methods of detection are satisfactory.

The geochemistry of urainum has been extensively studied. Eh-pH and solubility diagrams published by Garrels and Christ (1964, p. 254-256) show that reduced species, where the oxidation state is U^{+4}, are relatively insoluble, but more highly oxidized forms such as the uranyl ion, UO_2^{+2}, or the anionic species present at high pH are more soluble. Uranyl complexes with carbonate and sulfate are known (Sillén and Martell, 1964) and may influence the behavior of dissolved uranium. The chemical properties of the U^{+6} state favor the wide dispersion of uranium in the oxidized portion of the earth's crust. The element, however, is not soluble enough to be present in large amounts in the ocean.

Uranium is present in amounts between 0.1 and 10 $\mu g/l$ in most natural water. Amounts greater than this are somewhat unusual.

RADIUM

Four isotopes of radium naturally occur: radium-223, radium-224, radium-226, and radium-228. Two of these, radium-228 and radium-224, are disintegration products of thorium. Radium-223 is a disintegration product of uranium-235, and radium-226 is a disintegration product of uranium-238. The half life of radium-226 is 1,620 years, much longer than any of the others, and it seems safe to assume that radium-226 is the dominant form in natural water. All radium isotopes are strongly radioactive and can be detected in small amounts. The data obtained by the U.S. Geological Survey's analytical procedure, however, are for the radium-226 isotope alone.

Radium is an alkaline-earth metal and behaves chemically somewhat like barium. The solubility of $RaSO_4$ at 25°C appears from data quoted by Sillén and Martell (1964, p. 237) to be somewhat less than the solubility of barium sulfate under the same conditions.

Radium that is produced in the disintegration of uranium-238 will reach equilibrium with the parent material in time if the system is not disturbed. It seems likely, however, that few aqueous environments could provide sufficient time for this equilibrium to become established, and because the chemistries of the two elements are so different the likelihood of separation is further increased.

The concentration of radium in most natural waters is far less than that of uranium and is usually below 1.0 picocurie per liter. The highest concentration reported by Scott and Barker (1962) was 720 picocuries per liter in water from the brine well represented by analysis 4, table 15. Values for radium of several thousand picocuries per liter have been reported in the literature, but it seems possible that some of these data were obtained by methods that did not distinguish between radium and radon; thus they should be interpreted accordingly (Stehney, 1955).

Radium concentrations amounting to more than 3.3 picocuries per liter have been found in potable water from deep aquifers in a considerable area of Iowa, Illinois, and Wisconsin (Scott and Barker, 1961). The upper limit recommended for drinking water is 3 picocuries per liter (U.S. Pub. Health Service, 1962).

RADON

Radium-226 decays to radon-222. This element is a gas and is strongly radioactive with a half life of 3.82 days. Radon is soluble in water, although it can move through openings in rocks without water being present. Small amounts are present in rainwater, and larger quantities are associated with ground water from certain localities. Obviously, the amount of radon present in a freshly collected sample may be much greater than could be supported by radioactive equilibrium with dissolved radium-226. The radon, however, is lost to the air or by radioactive decay so rapidly that measurements must be made at the sample-collection site. Even if this is done, the significance to be attached to the result is uncertain. Kuroda and Yokoyama (1954) and Rogers (1958) have described methods for field measurements. It is interesting to note that field measurements of radioactivity of spring waters in Colorado were made, as early as 1914, with equipment that was transported around the State in an automobile (George and others, 1920).

Radon-220 (thoron) is produced by the radioactive disintegration of thorium. This nuclide may occur in water, but it has a half life of less than a minute.

THORIUM

Although thorium is more abundant than uranium, the chemistry of thorium seems likely to cause the element to be deposited in hydrolyzate sediments and not to be carried extensively in solution in water. The methods available for determining thorium or separating daughter products of thorium from those of uranium are inadequate to permit any discussion of their relative importance in naturally radioactive solutions.

OTHER RADIOACTIVE NUCLIDES

Some of the fission products that have been released into the atmosphere and other parts of the environment have appeared in water in significant amounts. Presumably, fallout of fission products from the atmosphere was responsible for some of the otherwise unexplained high values of beta activity that have been reported in rivers and public water supplies (Setter and others, 1959). Amounts of radioactive fallout have trended downward after most of the atmospheric testing of nuclear devices was stopped in the early 1960's.

Tritium, H^3, is produced naturally in small quantities in the upper atmosphere. Tritium so produced is incorporated into water molecules and is, therefore, present in rainwater. Concentrations of tritium in water can be determined, but extensive concentration and highly sensitive detection procedures are required, because the tritium concentration in normal rainwater is only about 10 tritium atoms per 10^{18} normal hydrogen atoms. Stewart and Farnsworth (1968) presented detailed data on tritium in rainfall in the United States in the period 1963–65. The half life of tritium is 12.3 years, and it can, therefore, be used as a means of determining how long a particular water may have been stored out of contact with fresh supplies of tritium. The amount of tritium in the atmosphere was greatly increased by the testing of nuclear weapons, and this caused all the precipitation especially in the Northern Hemisphere for a long period, beginning in 1954, to be "tagged" with excess tritium. Various studies of the tritium contents of water have been made and used for such purposes as tracing and dating of ground water (von Buttlar and Wendt, 1958) and calculating rates of water circulation in the hydrologic cycle (Begeman and Libby, 1957).

Another radioactive nuclide formed by cosmic-ray bombardment in the atmosphere is carbon-14. This carbon isotope occurs naturally in atmospheric carbon dioxide and is incorporated into organic material synthesized by growing plants. Carbon-14 has a half life of 5,730 years and can be used as a means of determining the age of plant material, such as wood, or other substances containing carbon that originated from the atmosphere at a specific time and then was cut off from further carbon-14 supplies. Ground-water movement rates can be estimated. Hanshaw, Back, and Rubin (1965), using carbon-14 measurements, calculated that the movement rate in the principal artesian aquifer of central Florida was 23 feet a year on the average over a distance of 85 miles. Pearson and White (1967) found movement rates in the Carrizo sand aquifer in Atascosa County, Tex., were 8 feet a year 10 miles down dip from the outcrop of the formation and 5.3 feet a year at a distance of 31 miles. The rates obtained in both studies by the carbon-14 technique agree with values calculated from hydrologic measurements.

The possible uses of radioactive nuclides in hydrology have been the subject of many papers, but descriptions of actual applications have been fewer. Besides the dating studies mentioned above, some experiments have been made by adding radioactive material to water and measuring recovery at other points. An indication of the nature of pilot-type studies can be gained from the work of Kaufman and Orlob (1956). Several papers describing work on tritium as a hydrologic tool (Carlston and Thatcher, 1962; Carlston, 1964) were issued as a result of U.S. Geological Survey research.

Radioactive tracer techniques may make possible some types of studies that otherwise could not be attempted and offer a possibility of simple and sensitive detection of the tracer. As with all tracing studies, however, the results must be interpreted in the light of other important hydrologic information.

ORGANIC CONSTITUENTS

The determination of the content of specific organic compounds in water has gained considerable impetus from the need to have more information about potentially or actually dangerous organic pollutants that might enter water supplies. Development of instrumentation for organic analyses has progressed rapidly, and some classes of organic compounds, notably chlorinated hydrocarbons, can be detected at extremely low concentrations. It is to be expected that technology will be available in time whereby a good many individual constituents can be determined.

Some water pollution studies have utilized a carbon-filtration technique for extracting traces of organic compounds from large volumes of water. The procedure involves filtration of a large volume of water, up to thousands of gallons, through activated carbon, followed by extraction of the filter with chloroform (Middleton and others, 1962). The possibility exists that organic compounds might be altered by reactions on the filter during and after extraction and recovery, and the efficiency of the method is probably not the same for all compounds. A standard carbon-chloroform extract procedure is included in the 12th edition of "Standard Methods for the Examination of Water and Waste" (American Public Health Association, 1962).

As noted earlier, Lamar and Goerlitz (1966) identified various organic acids with chromatographic techniques and infrared spectroscopy. Except in highly polluted or strongly colored waters, the organic-carbon content of ordinary river or lake water is probably not over the amount common in sea water, or some 10 mg/l (Lee and Hoadley, 1967). This includes living organisms, which, however, makeup only about 2 percent of the total. Until more study of the composition of this material has been made, little more can be said about concentration of individual components.

DETERGENTS

Surface-active agents present in commonly used household detergents include several types of synthetic organic compounds. Prior to 1965, the most widely used active agent in household detergents was alkyl benzene sulfonate, sometimes called ABS. This material is resistant to digestion by organisms in sewage treatment and tends to persist for long periods in water that has received sewage effluents. Concentrations of ABS have been determined in many waters. One mg/l of ABS is sufficient to cause a noticeable froth, and the foam produced with higher concentrations can be a serious problem. The household detergents sold in the United States since 1965 contain other surface-active agents which are more readily biodegradable than ABS in sewage-digestion processes.

COLOR

The determination of color that is sometimes included in water analyses represents an evaluation of a physical property and has no direct chemical significance. The color of natural stream water usually results from leaching of organic debris. The natural color is a yellowish brown which can generally be matched fairly well by dilutions of a mixture of cobalt chloride and potassium chloroplatinate solutions. An arbitrary standard solution containing 1 gram of cobalt chloride, 1.245 grams of potassium chloroplatinate, and 100 milliliters of concentrated hydrochloric acid in a total volume of 1 liter has a color rating of 500. Usually, permanently colored glass discs are used as standards with which tubes of a specific depth, containing samples of water, are compared in a color-matching device. The intensity of color is rated numerically, a color of 5 being equivalent to $\frac{1}{100}$ that of the standard. The color number has no direct connection with the actual amount of organic material causing the color and is purely empirical.

Nature of Colored Material in Water

Some water appears to be colored because of relatively coarse particulate material in suspension, which is readily removed by filtration. Some color may be due to inorganic or organic waste. The brown color of clear unpolluted water, however, is not completely removable by filtration, even through filters having pores 0.10 micron in diameter. The chemical composition of this material is not well known, but commonly is supposed to be similar to substances that can be extracted from soil. The organic fraction of soils can be partly brought into solution by treating the soil with a sodium hydroxide solution. The colored extract is commonly further treated by acidification, which causes part of the organic material to precipitate. This acid-insoluble fraction is termed "humic acid." The acid-soluble organic matter is termed "fulvic acid" (Black and others, 1965, p. 1414-1416). These terms do not denote definite chemical compounds, but are operational definitions which might imply that the materials obtained from different soils by one kind of treatment have some chemical similarity. Other names are applied by soil chemists to certain other fractions of the organic extract which can be separated by using organic solvents.

Black and Christman (1963a, p. 897) reported the colored materials in water they studied were of the fulvic-acid type, and they described them as polyhydroxy aromatic methoxy carboxylic acids. They found most of the colored material was particulate (Black and Christman, 1963b, p. 751) and that diameters were between 4.8 and 10 millimicrons.

Lamar and Goerlitz (1966) identified 13 organic acids in colored water by gas chromatography. Most of the colored material, however, was not sufficiently volatile to be identifiable by this technique.

The colored material in natural water is evidently a mixture of high molecular weight organic substances. The material extracted from lake water by Shapiro (1957) was stated by him to have a molecular weight near 450. This material might be capable of forming complexes with metal ions. Its behavior toward iron was studied extensively by Shapiro (1964). Szalay (1964, p. 1605) reported that humic acid concentrated uranium and might influence other metals. Pommer and Breger (1960) reported an equivalent weight of 144 for humic acid, but noted this value increased with time. Stability constants for six metal-ion complexes with fulvic acid were determined by Schnitzer and Skinner (1967). These investigators also determined a molecular weight of 670 for fulvic acid, which was needed to express concentrations of the material in solution in molar terms. It should be noted, however, that the stability constants reported by Schnitzer and Skinner are applicable only to solutions having certain specified pH values. Also, the composition and behavior of fulvic or humic acid extracted from different soils may be different Consequently, the reporting of stability constants for combinations of metal ions with these rather ill-defined organic solutes does not provide for a complete understanding of the behavior of organic solutes in natural water toward metal cations. More information on the chemical composition and structure of the organic material as well as the complexes must eventually be obtained to accomplish an understanding comparable to that which has been attained for many inorganic complexes.

The color of some streams in the southeastern United States which drain swamps exceeds 200 units at times (U.S. Geological Survey, water-supply paper 1947). Color below 10 units is barely noticeable to the casual observer.

CHEMICALLY RELATED PROPERTIES

The kinds of data discussed here are included in many water analyses, but are not specific chemical components. They may be properties resulting from the combined effects of several constituents or they may be general evaluations of water quality that have been developed as empirical indexes for certain purposes.

RESIDUE ON EVAPORATION

The total concentration of dissolved material in water is ordinarily determined from the weight of the dry residue remaining after evaporation of the volatile portion of an aliquot of the water sample. Total solids, total dissolved solids, or dissolved solids are terms used more or less synonymously for this value. The dissolved solids also may be calculated if the concentrations of major ions are known, and it is not always possible to tell from an analysis whether the value given was determined or calculated.

SIGNIFICANCE OF DETERMINATION

Although evaporation of the water from an aliquot and weighing the residue seems about as simple and direct a way of measuring the solute content as could be devised, the results of such a determination can be difficult to interpret. Water will be strongly retained by some types of residue. Some solutes are volatile or partly volatile at the drying temperature. Furthermore, comparison among analyses from different sources may be troublesome because different drying temperatures have been used. The American Society for Testing and Materials' (1964) procedure specifies final drying shall be for 1 hour at either 103°C or 180°C (the temperature must be reported in the analysis), and the American Public Health Assoc. (1965) specifies 1 hour at 103°C–105°C. The U.S. Geological Survey's procedure, however, calls for drying at 180°C for 1 hour (Rainwater and Thatcher, 1960). The higher drying temperature is mainly intended to remove a higher proportion of the water of crystallization which some residues contain. The different drying temperatures, however, do not produce significantly different results for most natural waters compared with the other factors that may influence this determination.

At 100°C bicarbonate ions are unstable. Half are converted to water and carbon dioxide, and the other half to carbonate ions:

$$2HCO_3^- = CO_3^{-2} + H_2O(g) + CO_2(g).$$

The bicarbonate-ion content of a solution, therefore, is partly volatile, and that part does not appear in the dissolved-solids determination.

Organic matter if present may be partly volatile, but is not usually completely removed unless the residue is strongly ignited. Inorganic constituents such as nitrate and boron are partly volatile, and water that has a low pH will generally lose a considerable amount of its anion content when evaporated to dryness. On the other hand, waters high in sulfate, especially those from which crystals of gypsum are deposited, will give residues containing water of crystallization that may be very difficult to remove, even at 180°C.

Dissolved-solids values often are used in computing rates at which rivers transport weathering products to the ocean and to compute the rate at which rock weathering is lowering the land surface. It is an interesting fact, however, that a considerable part of the dissolved-solids load of a stream draining an igneous terrain represents bicarbonate ions which were derived from the atmosphere rather than from the rocks.

Dissolved-solids values, in spite of the handicaps noted above, are widely used in evaluating water quality and are a convenient means of comparing waters with one another. The residue on evaporation can be used as an approximate check on the general accuracy of an analysis when compared with the computed dissolved-solids value.

In regions of high rainfall and relatively insoluble rocks, dissolved-solids concentrations of runoff may be as low as 25 mg/l. A saturated sodium chloride brine, on the other hand, may contain more than 300,000 mg/l. The U.S. Geological Survey has assigned terms for waters of high dissolved solids as follows (Robinove and others, 1958):

	Dissolved solids (mg/l)
Slightly saline	1,000–3,000.
Moderately saline	3,000–10,000.
Very saline	10,000–35,000.
Briny	More than 35,000.

TOTAL DISSOLVED SOLIDS—COMPUTED

The measurement of the total dissolved mineral matter in water by evaporating an aliquot to dryness and weighing the residue has been discussed. The specific conductance of the water provides another general indication of the content of dissolved matter for water that is not too saline or too dilute. An approximate measure for water very high in dissolved solids can be obtained from its specific gravity. A fourth procedure for measuring dissolved solids is to sum up the concentrations reported for the various dissolved constituents. This computed value may give a more useful indication of dissolved-solids content than the residue on evaporation for certain types of water. A rather complete analysis is required, however, to obtain an accurate value.

CHEMICAL FACTORS IN DISSOLVED-SOLIDS DETERMINATION

In developing standard procedures for computing the total-dissolved constituents, the assumption has generally been made that the result will be used either as a substitute for the determined residue on evaporation or as a means of checking the analysis by comparing the computed value with the corresponding determined value. In the determination of dissolved solids, the bicarbonate ions present in solution are converted to carbonate in the solid phase. Therefore, the bicarbonate is generally converted by a gravimetric factor (mg/l $HCO_3 \times 0.4917 = $ mg/l CO_3) which assumes half the bicarbonate is volatilized as CO_2, and this carbonate value is used in the summation. The value obtained is thus supposed to correspond to the conditions which would exist in dry residue. The assumption that titrated alkalinity represents only OH^-, CO_3^{-2}, and HCO_3^- ions is inherent in this computation. As pointed out elsewhere, this assumption is not always correct, but it is usually a good approximation.

Even though dehydration of gypsum is supposed to be complete at 180°C, it is not uncommon for water high in calcium and sulfate concentrations to yield a residue after drying for an hour at 180°C that exceeds the computed dissolved solids by several hundred milligrams per liter. On the other hand, some waters give residues which are partly decomposed or volatilized at the drying temperature. Such effects can be observed in some waters high in magnesium, chloride, and nitrate. Other waters may yield residues that are hygroscopic and difficult to weigh, and if the water is acid (pH<4.5), some of the solutes may form liquids such as H_2SO_4; thus a meaningful dissolved-solids determination is perhaps impossible to obtain. In many instances the calculated dissolved solids value may be preferable to the determined residue on evaporation value especially if the concentration is greater than 1,000 mg/l.

The computed dissolved solids may differ from the residue on evaporation by 10–20 mg/l in either a positive or negative direction where the total solids content is on the order of 100–500 mg/l. It is evident that only major analytical or computation errors can be detected by comparing these values.

In the sixth edition of Data of Geochemistry (White and others, 1963; Livingstone, 1963) water-analysis tables include a value for "total, as reported" applicable to dissolved components in which no adjustment of bicarbonate to carbonate ions was made. These are not comparable with usual dissolved-solids values, but were believed to represent a more useful total ion content for most geochemical purposes.

DISSOLVED OXYGEN

The solubility of oxygen in water is mainly a function of temperature and pressure. At the normal atmospheric partial pressure of oxygen the solubility at 10°C is 11.33 mg/l (American Society for Testing and Materials, 1964, p. 820), and at 30°C, 7.63 mg/l. The dissolved-oxygen content may be reported in terms of actual concentration or as a percentage of saturation.

The chemical state of the dissolved oxygen seems not to be fully understood. Although the dissolved oxygen is capable of entering into oxidation reactions, the potential that can be measured in an oxygenated solution by a standard reference electrode and a noble-metal electrode is considerably below the theoretical value for the direct couple

$$O_2(aq)+4e^-=2O^{-2},$$

and the reactions are typically slow. A relatively complicated reaction route involving peroxide species is usually assumed (Latimer, 1952, p. 39-44).

The ultimate source of oxygen in water exposed to air is the atmosphere, but some oxygen is contributed by an indirect route as a byproduct of photosynthesis. Water bodies in which there is much organic productivity often display wide fluctuations of dissolved oxygen in response to the biological activity. The relative importance of organic sources and atmospheric ones perhaps can be evaluated by comparing productivity with the hydraulic parameters which essentially control absorption of oxygen from the air at a water surface (Langbein and Durum, 1967), but few such comparisons have yet been made. Oxygen may be retained in solution for a long time in ground water circulating far out of contact with air if there are no oxidizable materials of consequence in contact with the water. If organic material and other oxygen-consuming substances are present, however, the dissolved oxygen may be depleted to very low concentrations after the water has moved only a short distance in the saturated zone.

Owing to the rapidly changing input and consumption rates, the oxygen content of a surface-water body or stream is a highly transient property. A measurement is meaningful only for the spot of sampling and a brief time period. The oxygen content of a sample of water can readily change after collection and thus must be chemically preserved and determined quickly. The development of electrodes for sensing dissolved oxygen has greatly simplified the sampling and determination problem. This kind of instrumentation was very much needed, and its development represents a major advance in the technology of water-quality studies. With it, continuous records of dissolved-oxygen content of water at designated sampling points in water bodies can be obtained.

The dissolved-oxygen content of a water is an indicator of the biochemical condition of the water at that time and place. Fish and other desirable clean-water biota require relatively high dissolved-oxygen levels at all times. Streams with large loads of organic material, however, may have oxygen-consuming organic and inorganic reactions that deplete oxygen to levels unfavorable for the clean-water species. The dissolved-oxygen content is an indication of the status of the water with respect to balance between oxygen-consuming and oxygen-producing processes at the moment of sampling.

The subject of organic pollution and its relationship to dissolved oxygen is not considered extensively in this paper. A large literature on the subject exists. An introduction to this literature can be obtained from a compilation by Ingram, Mackenthun, and Bartsch (1966).

Dissolved-oxygen determinations usually have not been made in geochemical studies of ground water, but they may have some value for such purposes and are being increasingly utilized. Heidel (1965) showed that dissolved-oxygen content was related to the behavior of iron in ground water. The concentration of oxygen in some respects is a better indication of oxidation potential than the measurement of electrical potential, which is an uncertain process in aerated solutions. The range of measurable dissolved-oxygen content, however, covers only the oxidizing part of the Eh range, and highly reducing potentials imply oxygen concentrations much below feasible measurement by chemical methods.

OXYGEN DEMAND AND OTHER EVALUATIONS OF ORGANIC POLLUTION LOAD

The dissolved-oxygen concentration of a water body represents the status of the system at the point and moment of sampling. The processes by which organic debris, or other kinds of material in the water, react with oxygen are relatively slow. The processes generally are biological, which means a suitable incubation and growth period is required for the organisms involved. As the organisms multiply, the rate at which they use oxygen may exceed the rate at which the dissolved oxygen in the water can be replenished from the atmosphere, and the dissolved-oxygen content may decrease—perhaps drastically. After the processes have attained essential completion, the normal oxygen level of the water can be reestablished, and if the oxidizable load is light, the oxygen level may not be depleted much at any time. If considerable oxidizable material is present, however, the oxygen may be considerably depleted while the digestive processes are going on. A system may be so overloaded by pollutants that oxygen levels fall practically to zero, and aerobic organisms are destroyed. Because considerable time may be required for the natural purification processes to become effective again, the polluted water may move through many miles of river channel with very low oxygen contents.

Various methods have been used to estimate the requirement of a given water for oxygen or to evaluate the pollution load in a quantitative way. These include measurement of biochemical oxygen demand (BOD), chemical oxygen demand (COD), and total organic carbon.

BIOCHEMICAL OYYGEN DEMAND

The BOD determination is commonly made by diluting portions of a sample with oxygenated water and measuring the residual dissolved oxygen after a period of incubation (usually 5 days at 20°C). The results are commonly expressed in terms of weight of oxygen required per unit volume of the initial sample. Sometimes the pollution load of a given waste stream is expressed in terms of the population level whose normal domestic sewage production would equal the BOD of the stream. The determination is slow to make and has no particularly direct geochemical significance, but it is extensively applied in pollution studies. It is generally considered to be the most useful way of expressing stream-pollution loads and for comparing one set of conditions with another.

CHEMICAL OYYGEN DEMAND

To determine pollution or oxidizable material loads more quickly, wet oxidations with strong oxidizing agents have been extensively used. The results can be expressed in terms of oxygen equivalent. Heating of the sample with permanganate or dichromate is one such procedure. The results of COD determinations obviously may not correspond to values obtained by BOD determinations, but may give some aid in estimating organic pollution loads.

TOTAL ORGANIC CARBON

In recent years, instruments have been developed that are capable of determining the total carbon content of water, and although this development is still in fairly early stages, it seems likely a more comprehensive indication of organic pollution loads can be obtained in this way than in any other. Several papers relating to measurement of pollution loads by converting all carbon to carbon dioxide and measuring the CO_2 by infrared spectrometry were included in the review by Fishman, Robinson, and Midgett (1967, p. 278R).

HARDNESS

The concept of hardness as an evaluation of certain chemical properties of water has become deeply imbedded in the literature of water analysis and in the habits of thought of almost everyone who is concerned with water quality. In spite of wide usage, however, the property of hardness is difficult to define exactly, and several definitions are in common use.

The terms "hard" and "soft" are contained in a discourse on water quality by Hippocrates (460–354 B.C.) quoted as follows by Baker (1949): "Consider the waters which the inhabitants use, whether they be marshy and soft, or hard and running from elevated and rocky situations * * *." The use of the terms here could have some of the modern meaning; at least limestone was probably present in many of the upland regions familiar to Hippocrates. Over the years, the property of hardness has been associated with effects observed in the use of soap or with the encrustations left by some types of water when they are heated. If the reactions with soap are the only ones considered, one might say that hardness represents the soap-consuming capacity of a water. The effect results from cations that form insoluble compounds with soap. More than 100 years ago, a procedure was developed for titration of hardness with standard soap solution.

Because most of the effect observed with soap results from the presence of calcium and magnesium, hardness is now generally defined in terms of these constituents alone with some rather indefinite reservations about interferences (American Society for Testing and Materials, 1964, p. 391). The other ions which might precipitate soap include H^+ and all polyvalent metals, but usually they are present in insignificant amounts in waters of the type that are usable domestically and for which hardness data might be obtained.

Because hardness is a property not attributable to a single constituent, some convention has to be used for expressing concentrations in quantitative terms. Usually, this consists of reporting hardness in terms of an equivalent concentration of calcium carbonate. In practical water analysis, the hardness is computed by multiplying the sum of milliequivalents per liter of calcium and magnesium by 50. The hardness value resulting is generally entitled "hardness as $CaCO_3$" in tabulated data. The same quantity is sometimes referred to as "calcium+magnesium hardness" or "total hardness." The usual procedures for determining calcium and magnesium will include an amount of these metals approximtely equivalent to any other alkaline-earth metals; hence a reasonable practical definition of hardness is "the effect of alkaline-earth cations."

Carbonate hardness, where reported, includes that portion of the hardness equivalent to the bicarbonate+carbonate (or alkalinity). If the hardness exceeds the alkalinity (in milligrams per liter of $CaCO_3$ or other equivalent units), the excess is termed "noncarbonate hardness" and frequently is reported in water analyses. In some older reports, the terms "temporary" and "permanent" are used instead of "carbonate" and "noncarbonate" in speaking of these subdivisions of hardness.

Hardness values are reported in some European countries in terms of "degrees." One French degree is equivalent to 10 mg/l, one German degree to 17.8 mg/l, and one English or Clark degree to 14.3 mg/l, all in terms of calcium carbonate.

The soap procedure for titration of hardness has been supplanted by chelation methods for titration of the alkaline-earth metals, and analyses made since the late 1940's can generally be assumed not to have used the soap procedure. The chelation method gives results that are far more reliable than the old procedure.

Hardness determinations have a very limited value in geochemical studies. Modern analytical procedures can provide separate calcium and magnesium values practically as easily as combined ones, and the increase in usefulness of the results is well worth the trouble.

RANGE OF HARDNESS CONCENTRATION

The adjectives "hard" and "soft" as applied to water are inexact, and some writers have tried to improve on this situation by adding qualifying adverbs. Durfor and Becker (1964, p. 27) use the following classification:

Hardness range (mg/l of $CaCO_3$)	Description
0–60	Soft.
61–120	Moderately hard.
121–180	Hard.
More than 180	Very hard.

In some areas of the United States, however, where most water has a low dissolved-solids content, a water containing 50 mg/l of hardness would be considered hard by most residents.

The standards by which water hardness is judged have tended to become more rigorous over the years. Many public water supplies now are softened to less than 100 mg/l of hardness. The U.S. Public Health Service standards (1962) do not specify any value for hardness. According to the American Water Works Association (Bean, 1962), however, "ideal" quality water should not contain more than 80 mg/l of hardness.

Hardness in water used for ordinary domestic purposes does not become particularly objectionable until it reaches a level of 100 mg/l or so. Hardness can greatly exceed this level, and in many places— especially where waters have been in contact with limestone or gypsum—200 or 300 mg/l or more of hardness will be common. At these levels, hardness becomes noticeable in all uses and becomes increasingly troublesome as the concentration rises. In water from gypsiferous formations 1,000 mg/l of hardness is not uncommon.

In recent years, some authors have reported apparent correlations between the hardness or other properties of drinking-water supplies and the death rates from cardiovascular diseases. Muss (1962) reviewed literature on this subject and expressed the belief that in a very general way the lower death rates from heart and circulatory diseases occurred in States where the public water supplies are highest in hardness. Kobayashi (1957) reported that the geographical distribution of high death rates from apoplexy in Japan seemed to suggest the high rates occurred in areas where the river waters had a low alkalinity and relatively high sulfate content. All these Japanese waters, however, were relatively soft. Because many other factors affect these apparent correlations, their significance is uncertain.

REDOX POTENTIAL

The terms "redox potential (Eh)," "oxidation potential," and "oxidation-reduction potential" are interchangeably used in the literature of water chemistry to represent the relative intensity of oxidizing or reducing conditions in solutions. The sign convention used for these potentials in natural-water chemistry results in relatively oxidizing systems being assigned positive potentials and reducing systems relatively negative ones. The relation of Eh values to standard potentials by means of the Nernst equation has been discussed earlier in this paper under the heading "Electrochemical equilibrium." For solutions containing oxidized and reduced ionic species at equilibrium, the Eh will be fixed by the activity of pairs of those species which are dominant. Usually, the dominant species are the ones present in highest concentration. In many natural waters, however, the oxidation-reduction reactions are not at a state of equilibrium, and the Eh of the system may be controlled by the most rapid reaction of the several possible ones The stabilization of Eh brought about by equilibration of large activities or reserves of participating species is parallel to the buffering effect observed in measuring pH. In connection with Eh, the effect is referred to as "poising." Thus, a solution which resists change in Eh is said to be highly poised.

Natural water in contact with air should be poised, supposedly by the large supply of available oxygen in the atmosphere, and potentials between 0.35 volts (350 millivolts) and 0 55 volts are usually reported for oxygenated water in the range of pH from 6 to 8. This is considerably below the theoretical potential for direct conversion of dissolved-oxygen gas at atmospheric partial pressure to O^{-2} or OH^- ions, and as noted earlier in this paper, there remains some disagreement as to whether the deviation of measured potentials from theoretical ones is related to measurement difficulties or to complications in the chemical behavior of dissolved oxygen.

The observed potential is that developed between a bright, inert metal electrode, usually platinum or gold, and a standard reference electrode. Most pH meters can be used for measuring these potentials. Results are expressed in volts or millivolts. ZoBell (1946) was one of the first investigators to make measurements in materials of geologic interest. A very large number of published measurements on various kinds of water and sediment were compiled by Baas-Becking, Kaplan, and Moore (1960). Measurements have been made in connection with studies of ground-water composition, especially with respect to iron chemistry (Barnes and Back, 1964b; Back and Barnes, 1965), in connection with water-treatment plant operations (Weart and Margrave, 1957), as an indication of the condition of sewage digestors, and in numerous studies of soils, biochemical systems, and lake and ocean sediments. Redox data also were used by Clarke (1966) and by Barnes and Clarke (1969) in describing and studying causes of corrosion and encrustation of well casings. Measurements in ground water in the U.S.S.R. were described by Germanov and others (1959).

The practical usefulness of these measurements seems well established for a number of different applications, where the potential is ascribed to oxidized and reduced species of iron that are present in moderate concentrations. Questions have been raised, however, as to the relationship of the measured potentials to the theoretical Eh computed from the Nernst equation. As Stumm (1961) and Morris and Stumm (1967) pointed out, it is necessary to be cautious in interpreting measured potentials under the best of circumstances. Many measurements reported in the literature probably give potentials that are not closely related to the theoretical redox potential.

Solutions in contact with air can be expected to show effects from oxygen. Measurements on reducing systems such as ground water containing ferrous and ferric iron can be made to give a meaningful answer only if oxygen is very carefully excluded from all parts of the sampling and measuring system Back and Barnes (1965) described equipment and techniques they used to avoid the effect of oxygen.

Whatever may be the experimental difficulties in measurement of Eh values, the usefulness of the Nernst equation and the Eh-pH diagram in theoretical studies of redox equilibria in natural water is well established The diagram is an indication of limiting conditions and is a means of evaluating the chemical behavior of multivalent elements that is clear, simple, and convenient.

RANGE OF REDOX POTENTIAL

Although the data collected by Baas-Becking, Kaplan, and Moore (1960) probably contain many measurements of doubtful significance, the observed values are well clustered within the stability limits for water. At a pH of 7.0 this is a maximum range of $+0.82$ to -0.42 volts. Values reported for ground water in Maryland (Back and Barnes, 1965) ranged from 0.471 to -0.020 volts, and a value of -0.103 volts was measured by Barnes, Stuart, and Fisher (1964) in a flooded coal mine in Pennsylvania. The latter water had a pH of 3.92.

OTHER UNITS OF EXPRESSION

Instead of expressing redox potentials in volts, some writers have preferred to treat the symbol "e^-" in a redox couple such as

$$Fe^{+3} + e^- = Fe^{+2}$$

as an ion so that it can be assigned an activity. At equilibrium, the negative logarithm of the electron activity, or pE, will represent the intensity of oxidizing or reducing conditions in the same way as Eh but not in conventional units. The value of pE at 25°C and 1 atmosphere is Eh/0.0592 volts.

As demonstrated by Sillén and Martell (1964, p. XV–XVII), the arithmetic of some types of equilibrium calculations can be simplified by using pE in place of Eh values in volts.

SODIUM-ADSORPTION-RATIO (SAR)

The U.S. Department of Agriculture Salinity Laboratory (1954) defined sodium-adsorption-ratio (SAR) of a water as

$$SAR = \frac{(Na^+)}{\sqrt{\frac{(Ca^{+2}) + (Mg^{+2})}{2}}},$$

where ion concentrations are expressed in milliequivalents per liter. The experiments cited by the Salinity Laboratory show that the SAR predicts reasonably well the degree to which irrigation water tends to enter into cation-exchange reactions in soil. High values for SAR imply a hazard of sodium replacing adsorbed calcium and magnesium, and this replacement is damaging to soil structure.

Values for SAR are included in chemical analyses for irrigation water or water which might be considered for that use. The value is empirical and of otherwise limited geochemical significance.

In older reports, the tendency for a water to enter into cation-exchange reactions was commonly evaluated in terms of the sodium percentage. The sodium percentage is the percentage of total cations made up by sodium (concentrations expressed in milliequivalents per liter). Because divalent cations are usually preferentially held in exchange positions on clay minerals, the displacement of Ca^{+2} and Mg^{+2} by Na^+ is unlikely unless the sodium percentage is considerably higher than 50 or the total concentration of solutes is very large.

The correlation of SAR with observable ion-exchange effects is superior to that obtained by using sodium percentage. This is probably related to the fact that the equation used for calculating SAR has the form of a mass-law equilibrium relationship, whereas that for sodium percentage does not.

DENSITY

In water analyses made by the U.S. Geological Survey, the density of water is considered to be significantly different from 1 gram per milliliter only when the dissolved-solids concentration exceeds about 7,000 mg/l or when the specific conductance exceeds 10,000 micromhos per centimeter. "Significantly" here means that the difference between concentration expressed in milligrams per liter and in parts per million is greater than ordinary analytical error in the determination of major constituents. Obviously, the density of a water is a function of temperature as well as the nature and quantity of dissolved impurities. The density is usually determined to three decimal places, and the temperature at which it was made (usually 20°C) is given. The density of a water is a significant physical property that affects its behavior in natural systems and may influence its chemical composition in an indirect way. The density value is required to convert concentrations in weight per volume to weight per weight units.

STABLE NUCLIDES

The hydrologic significance and utilization of radioactive nuclides has already been noted. Many of the elements occur in the form of mixtures of two or more isotopes, mostly stable rather than radioactive. The stable isotopes of an element behave essentially the same chemically, and usually the isotopic composition of an element is nearly constant no matter where it may be found. Some types of processes, however, tend to occur more rapidly or easily with lighter or heavier isotopes, and the isotopic composition of an element may, therefore, provide an indication of its past environment, owing to enrichment or impoverishment of light or heavy forms.

The oxygen isotope O^{18} and the hydrogen isotope H^2 (deuterium) constitute about 0.2 percent and 0.016 percent, respectively, of normal oxygen and hydrogen, and O^{17} constitutes about 0.04 percent of normal oxygen. The radioactive isotope H^3 (tritium) is extremely rare by comparison. Even in the higher concentrations observed in rainfall in the United States from 1963 to 1965 (Stewart and Farnsworth, 1968), the abundance of tritium seldom reached as much as one tritium atom for each 10^{14} atoms of hydrogen. In terms of percentage, this would be some ten orders of magnitude below the abundance of deuterium. These isotopes are of particular hydrologic significance because they are present in water and produce a significant proportion of molecules of H_2O that are heavier than normal water. In the process of evaporation, the heavier molecules tend to be enriched in residual water, and thus the lighter species are more abundant in water vapor, rain and snow, and most fresh water of the hydrologic cycle; the heavier forms are more abundant in the ocean.

Some of the first studies of deuterium and oxygen-18 contents of water from various sources were made by Friedman (1953) and by Epstein and Mayeda (1953), and the usefulness of isotopic-abundance data in studies of water circulation has been amply demonstrated by continuing studies. The abundance of the hydrogen isotopic species has been considered a useful key to deciding whether a water from a thermal spring contains a significant fraction of water of magmatic or juvenile origin that has not been in the hydrologic cycle previously (Craig, 1963).

Biological processes tend to produce some fractionation of isotopes. Among the studies of these effects is the paper by Kaplan, Rafter, and Hulston (1960) relating to enrichment of sulfur-32 over sulfur-34 in bacterially reduced forms of the element and the papers on fractionation of carbon-12 and carbon-13, as in fermentation (Nakai, 1960) or in processes related to calcite deposition (Cheney and Jensen, 1965). The dating of rocks by using isotopic composition is extensively practiced, and the isotopic composition of lead may give valuable information as to its origin. The natural radioactive decay series each produce a single stable and distinctive lead isotope.

Isotopic separations and analyses are generally made by mass spectrometry, and a relatively large amount of research is being conducted on the utilization of isotopic variations in geochemistry. The importance and potential future value of this work is obvious even from the very brief mention that is all that can be given it here.

ORGANIZATION AND STUDY OF WATER-ANALYSIS DATA

Hydrologists and others who use water analyses must make interpretations of individual analyses or of large numbers of analyses at a time. From these interpretations final decisions regarding water utilization and development are made. Although the details of water chemistry often must play an important part in water-analysis interpretation, a fundamental need is for means of correlating analyses with each other and with hydrologic or other kinds of information that are relatively simple, as well as scientifically reasonable and correct. It may be necessary, for example, to make an organized evaluation in a summary report of the water resources of a region to correlate water quality with environmental influences, to develop plans for management of water quality, control of pollution, setting of water-quality standards, or selecting and treating public or industrial water supplies.

The interpretive techniques and correlation procedures described here do not require extensive application of chemical principles. The procedures range from simple comparisons and inspection of analytical data to more extensive statistical analyses and the preparation of of graphs and maps that show significant relationships and allow for extrapolation of available data to an extent sufficient to be most practical and useful.

The use of water-quality data as a tool in hydrologic investigations of surface- and ground-water systems often has been neglected. In appropriate circumstances, chemical data may rank with geologic, engineering, and geophysical data in usefulness in solution of hydrologic problems. Arraying and manipulating the data, as suggested in the following pages, may lead the hydrologist to insight into a problem that appears from other available information to be insoluble.

EVALUATION OF WATER ANALYSIS

The chemical analysis, with its columns of concentration values reported to 2 or 3 significant figures, accompanied by descriptive material related to the source and the sampling and preservation techniques, has an authoritative appearance, which unfortunately can be misleading. Although mention has already been made of some of the effects of sampling techniques, preservation methods, and the length of storage time before analysis on the accuracy of results, it should be noted again that many completed analyses include values for constituents and properties that may be different from the values in the original water body. Most analyses, for example, report a pH determined in the laboratory; almost certainly such pH values deviate from the pH at time of sampling. Roberson, Feth, Seaber, and Anderson (1963) published some data on the extent of such deviations. The user of the analysis also should be concerned, however, with the general reliability of all the analytical values, including those for constituents generally assumed to be stable.

ACCURACY AND REPRODUCIBILITY

Under optimum conditions, the analytical results for major constituents of water have an accuracy of $\pm 2-\pm 10$ percent. That is, the difference between the reported result and the actual concentration in the sample at time of analysis should be between 2 and 10 percent of the actual correct value. Solutes that are present in concentrations above 100 mg/l can generally be determined with an accuracy of better than ± 5 percent. Limits of precision (reproducibility) are similar. For solutes present in concentrations below 1 mg/l the accuracy is generally not better than ± 10 percent and can be poorer than this. Specific statements about accuracy and reproducibility have not been made for most of the analytical determinations discussed here because such statements are subject to many uncertainties and can be misleading.

Analytical errors are at least partly within control of the chemist, and for many years efforts have been made to improve the reliability of analytical methods and instruments and to bring about uniformity in procedures. The majority of laboratories active in the water-analysis field in the United States use procedures described in "Standard Methods for Analysis of Water and Waste" (American Public Health Association, 1962) which is kept up to date by frequent revisions. Other manuals such as those of the American Society for Testing and Materials (1964) and the U.S. Geological Survey (Rainwater and Thatcher, 1960) specify much the same set of procedures. The Analytical Reference Service operated by the U.S. Public Health Service at its Taft Center laboratory in Cincinnati has circulated many standard samples among cooperating testing laboratories that use these methods and has observed many interesting and, at times, somewhat disconcerting results. The results which have been published show that when the same sample is analyzed by different laboratories, a spread of analytical values is obtained that considerably exceeds the degree of precision most analytical chemists hope to attain.

Lishka, Kelso, and Kramer (1963) summarized results of several Analytical Reference Service studies, where the spread in analytical values reported by different laboratories for the same sample are pointed out. For example, of 182 reported results for a standard water sample, 50 percent were within ± 6 mg/l of the correct value for chloride (241 mg/l). The standard deviation reported for this set of determinations was 9.632 mg/l. If these results are evaluated in terms of confidence limits, or probability, the conclusion may be drawn that a single determination for chloride in a sample having around 250 mg/l has an even chance of being within ± 6 mg/l of the correct value, and assuming a near-normal distribution, the probability is 68 percent that the result will be within ±9.6 mg/l of the correct value. The probability of the result being within ±20 mg/l of the correct value is 95 percent. The results for other determinations reported by Lishka, Kelso, and Kramer were in the same general range of accuracy. Before the statistical analysis was made, however, they rejected values that were grossly in error. The rejected determinations amounted to about 8 percent of the total number of reported values. The probable accuracy would have been poorer had the rejected data been used, but it may not be too unrealistic to reject the grossly erroneous results. In practice, the analytical laboratory and the user of the results will often be able to detect major errors in concentration values and reject such results if there is some prior knowledge of the composition of the water or if the analysis is reasonably complete. Methods for detecting major errors will be described presently.

The results of a single analyst or of one laboratory should have somewhat lower deviations than the data cited above. It would appear, however, that the third significant figure reported in water analysis determinations is usually not really meaningful and that the second figure may sometimes have a fairly low confidence limit. In the data cited here, there is only a 5 percent chance that a chloride determination in this range of concentration would be as much as 10 percent in error. The effects of sampling and other nonanalytical error are excluded from this consideration.

Organizations in Federal and State governments which publish analyses intended for general purposes use accuracy standards that are generally adequate for the types of interpretation to be discussed in this section of this report. All such organizations, however, share human tendencies toward occasional error. A first step in analysis acceptance often is the opinion of the user of the data about the originating laboratory's reputation for accuracy of results and perhaps its motivation for obtaining maximum accuracy. Data obtained for some special purposes may not be satisfactory for other uses. For example, some laboratories are concerned with evaluating water for conformity to selected standards and may not determine concentrations closely if they are far above or below some limiting value.

ACCURACY CHECKS

The accuracy of a complete chemical analysis of a water sample can be checked by calculating the cation-anion balance. When all the major anions and cations have been determined, the sum of the cations in milliequivalents per liter should equal the sum of the anions expressed in the same units The difference between the two sums will generally not exceed 1 or 2 percent of the total of cations and anions in waters of moderate concentration (250–1,000 mg/l), if the analytical work has been done carefully. If the total of anions and cations is less than about 5.00 meq/l, larger percentage errors are sometimes unavoidable. If an analysis is found acceptable on the basis of this check, it can be assumed there are no important errors in concentrations reported for major constituents.

Water with dissolved-solids concentrations much greater than 1,000 mg/l tends to have large amounts of a few constituents. In such waters, the test of anion-cation balance does not adequately evaluate the accuracy of the values of the lesser constituents.

The concept of equivalence of cations to anions is chemically sound, but in some waters it may be difficult to ascertain the forms of some of the ions reported in the analysis. To check the ionic balance, it must be assumed that the water does not contain undetermined species participating in the balance and that the formula and charge of all the anions or cations reported in the analysis are known. Solutions that are strongly colored, for example, often have organic anions that form complexes with metals, and the usual analytical procedures will not give results that can be balanced staisfactorily.

Another source of possible difficulty is in the assumptions underlying the alkalinity and acidity titrations. If species contributing to the alkalinity are present and are not fully titrated at the methyl-orange end point (or at a pH of 4.50 if a fixed-pH end point is used), they will not be accounted for in the analysis, unless determined by other procedures. If determined in other ways besides titration, such species may be represented in the analysis twice. Forms of orthophosphate, for example, influence alkalinity and are usually determined by procedures other than titration. Water having a pH below 4.50 presents more serious problems because, as noted in the discussion of acidity earlier in this paper, the titration is affected by several kinds of reactions and may not provide a value that can be used in reaching an ionic balance. For some waters the alkalinity-titration end point pH should be higher than 4.50, and for others, lower, as noted by Barnes (1964).

In the published literature, many analyses report computed values for sodium or sodium plus potassium. These values were obtained by assigning the difference between milliequivalents per liter of total anions determined and the sum of milliequivalents per liter of calcium and magnesium. Obviously, such an analysis cannot be readily checked for accuracy by cation-anion balance. Although calculated sodium concentrations are not always identified specifically, exact or nearly exact agreement between cation and anion totals for a series of analyses is a good indication that sodium concentrations were calculated. The improvement in methods of analysis since the 1940's has made it easy to determine sodium, and analyses with calculated values are no longer common.

Another procedure for checking analytical accuracy that is sometimes useful is to compare determined and calculated values for total dissolved solids. The two values should agree within a few milligrams per liter unless the water is of exceptional composition, as noted in the discussion of these determinations earlier in this report. The comparison is often helpful in identifying major analytical or transcribing errors.

An approximate accuracy check is possible with the conductivity and dissolved-solids determinations. The dissolved-solids value in milligrams per liter should generally be from 0.55 to 0.75 times the specific conductance in micromhos per centimeter for waters of ordinary composition, up to dissolved-solids concentrations as high as a few thousand milligrams per liter. Water in which anions are mostly bicarbonate and chloride will have a factor near the lower end of this range and waters high in sulfate may reach or even exceed the upper end. Waters saturated with respect to gypsum (analysis 3, table 15, for example) or containing large amounts of silica may have factors as high as 1.0. For repeated samples from the same source a well-defined relationship of conductivity to dissolved solids often can be established, and this can afford a good general accuracy check for analyses of these samples. The total of milliequivalents per liter for either anions or cations multiplied by 100 usually agrees approximately with the conductivity in micromhos per centimeter. This relationship is not exact, but is somewhat less variable than the relationship between conductivity and dissolved solids in milligrams per liter. The relationship of dissolved solids to conductance becomes indefinite for waters very high in dissolved solids (over about 30,000 mg/l) and for very dilute solutions, such as rainwater, and thus is not useful as a check of accuracy of analysis for such waters.

One of the most useful procedures when a number of analyses for the same or similar sources is available is to compare the results with one another. Errors of transcribing or analytical error in minor constituents containing factors of 2 or 10 sometimes become evident when this is done. It is common practice, however, to make this type of scrutiny before data are released from the laboratory, and it is most useful to do it before the analytical work is completed so that any suspected values can be redetermined.

Analyses reporting calculated zero values for sodium, or indicating sodium concentration as less than some round number, commonly result from analytical errors causing milliequivalents per liter for calcium and magnesium to equal or exceed the total milliequivalents per liter reported for anions; thus, there is nothing to assign to sodium for calculation purposes. A zero concentration for sodium is very rarely found if the element is determined.

Certain unusual concentration relationships for major cations can be considered as grounds for suspicion of the validity of the analysis. A zero value for calcium, where more than a few milligrams per liter are reported for magnesium, or a potassium concentration substantially exceeding that of sodium, unless both are below about 5 mg/l, are examples.

Groups of analyses from the same or similar sources in which magnesium concentrations are all similar but calcium concentrations have a rather wide range may indicate that calcium and bicarbonate were lost by precipitation of calcium carbonate. This can occur in water-sample bottles during storage and also can occur in the water-circulation system before sampling.

SIGNIFICANT FIGURES

Water analyses in which concentrations in milligrams per liter are reported to four or five significant figures are commonly seen. The notion of high accuracy and precision conveyed by such figures is misleading, as ordinary chemical analytical procedures rarely give better than two-place accuracy. Usually, the third significant figure is in doubt, and more than three is entirely superfluous. Analytical data in terms of milliequivalents per liter in U.S. Geological Survey reports are commonly reported to two decimal places, without rounding. These values can be no more reliable than the milligrams per liter values from which they were calculated.

A concentration of 0 mg/l reported in a chemical analysis should be interpreted as meaning the amount present was less than 0.5 mg/l and the procedure used could not detect concentrations less than 0.5 mg/l. Concentrations of 0.0 or 0.00 mg/l imply lower detection limits. Some analysts report such findings in terms such as "<0.05 mg/l" where the figure given is the detection limit. These values should not be interpreted as indicating that any specific amount of the element sought was present.

GENERAL EVALUATIONS OF AREAL WATER QUALITY

The type of water-analysis interpretation most commonly required of hydrologists is the preparation of a report summarizing the water quality in a river, a drainage basin, or some other areal unit which is under study. The writer of such a report is confronted with many difficulties. The chemical analyses with which the writer must work usually represent only a few of the water sources in the area and must be extrapolated. The finished report must convey water-quality information in ways in which it will be understandable both to technically trained readers and to those whose interests are more general. Conclusions or recommendations are usually required, relating to existing conditions or to conditions expected in the future.

As an aid to interpreting groups of chemical analyses, several approaches will be cited that can serve to relate analyses to each other and to provide means of extrapolating data areally and in time. Different types of visual aids which often are useful in reports will be described. The basic methods considered are inspection and simple mathematical or statistical treatment to bring out resemblances among chemical analyses, procedures for extrapolation of data in space and time, and preparation of graphs, maps, and diagrams to show the relationships developed.

INSPECTION AND COMPARISON

A simple inspection of a group of chemical analyses generally will make possible a separation into obviously interrelated subgroups. For example, it is easy to group together the waters that have dissolved-solids concentrations falling within certain ranges. The consideration of dissolved solids, however, should be accompanied by consideration of the kinds of ions present as well.

A common practice in literature on water quality is to refer to or classify waters by such terms as "calcium bicarbonate water" or "sodium chloride water." These classifications are derived from inspection of the analysis and represent the predominant cation and anion expressed in milliequivalents per liter. These classifications are meant only to convey general information and cannot be expected to be exact or precise. They may, however, be somewhat misleading if carelessly applied. For example, a water ought not be classed as a sodium chloride water if the sodium and chloride concentrations constitute less than half the total of anions and cations, even though no other ions exceed them. Water in which no one cation or anion constitutes as much as 50 percent of the totals should be recognized as a mixed type and identified by the names of all the important cations and anions.

ION RATIOS AND WATER TYPES

Classifications of the type just described are only rough approximations, and for most purposes in the study of chemical analyses a more exact and quantitative procedure is required. Expression of the relationships among ions, or of one constituent to the total concentration in terms of mathematical ratios, is often helpful in making resemblances and differences among waters stand out clearly. An example of the use of ratios is given in table 21, where three hypothetical analyses are compared. All three of these chemical analyses

TABLE 21.—*Hypothetical chemical analyses compared by means of ratios*

[Date below sample letter is date of collection]

Constituent	(A) Jan. 11, 1950		(B) Feb. 20, 1950		(C) Mar. 5, 1950	
	mg/l	meq/l	mg/l	meq/l	mg/l	meq/l
Silica (SiO_2)	12		33		30	
Iron (Fe)						
Calcium (Ca)	26	1.30	12	0.60	11	0.55
Magnesium (Mg)	8.8	.72	10	.82	9.2	.76
Sodium (Na)	} 73	3.16	89	3.85	80	3.50
Potassium (K)						
Bicarbonate (HCO_3)	156	2.56	275	4.51	250	4.10
Sulfate (SO_4)	92	1.92	16	.33	15	.31
Chloride (Cl)	24	.68	12	.34	12	.34
Fluoride (F)	.2	.01	1.5	.08	1.2	.06
Nitrate (NO_3)	.4	.01	.5	.01	.2	.00
Dissolved solids:						
Calculated (mg/l)	313		309		282	
Calculated (tons per acre-ft)	.42		.42		.38	
Hardness as $CaCO_3$:						
Ca and Mg	101		71		66	
Noncarbonate	0		0		0	
Specific conductance (micromhos at 25° C)	475		468		427	
pH	7.7		8.0		8.1	

Note. Comparison of analyses of the samples.

	$\dfrac{SiO_2\ mg/l}{Sum\ mg/l}$	$\dfrac{Ca\ mg/l}{Mg\ mg/l}$	$\dfrac{Na\ mg/l}{Cl\ mg/l}$	$\dfrac{SO_4\ mg/l}{Total\ anions\ mg/l}$
A	0.038	1.8	4.6	0.37
B	.11	.73	11.3	.063
C	.11	.72	10.3	.064

could be considered to represent sodium bicarbonate waters, and they do not differ greatly in total concentration. The high proportion of silica in waters B and C, their similar Ca:Mg and Na:Cl ratios and their similar proportions of SO_4 to total anions establish the close similarity of B to C and the dissimilarity of both to A. For most comparisons of this type, concentration values expressed in terms of milliequivalents per liter or moles per liter are the most useful.

The data in table 21 are synthetic, and actual analyses often do not show such well-defined relationships. Ratios are obviously useful, however, to establish chemical similarities among waters, for example, in grouping analyses representing a single geologic terrane, or a single aquifer, or a water-bearing zone. Fixed rules regarding selection of the most significant values to compare by ratios cannot be given, but some thought as to the sources of ions and the chemical behavior which might be expected can aid in this selection. The ratio of silica to dissolved solids may aid in identifying water influenced by solution of silicate minerals and the type of mineral itself may be indicated in some instances by ratios among major cations (Garrels and MacKenzie, 1967). The ratio of calcium to magnesium may be useful in studying water from limestone and dolomite (Meisler and Becker, 1967) and may help in tracing sea-water contamination. The ratio of sodium to total cations is useful in areas of natural cation exchange. The ratio of chloride to other ions also may be useful in studies of water contaminated with common salt (sodium chloride).

The study of analyses using ratios has many undeveloped possibilities. Schoeller (1955) made some suggestions for the use of ratios in connection with water associated with petroleum, and White (1960) published a set of median ratios of ion contents in parts per million which he believed to be characteristic of water of different origins. Ratios are included for all the analyses of ground waters given by White, Hem, and Waring (1963).

Extensions of the concept of ion ratios to provide comprehensive classification schemes have appeared in the literature from time to time. Most of these schemes are not simple two-ion ratios, but are attempts to express proportions of all the major ions within the total concentration of solutes.

The principal classification schemes that have been proposed in the literature have been well reviewed by Konzewitsch (1967), who also described the principal graphical methods that have been used. Some classification methods and graphs also are described by Schoeller (1962). Some of the classification schemes are very elaborate, recognizing more than 40 types of water.

Although classification procedures provide a basis for grouping waters that are closely related to each other and can be a starting point for deciding what brought about the resemblance, the process of classifying waters into chemical types is of little inherent value, and in the United States literature relatively minor attention has been paid to this activity.

STATISTICAL TREATMENT OF WATER-QUALITY DATA

Various simple procedures such as averaging, determining frequency distributions, and making simple or multiple correlations are widely used in water-analysis interpretation. The more sophisticated applications of statistical methods and particularly procedures which utilize analog or digital computers are being more and more widely applied. Some potential applications of these techniques will be suggested here. It is essential that proper consideration be given to chemical principles when such applications are made.

USE OF AVERAGES

A detailed record of the composition of water passing a river sampling point may be conveniently summarized by computing an average. The usefulness of the average and its significance, however, will be considerably influenced by the nature of the observed data as well as by the method used in computing the average.

Many published water-quality records for streams consist of chemical analyses of successive single samples taken at about equal intervals of time during a year or series of years. Others consist of analyses of composites of daily samples, prepared by combining equal volumes of each daily sample. These kinds of data can logically be averaged on a time basis either by assigning equal weight to each analysis (if the sample intervals or composite periods are of equal length) or by weighting each analysis by the time period during which it is assumed to represent the river water. Frequent sampling decreases the possible error of this extrapolation.

The time-based average of water composition is helpful for the user of water who may wish to know what composition can be expected from day to day and may have some general hydrologic significance as well. For example, the annual averages may be considered as historical reference data showing the chemical composition of water in the past for comparison with later observations under changed conditions.

A statistical treatment of the individual observations also can be made, showing for example the frequency of occurence of high or low concentrations, and a crude predictive model eventually might be established based on statistical treatment of recurrence patterns revealed in past records. The chemical composition of river water, however, is not a random variable and can be studied by much more refined methods. The simple time distribution study of the data has value only to the degree that it organizes more effectively the historical base with which other observations may be compared.

The time-based average concentration is not mathematically compatible with the mean water discharge for the same period because of differences in the way the two quantities are computed. Hence, any correlation between the two will be ill defined and without much hydrologic meaning. In a wet year the mean water discharge will tend to be high, and if runoff occurs over longer time periods than usual the time-based average concentration may be notably lower than in a dry year. The spacing of runoff events in time will tend to influence the record computed from time-based averages and may obscure any relationship. It should also be noted that time-based average chemical analyses cannot be used with mean water discharges to compute total loads of dissolved material carried by a stream, unless the discharge remains constant through the period of the average. This condition generally cannot be expected

The relationship of dissolved-solids concentrations to stream discharge has generally been recognized in the design of sampling programs and the treatment of data obtained by the U.S. Geological Survey. Various ways of integrating these two kinds of data have been used, and some of the procedures influence the water-quality record considerably. A large proportion of published average analyses are weighted by discharge. Instead of assuming an analysis represents a time period, it can be assumed to represent a volume of water, part of the total volume of water delivered by the stream during the period of the average. In computing the average, the results in each analysis are multiplied by the appropriate water volume, and the total of these products is finally divided by the total water volume discharged during the period of the average. This average is mathematically compatible with the mean water discharge.

A discharge-weighted average tends to emphasize strongly the composition of water during periods of high flow rate. A reliable record of flow must be available for such periods, with sufficient frequency of sampling to cover the possible changes of water composition. Many discharge-weighted averages have been estimated for chemical analyses of composites of equal volumes of daily samples. The composition of a composite made in this way is not directly related to the water discharge during the composite period. However, if either discharge or concentration does not change greatly during the period of the composite, the analysis can be used to compute a weighted average without serious distortion of the final result.

As noted earlier, a composite sample made up by using volumes of the daily samples proportional to water discharge at the time of sampling is more appropriate for analyses that will be used to compute discharge-weighted averages. The risk remains for some streams that both the sample and the instantaneous discharge may fail to represent the true average for the day they are collected. The chemical analyses of composites made in this way have a built-in component of stream discharge which may decrease their usefulness for some purposes.

The discharge-weighted average may be thought of as representing the composition which the water passing the sampling point during the period of the average would have if it had been collected and mixed in a large reservoir. Actually, reservoir storage would bring about changes in the composition of the water owing to evaporation and other complications such as precipitation of some components; however, the water discharge-weighted average does give a reasonably good indication of the composition of water likely to be available from a proposed storage reservoir and is useful in preconstruction investigations for water-development projects.

The discharge-weighted average also is useful in geochemical studies, because it can be used directly with annual discharge values to compute total quantities of the various solutes transported by the stream during a year. These quantities are used to compute erosion rates, and the inflow-outflow balance of solutes.

Averages weighted by time are useful to water users or potential users who do not have storage facilities and must use the water available in the river. The discharge-weighted average is strongly affected by comparatively short periods of very high discharge, but the influence of high flow rate on the chemistry of the river flow observable at a point is quickly dissipated when discharge returns to normal.

Important facts relating the composition of river water to environmental influences may be brought out by means of averages of several kinds, by using time periods that have been judiciously selected and with some knowledge of the important factors involved. Interpretation of water-quality records for an actual stream may help demonstrate the processes which can be used.

From 1940 to 1956, the U.S. Geological Survey conducted a continuous study of the quality of the water of the Rio Grande at the San Acacia diversion dam a few miles north of Socorro, N. Mex. In accordance with standard practices used in that period, a water sample was obtained once a day by a local observer. The samples were analyzed by the New Mexico district laboratory of the U.S. Geological Survey. Specific conductance was determined for each daily sample, and the samples were then combined into 10-day composites (three per month), made from equal volumes of each daily sample. During periods of fluctuating water composition, the number of days in composite periods was sometimes decreased so as to include only samples having similar conductances in any one composite.

The sampling station is upstream from irrigated land in the vicinity of Socorro, N. Mex , and downstream from two major tributaries, the Rio Puerco and the Rio Salado, that drain the arid plains and adjacent highlands of west central New Mexico. The drainage basin boundary encloses a total area of 26,770 square miles upstream from San Acacia, but some of this area is noncontributing. The basin includes mountainous areas of high precipitation and a wide variety of landforms and geologic terranes, as well as much arid land with low runoff. A large upstream acreage is irrigated with water diverted from the river and its tributaries, and residual water is returned to the stream through a network of drains, to be rediverted at downstream points. The discharge of the river and the chemical composition of the water fluctuate widely in response to these and other related influences.

The discharge of the river at San Acacia was measured at a gaging station 0.2 mile downstream from the sampling point and thus downstream from the diversion structure that took water from the river for irrigation in the vicinity of Socorro. The amount of water diverted ranged from about 30,000 to about 90,000 acre-feet per year during the period of record. In some years, this amounted to 20 percent of the flow measured in the river at the gaging station, but usually the amount diverted was a smaller proportion of the river flow. On many individual days, however, during the irrigation season, a high percentage of the flow was diverted. The water samples presumably represented flow at the gaging station as well as at the diversion point, as there was no inflow between these locations.

In normal years, the flood runoff at San Acacia occurs in two rather well-defined periods: the snowmelt period in the spring, when the flow originates largely from the melting of winter accumulation of the snow in the high mountains, and the summer storm period, when the flow originates largely from violent local rainstorms mostly centered in the parts of the basin that are at lower altitudes. The snowmelt water is characteristically low in dissolved solids. The runoff from storms in the lower parts of the basin may be comparatively high in dissolved solids and carries a heavy load of sediment, owing to the nature of the rock and the soil in parts of the area. The amount of runoff during the snowmelt period is normally a large part of the total for the year.

In table 22, the results of two methods of averaging the analytical

TABLE 22.—*Average of chemical analyses computed by different methods for Rio Grande at San Acacia, N. Mex., for 1941-42 and 1945-46 water years*

[Analyses by U S Geological Survey. Date below sample number is date of collection. Source of data U S. Geol. Survey Water-Supply Papers 950 and 1050]

Constituent	(1) Oct 1, 1941– Sept 30 1942		(2) Oct. 1, 1941– Sept 30, 1942		(3) Oct 1, 1945– Sept 30, 1946		(4) Oct. 1, 1945– Sept 30, 1946	
	mg/l	meq/l	mg/l	meq/l	mg/l	meq/l	mg/l	meq/l
Silica (SiO$_2$)	23		26		30		31	
Iron (Fe)	.05		.05		.06		.06	
Calcium (Ca)	45	2.246	53	2.64	76	3.79	84	4.19
Magnesium (Ng)	9.1	.748	10	.82	15	1.23	17	1.40
Sodium (Na)	34	1.478	47	2.04	78	3.39	98	4.26
Potassium (K)	3.7	.095	4.2	.11	5.9	.15	6.0	.15
Bicarbonate (HCO$_3$)	129	2.114	154	2.52	200	3.28	222	3.64
Sulfate (SO$_4$)	93	1.936	120	2.50	203	4.23	243	5.06
Chloride (Cl)	16	.451	23	.65	37	1.04	48	1.35
Fluoride (F)	.4	.021	.4	.02	.5	.03	.6	.03
Nitrate (NO$_3$)	1.8	.029	1.6	.03	1.6	.03	1.6	.03
Dissolved solids								
Calculated	289		361		546		639	
Hardness as CaCO$_3$	150		173		251		280	
Noncarbonate	44		47		87		98	
Specific conductance (micromhos per cm at 25°C)	435		550		800		926	
Mean discharge (cfs)	3,255		3,255		345		345	

[1] Average of analyses of composites of equal volumes of daily samples weighted by discharge for composite period
[2] Average of analyses of composites of equal volumes of daily samples weighted by number of days in composite period
[3] Average of analyses of composites of equal volumes of daily samples weighted by discharge for composite period
[4] Average of analyses of composites of equal volumes of daily samples weighted by number of days in composite period

data for the Rio Grande at San Acacia are given for 2 years. The 1941–42 water year was the year of highest runoff in the 16-year period of record, and 1945–46 was one of the 3 years of lowest runoff in the period. The discharge-weighted average for the wet year is strongly influenced by the low concentrations experienced during the snowmelt period, when the discharge was greatest. As a result, the average dissolved-solids value calculated by the discharge-weighting procedure is about 80 percent of the value calculated by weighting each composite analysis by the number of days represented. The amount of difference between the averages would depend on the fluctuation in dissolved solids in relation to the discharge of the stream and could be greater or smaller for other years.

The 1945–46 water year was in a drought period, and the runoff from snowmelt was so small that year that it hardly affected the discharge of the river at San Acacia. The difference between the two averages for 1945–46 is less than for the year of higher flow. The dissolved solids computed by the discharge-weighting procedure is about 85 percent of the time-weighted value.

The period October 1 to the succeeding September 30, which is the standard water year for streamflow, is not necessarily the best period to use for averages to show stream-water composition. The water-quality characteristics of a river can often be explored and demonstrated by using averages for different time periods. This approach is demonstrated for the Rio Grande data in table 23, which covers nearly the same total time period as table 22 but is divided into shorter time periods for computing averages.

The runoff at San Acacia comes from three major sources, each of which may be considered as normally dominant during a certain period of the year. The snowmelt period commonly begins in March and usually is over by the end of June. The period March 1 to June 30 was, therefore, selected to represent the snowmelt season. From July to the end of October is the season of summer thunderstorms, which normally cause a major part of the runoff during this period. The period July 1 to October 31, therefore, was selected as the summer storm-runoff period. These two periods also represent the irrigation season in the area upstream from the sampling point. From November 1 to February 28 there is usually not much runoff from rain or snow, and the streamflow is sustained by ground-water inflows. The water table has been raised in the irrigated area during the other two periods as a result of water applications, and the ground-water inflows to the river during the winter are thus essentially return flow from irrigation. This period may be called the winter base-flow period.

In table 23, discharge-weighted averages for each of the three periods are shown for 2 years, approximately coinciding with the 1942

TABLE 23.—*Discharge-weighted average of chemical analyses for Rio Grande at San Acacia, N. Mex., representing periods in the 1941–42 and 1945–46 water years in which different sources of runoff predominated*

[Analyses by U.S. Geological Survey. Date below sample number is date of collection. Source of data: U.S. Geol. Survey Water-Supply Papers 950, 970, 1050, and 1102]

Constituent	(1) Nov. 1, 1941–Feb. 28, 1942		(2) Mar. 1–June 30, 1942		(3) July 1–Oct. 31, 1942		(4) Nov. 1, 1945–Feb. 28, 1946		(5) Mar. 1–June 30, 1946		(6) July 1–Oct. 31, 1946	
	mg/l	meq/l	mg/l	meq/l	mg/l	meq/l	mg/l	meq/l	mg/l	meq/l	mg/l	meq/l
Silica (SiO$_2$)	27		22		26		32		33		21	
Iron (Fe)	.04		.06		.08		.04		.07		.04	
Calcium (Ca)	51	2.55	39	1.946	65	3.24	63	3.14	66	3.29	139	6.94
Magnesium (Mg)	10	.82	8.2	.674	12	.99	13	1.07	14	1.15	28	2.30
Sodium (Na)	46	2.00	25	1.087	63	2.74	60	2.61	68	2.96	161	7.00
Potassium (K)	3.7	.09	3.7	.095	4.3	.11	6.5	.14	5.5	.14	7.0	.18
Bicarbonate (HCO$_3$)	161	2.64	115	1.885	176	2.88	197	3.23	201	3.29	209	3.42
Sulfate (SO$_4$)	102	2.12	74	1.541	161	3.35	133	2.77	149	3.10	541	11.26
Chloride (Cl)	26	.73	11	.310	28	.79	32	.90	38	1.07	59	1.66
Fluoride (F)	.4	.02	.3	.016	.4	.02	.4	.02	.5	.03	.7	.04
Nitrate (NO$_3$)	1.3	.02	1.9	.031	1.7	.03	1.5	.02	1.4	.02	1.9	.03
Dissolved solids												
Calculated	347		242		448		437		475		1,060	
Hardness as CaCO$_3$	168	10.99	131	7.585	211	14.15	210	13.90	222	16.05	462	32.83
Noncarbonate	36		37		68		49		58		291	
Specific conductance (micromhos per cm at 25°C)	526		366		665		664		715		1,460	
Mean discharge (cfs)	1,610		6,620		692		633		149		189	

1. Winter base-flow period, 1941–42. Large part of flow from ground-water inflow and irrigation return, following wet year.
2. Snowmelt runoff period 1942. Most of flow from melting of above-normal mountain snow pack.
3. Summer runoff period, 1942. Most of flow resulted from summer rainstorms.
4. Winter base-flow period 1945–46. Flow from irrigation return and ground-water inflows, following year of about normal runoff.
5. Snowmelt runoff period, 1946. Drought conditions; no snowmelt reached station. Flow continued to be largely irrigation return.
6. Summer runoff period, 1946. Most of flow resulted from summer rainstorms, curtailed by drought.

and 1946 water years of table 22. It will be noted that for 1942 the weighted average for the year approaches most closely the average for the snowmelt period, when most of the annual runoff occurred. In 1946, the runoff during the normal snowmelt period was mostly from ground-water inflows. The averages for the first two periods (4 and 5, table 23) resemble each other closely. The annual discharge-weighted average falls between the average of the first two and the third periods, but is nearer the value for the ground-water inflow period, when most of the runoff occurred.

Comparison of corresponding periods for the 2 years in table 23 demonstrates the effect of drought on water quality, especially for the periods of winter-base flow and snowmelt runoff. Some hydrologic effects observed during droughts in the southwestern United States were described by Gatewood, Wilson, Thomas, and Kister (1964). The quality of runoff in the summer storm period is partly controlled by the geographic distribution of the rainfall, as some tributaries produce more highly mineralized floodwaters than others, and comparison of data for this period can provide knowledge about some of these effects. It has already been shown that the river water at San Acacia changes rapidly in composition during summer floods.

As technology has improved, detailed continuous records of water quality have become easier to obtain and can serve as inputs to computers that can be programmed to analyze the data in many ways. The rather simple manipulations shown in tables 22 and 23 represent only a starting point. The examination and treatment of the water analyses, however, for streams must be done in the light of some knowledge of other conditions in the area of study so that a hydrologically coherent picture results.

In the early literature of stream-water quality, the influence of discharge rate was sometimes considered to be a simple dilution effect. If this premise is accepted, the composition of the water could be represented for all time by a single analysis in which the results are expressed in terms of percentages of the dry residue. Clarke (1924a, b) gave many analyses for river water expressed in this way. Although the assumption that the water at any other time could be duplicated by either dilution or concentration is an obvious, gross oversimplification, this way of expressing analyses makes it possible to compare the composition of streams and make broad generalizations from sketchy data. In Clarke's time, this approach was about the only one possible. It should not be necessary to belabor the point that the composition of the water of only relatively few rivers can be characterized satisfactorily solely on the basis of dilution effects. Geochemistry textbooks, however, still often quote analyses for river water in terms of percentages of dry residue and make this unwarranted implication.

Although the annual discharge-weighted average analysis gives a useful geochemical summary of river-water composition, the weighting procedure may tend to magnify some of the effects of water development and utilization by man. For example, during a period of flow where some fairly consistent correlation between discharge and conductance might happen, as in the Rio Grande during the snowmelt period, a diversion of water from the river would influence the discharge and thus would influence the weighted-average composition of water downstream even if the diversion had no effect on the chemical quality of the water there. Likewise, the introduction of highly mineralized waste might change composition without much changing the discharge. Few important streams in the United States are fully free from such effects; therefore, the annual weighted-average composition of a stream may be poorly correlated with annual-mean discharge and may not provide a reliable basis for extrapolating back to the composition of the water in the stream before the basin was developed by man. The subject of extrapolation of data and the ways in which flow regulation may influence the correlation of water composition and discharge will be considered somewhat further in a later section.

FREQUENCY DISTRIBUTIONS

A useful generalization about an array of data, such as a series of chemical analyses of a stream, often may be obtained by grouping them by frequency of occurrence. Figure 23 is a duration diagram showing the dissolved-solids content of water in the Colorado River at Grand Canyon, Ariz., compared with the dissolved solids in the outflow at Hoover Dam, the next downstream sampling point. The curves show that the water is much less variable in composition as a result of storage. The median point, represented by 50 percent on the abscissa for the water at Grand Canyon, is also at a much higher dissolved solids than that for virtually the same water after storage, mixing, and release at Hoover Dam. Daily conductivity values for a period of record may be summarized conveniently by a graph similar to this. Figure 24 shows this kind of information for the Ohio River and its two source streams, the Allegheny and Monongahela Rivers in the vicinity of Pittsburgh, Pa Figures 23 and 24 are time distributions.

A frequency distribution for percent sodium in ground waters from wells in the San Simon artesian basin of Arizona is shown in figure 25. The double maximum shown by these data indicates the waters were of two types, which were identified as occuring in separate areas of the basin. This distribution is not related to time. Other applications of frequency distributions are obviously possible. It

248　CHEMICAL CHARACTERISTICS OF NATURAL WATER

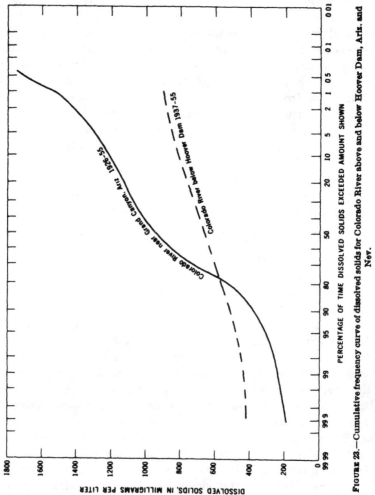

FIGURE 23.—Cumulative frequency curve of dissolved solids for Colorado River above and below Hoover Dam, Ariz. and Nev.

ORGANIZATION AND STUDY OF WATER-ANALYSIS DATA

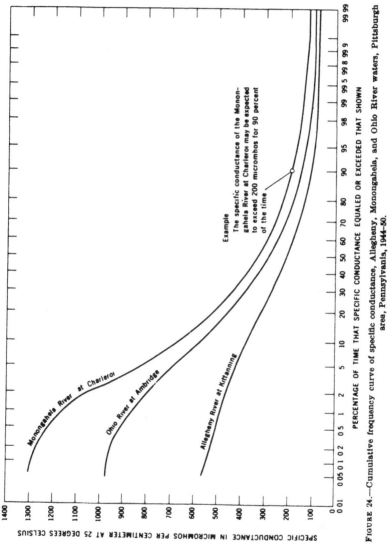

FIGURE 24.—Cumulative frequency curve of specific conductance, Allegheny, Monongahela, and Ohio River waters, Pittsburgh area, Pennsylvania, 1944-50.

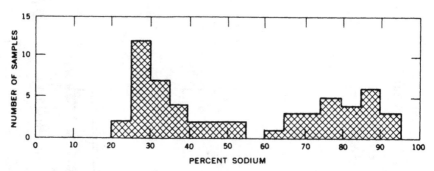

FIGURE 25.—Number of samples having percent-sodium values within ranges indicated, San Simon artesian basin, Arizona.

may be considered most useful as a means of summarizing a volume of data and often gives more information than a simple mean or median value alone would give.

CORRELATIONS

The examination of an array of water analyses frequently involves a search for relationships among constituents. Determining the existence of correlations that are sufficiently well defined to be of possible hydrologic or geochemical significance has sometimes been attempted in a random fashion by preparing many scatter diagrams with concentrations of one component as abscissa and another as ordinate. Commonly, one or more of the diagrams gives a pattern of points to which a regression line can be fitted. The statstical procedures for fitting the line, and the evaluation of goodness of fit by a correlation coefficient, provide a means of determining the apparent correlation rather closely in a numerical way. The significance of the correlation in a broader sense, for example, as the indicator of a particular geochemical relationship, is, however, likely to be a very different matter.

Random correlation procedures of the type suggested above are too laborious to do by hand if many analyses are utilized, but the electronic computer has made this type of correlation a simple process. Accordingly, a wide variety of calculations of this kind could be made, and their significance evaluated later should this procedure offer any reasonable promise of improving the investigator's understanding of hydrochemical systems.

The difficulties of interpreting random correlations are substantial. The chemical analyses of a series of more or less related samples of water have certain internal constraints that may result in correlations that have slight hydrologic or geochemical significance. For example, the total concentration of cations in milliequivalents per liter in each analysis must equal the total concentration of anions in the same units. Also, if many of the waters in the group being studied have one predominant anion and one predominant cation, these two constituents will have a good mutual correlation. Some of this effect may also be noticeable among less abundant constituents because their concentrations also represent a part of the ionic balance of each analysis. A correlation between calcium and bicarbonate concentrations commonly exists in many waters, but the ions may not result from calcite solution and this type of correlation is of little geochemical significance.

Before attempting to make correlations among ions in solution, the investigator should set up hypothetical relationships that might reasonably be expected to hold, and then proceed to test them with the data. Two kinds of relationships that might be anticipated will be described here. The first represents the simple solution of a rock mineral with no other chemical processes that could alter the proportions of dissolved ions. This relationship is based on simple stoichiometry, requires no chemical equilibria, and generally postulates a molar ratio between dissolved ions equal to the mole ratio of these ions in the source mineral. A more complex relationship is commonly found, the second type presented here, where concentrations of ions may rise freely until saturation with respect to one or more species is reached. From this point on, the concentrations are controlled by chemical equilibria.

Suppose it is desired to check the validity of assuming the sodium and chloride contents shown by a series of analyses of a river water or ground water body have been derived from solution of common salt, NaCl. These two ions will be brought into solution in equal quantity and retained in solution over a wide concentration range because of the rather high solubility of both ions. Over this range one might expect

$$C_{Na} = nC_{Cl},$$

where the C terms are ion concentrations and n is the conversion factor required to make the units in which C terms are reported chemically equivalent. If the values of C are in milliequivalents per liter, $n=1$. If data are in milligrams per liter, n is the ratio of combining weights of sodium to chloride or 0.65.

A plot of concentration of sodium versus concentration of chloride should give a straight line of slope 1.0 or 0.65, depending on units used, if the data fit the assumption. Curvature in the line or deviation from the two permissible slopes indicates the hypothesis is incorrect. If one plots log C_{Na} versus log C_{Cl}, a straight line also should be obtained.

Figure 26 is a plot of sodium versus chloride concentrations in milliequivalents for samples from the Gila River at Bylas, Ariz. The curved part of the regression line shows that sources of the ions other than common salt are involved, although at the higher concentrations the theoretical slope of 1.0 is closely approached. A similar correlation would be obtained by plotting log C_{Na} versus log C_{Cl}.

The correlation cited above does not involve much chemistry and is too simplified to have much practical value. If considerations of solubility are involved, the constituents may be correlated in a different way. For example, the solution of calcium and sulfate can only continue up to the solubility limit of gypsum. When this level is reached, at equilibrium the solubility-product relationship would hold. The two conditions would be

for dilute solutions
$$C_{Ca}^{+2} = nC_{SO_4}^{-2}$$
at saturation
$$[Ca][SO_4] = K_{sp}.$$

If the influences of ionic strength and ion-pair formation are ignored for the moment, one could represent the second equation in terms of concentrations rather than activities so that the same variables would be present in both expressions. Concentrations are all expressed in moles per liter.

The first equation would give a straight line if calcium concentrations were plotted against sulfate. The second equation, however, is for a hyperbola, and therefore an attempt to make a linear correlation of a group of analyses including concentrations affected by both relationships is likely to give ambiguous results. If both expressions are placed in logarithmic form, however, both give straight lines as

$$\log C_{Ca} = \log C_{SO_4}$$
and
$$\log C_{Ca} = \log K - \log C_{SO_4}.$$

The first relationship should hold up to the point where precipitation of gypsum begins and the second thereafter. A break in slope of the regression line would occur at that point. It might be noted that in this example the value of K for concentration data is not really constant, but tends to increase somewhat as concentrations rise. Therefore, the line might not reverse slope at the saturation point, but it should definitely change in slope there.

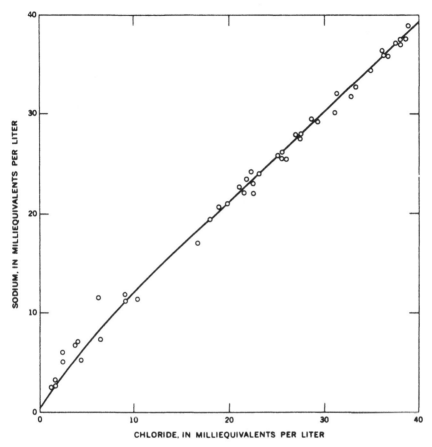

FIGURE 26.—Sodium-chloride relationship, Gila River at Bylas, Ariz., October 1, 1943 to September 30, 1944.

A more complicated equilibrium is that for calcite solution which involves three variables, as noted earlier:

$$\frac{[Ca^{+2}][HCO_3^-]}{[H^+]} = K_{eq}.$$

This could still be converted to a two-variable relationship by combining the bicarbonate and hydrogen ion into a composite variable:

$$[Ca^{+2}] \cdot \frac{[HCO_3^-]}{[H^+]} = K_{eq}$$

or

$$\log [Ca^{+2}] + \log \frac{[HCO_3^-]}{[H^+]} = \log K_{eq}.$$

The latter expression might be used as a basis for testing water for equilibrium with respect to calcium carbonate.

The degree to which correlations of the last type mentioned above can be obtained is of geochemical significance in many systems. Analyses may be checked for conformance to equilibrium with several solids at once. For example, one might want to examine data for equilibrium with gypsum and calcite. Figure 27 is one approach to this type of correlation. Several analyses from tables 11 and 15 are plotted in the diagram. This relationship also could be represented as

$$CaCO_3(c) + H^+ + SO_4^{-2} = CaSO_4(c) + HCO_3^-,$$

$$\frac{[HCO_3^-]}{[H^+]}[SO_4^{-2}] = K,$$

or

$$\log \frac{[HCO_3^-]}{[H^+]} + \log [SO_4^{-2}] = \log K,$$

where K is a combination of the two constants for the solubility equilibria for gypsum and calcite.

There are other combinations of equilibria that might be significant in ground-water systems and at low flow in some streams. For example,

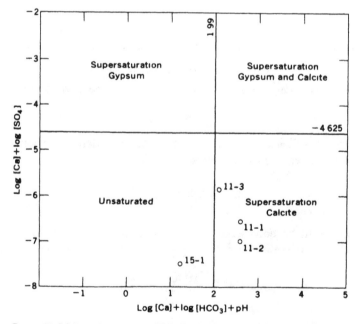

FIGURE 27 Calcite and gypsum equilibrium solubility limits, 25°C and 1 atmosphere pressure

ORGANIZATION AND STUDY OF WATER-ANALYSIS DATA 255

the combination of cation-exchange equilibria in soil and the solubility expression for calcite could be studied in this way. Garrels (1967) used a related procedure in studies of the composition of ground water associated with igneous rocks.

The abscissa and ordinate in figure 27 represent the ion-activity products for the calcite-solubility equilibrium and the solubility-product equation for gypsum. Values greater than the equilibrium constant indicate supersaturation with respect to these two solids. The activities of the ions must be calculated from analytical concentrations with due attention to the effects of ionic strength and complex or ion-pair formation. Irrigation drainage water in some areas may be near saturation with respect to both solids (Hem, 1966).

A study by Feth, Roberson, and Polzer (1964) of water composition associated with granitic rocks uses diagrams on which areas of stability for different rock minerals are shown in terms of activity of silica as abscissa and ratios of sodium or potassium to hydrogen activity as ordinate.

More sophisticated computer programs for examining data than have been used heretofore are certainly possible and hold much promise for better understanding of natural-water chemistry if they are carefully employed. Factor analysis as a means of studying water analyses from a hydrologic unit might be cited as an example (Dawdy and Feth, 1967), although it must be emphasized again that any such program must be set up with due reference to a hydrologic and chemical model for the area from which the data come and that the data must be used to test the model, rather than to establish apparent patterns of orgainzation.

ION AND SOLUTE INVENTORIES

In most hydrologic studies of water resources, the quantities of water present in, or recoverable from, different parts of the system are carefully measured or calculated. From quantities of water and adequate chemical analyses, it obviously is possible to calculate quantities of ions or solutes in storage or in transit. Usually, however, only the concentrations of dissolved matter in water in the different parts of the system are reported. Certain kinds of hydrologic studies, especially those where prediction of future water quality may be required, necessitate determining quantities of solutes in storage and in transit. The need for this kind of investigation likely will increase.

In any hydrologic system, the solute circulation can be represented by means of a simple equation analogous to the inflow-outflow equation of a water-resources inventory. The basic equation for the chemical system also contains inflow, outflow, and storage terms and in its most general form may be written

$$\Sigma W \text{ inflow } \pm \Delta W \text{ storage} = \Sigma W \text{ outflow},$$

where ΣW represents a total quantity of transported dissolved ions and ΔW represents a change in stored quantity of dissolved material. The water-quality information available for a section of river, for example, normally includes analyses and flow measurements sufficient to compute tonnages of inflow and outflow of ions in that section for any given time period. The change in the storage item for some solutes is minor and may be ignored, but for others it is very important. For example, some ions may become adsorbed on stream-bed sediments, be taken up by biota, enter into chemical reaction and precipitate, or otherwise be altered in the section of river involved. A particularly significant set of behavior patterns in river water is that associated with organic pollutants. The solute-balance equation is the basis for many quantitative interpretations of water chemistry, and several forms will be considered later in this paper in reviewing water discharge-quality relationships in streams and salt balance in irrigated areas.

The terms in the solute-balance equation may be subdivided into specific portions, some of which at least will be known or measurable and others of which can be approximately estimated. Computer programs which offer the means of considering simultaneously a great number of variables offer a technique for modeling hydrologic systems for water-quality studies. Ultimately, this approach will be very useful, and many studies involving this kind of modeling are already being made. The amount of fundamental knowledge concerning the behavior of ions in many hydrologic systems is not yet sufficient to provide movement rates for some parts of the model. The rates at which ions move through the soil and subsoil zones for example are not well known. A mathematical model for estimation of certain changes in water quality likely to be produced by irrigation development has been made by Tanji, Doneen, and Paul (1967).

GRAPHICAL METHODS FOR REPRESENTATION OF ANALYSES

Over the years, a considerable number of techniques for graphical representation of analyses have been proposed. Some of these are useful for display purposes—that is, to illustrate oral or written reports on water quality, to provide means for comparing the analyses with each other, or to emphasize differences and similarities. Graphical procedures do this much more effectively than numbers quoted in tables.

In addition to the types of graphs suitable for display and comparison of analyses, graphical procedures have been described that are intended to help detect and identify mixing of waters of different composition and to identify some of the chemical processes that may take place as natural waters circulate. Graphing techniques of the latter type may be useful in the study of data prior to preparing reports or arriving at conclusions. Some of the graphical techniques that appear to be useful are described here, but this discussion is not intended to include all the methods that have been suggested in the literature. Graphing of water analyses is a study technique and not an end in itself.

ION-CONCENTRATION DIAGRAMS

Most of the methods of graphing analyses are designed to represent simultaneously the total solute concentration and the proportions assigned to each ionic species for one analysis or a group of analyses. The units in which concentrations are expressed in these diagrams generally are milliequivalents per liter.

The ion-concentration graphing procedure most widely used in the United States probably is that originated by the late W. D. Collins (1923). In this system, each analysis is represented by a vertical bar graph whose total height is proportional to the total concentration of anions or cations, in milliequivalents per liter. The bar is divided by a vertical line with the left half representing cations and the right half anions. These segments are then divided by horizontal lines to show the concentrations of the major ions, which are identified by distinctive colors or patterns. Usually, six divisions are used, although more can be provided if necessary. The concentrations of closely related ions are often added together and represented by a single pattern.

An example of the Collins' diagram representing four analyses from tables in this paper is given in figure 28. The other graphing procedures discussed here are illustrated for the most part with the same four analyses. The analyses are identified by table number and analysis number in the table. Analysis 15–1 is analysis 1, table 15, for example.

A bar graph proposed by Reistle (1927) used ion concentrations in parts per million. It offers no advantages over the Collins' procedure and need not be discussed further. The Collins' system as described does not consider nonionic constituents, but they may be represented, if desired, by adding an extra bar or other indicating device with a supplementary scale. In figure 29, the hardness of two waters is shown. The hardness in milligrams per liter as $CaCO_3$ is equivalent to the height of the calcium plus the magnesium segments, in milliequivalents per liter, multiplied by 50. In figure 30, the concentration of silica is represented in millimoles per liter, because milliequivalents cannot be used for uncharged solute species or species whose form in solution is

258 CHEMICAL CHARACTERISTICS OF NATURAL WATER

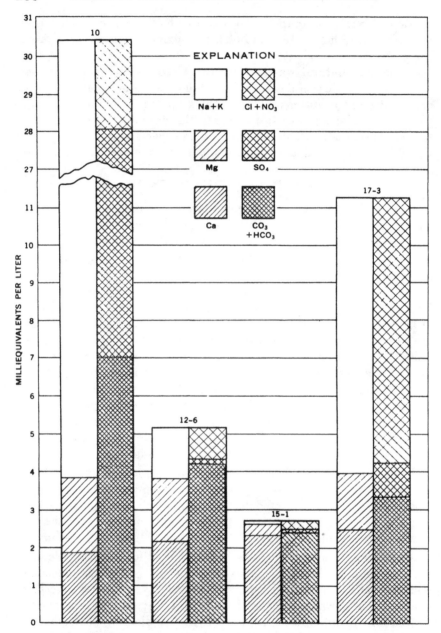

FIGURE 28.—Analyses represented by vertical bar graphs of milliequivalents per liter.

uncertain. The units are closely related to milliequivalents per liter, and this type of graph will be used in depicting analyses of water associated with different rock types later in this report. The pattern for silica occupies the full width of the bar graph, because the solute species H_4SiO_4 actually can be considered to include a silicate anion whose charge is balanced by H^+. The silica, therefore, includes in effect both anion and cation fractions. Furthermore, the contribution to the total solute content is in proper proportion. This modification is useful in showing results of geochemical processes. The concentration of silica in millimoles per liter, however, is not usually included in water-analysis tabulations.

A system of plotting analyses by radiating vectors was proposed by Maucha (1949) of Hungary and is illustrated in figure 31. The lengths of each of the six vectors from the center represents the concentration of one or more ions in milliequivalents per liter. This plotting system had little acceptance in the United States. It may have some potential, however, as a means of showing analytical values in a small space, for example, as a symbol on a map. This system suggested by Stiff (1951) seems to give a more distinctive pattern and has been used in many reports written by United States investigators. The Stiff method uses

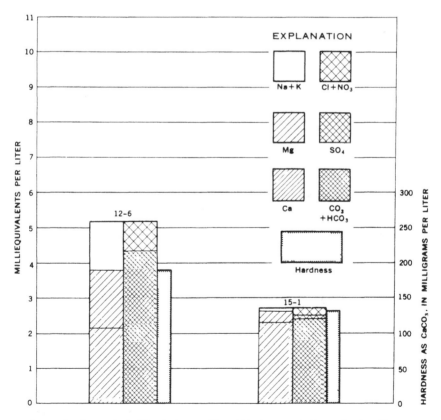

FIGURE 29 —Bar graph of milliequivalents per liter which also shows hardness values in milligrams per liter

260 CHEMICAL CHARACTERISTICS OF NATURAL WATER

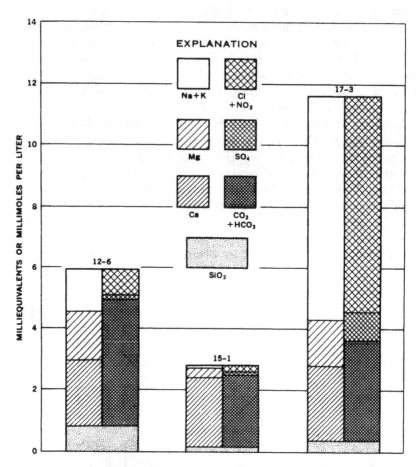

FIGURE 30.—Bar graph showing millimoles per liter of dissolved silica and milliequivalents per liter of anions and cations.

four parallel horizontal axes extending on each side of a vertical zero axis. Concentrations of four cations can be plotted, one on each axis to the left of zero, and likewise four anion concentrations may be plotted, one on each axis to the right of zero; the ions should always be plotted in the same sequence. The concentrations are in milliequivalents per liter. The resulting points are connected to give an irregular polygonal shape or pattern, as in figure 32. The Stiff patterns can be a relatively distinctive method of showing water composition differences or similarities. The width of the pattern is an approximate indication of total ionic content.

Two other procedures that have been used to prepare pattern diagrams are worthy of mention. The "pie" diagram can be drawn with a scale for the radii which makes the area of the circle represent

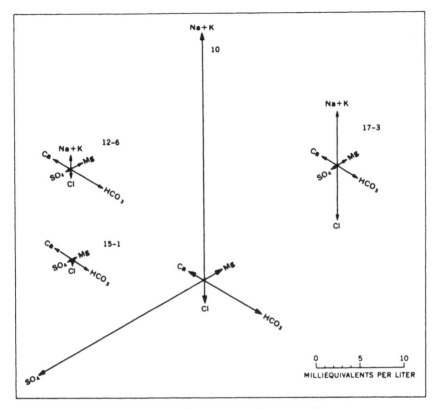

FIGURE 31.—Analyses in milliequivalents per liter represented by vectors.

the total ionic concentration (fig. 33) and subdivisions of the area represent proportions of the different ions. Colby, Hembree, and Rainwater (1956) used a pattern diagram in which four components, $Ca+Mg$, CO_3+HCO_3, $Na+K$, and $Cl+SO_4+NO_3$, were represented on rectangular coordinates. The kitelike figure resulting from connecting the four points made a convenient map symbol (fig. 34).

A nomograph proposed by Schoeller (1935) and modified by R. C. Vorhis of the U.S. Geological Survey is shown in figure 35. This diagram is a means of depicting a group of analyses that has the advantage of also showing relationships among milligrams per liter and milliequivalents per liter for the different ions. Waters of similar composition plot as near-parallel lines. This diagram, however, uses logarithmic scales which may complicate the interpretation for waters that differ greatly in concentration.

Figure 36 is a cumulative percentage plot of analyses in milligrams per liter. This method of graphing permits differentiating between types of water on the basis of the shape of the profile formed by joining the successive points.

262 CHEMICAL CHARACTERISTICS OF NATURAL WATER

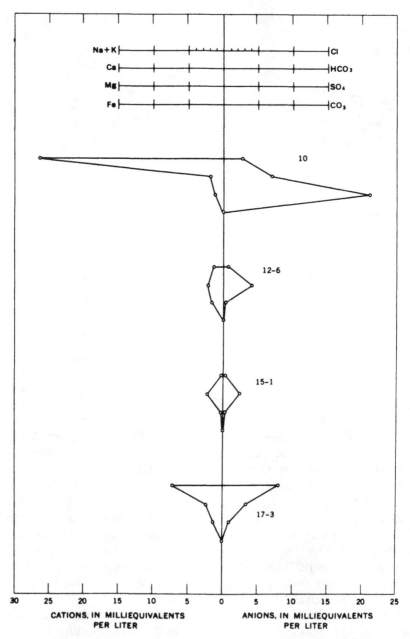

FIGURE 32.—Analyses represented by patterns based on milliequivalents per liter.

ORGANIZATION AND STUDY OF WATER-ANALYSIS DATA   263

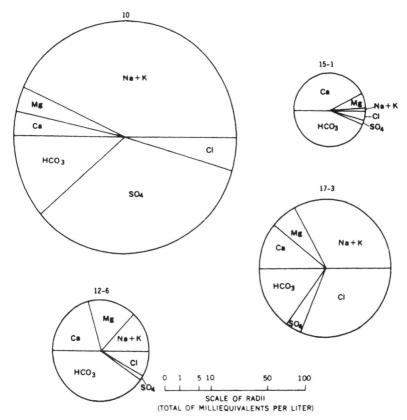

FIGURE 33.—Analyses represented by circular diagrams subdivided on the basis of percentage of total milliequivalents per liter.

Ion-concentration diagrams are useful for several purposes. They aid in correlating and studying analyses and are especially helpful to the novice in this field. They also aid in presenting summaries and conclusions about water quality in areal-evaluation reports. The Collins' diagram can form a very effective visual aid in oral presentations on water composition. For this purpose, the bar symbols can be made at a scale of about 10 centimeters=1 milliequivalent per liter, by using cards fastened together end to end with hinges of flexible tape to give a length sufficient to represent the total concentration of ions. Segments representing the ions are then drawn on the graph, and they can be colored distinctively to show the six species. As the lecturer discusses the analyses he can take out the appropriate set of cards and hang them up in view of his audience. The contrast between water of low concentration shown by only one or two cards and more concentrated solutions, occurring naturally or as a result of pollution,

264   CHEMICAL CHARACTERISTICS OF NATURAL WATER

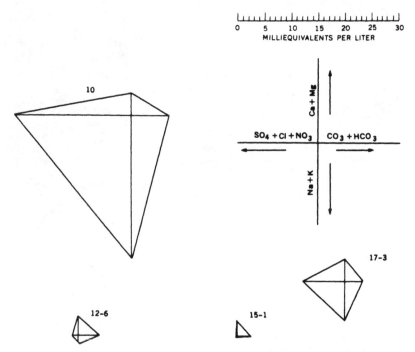

FIGURE 34.—Analyses represented by patterns based on combined anion and cation concentrations.

shown perhaps by a thick stack of cards that might reach a length of 20 feet when unfolded is very striking.

TRILINEAR PLOTTING SYSTEMS

If one considers only the major dissolved ionic constituents in milliequivalents per liter and lumps potassium and sodium together and floride and nitrate with chloride, the composition of most natural waters can be closely approximated in terms of three cationic and three anionic species. If the values are expressed as percentages of the total milliequivalents per liter of cations, and anions, the composition of the water can be represented conveniently by a trilinear plotting technique.

The simplest trilinear plots utilize two triangles, one for anions and one for cations. Each vertex represents 100 percent of a particular ion or group of ions. The composition of the water with respect to cations is indicated by a point plotted in the cation triangle, and the composition with respect to anions by a point plotted in the anion triangle. The coordinates at each point add to 100 percent.

Emmons and Harrington (1913) used trilinear plots in studies of mine-water composition. This application was the earliest found by

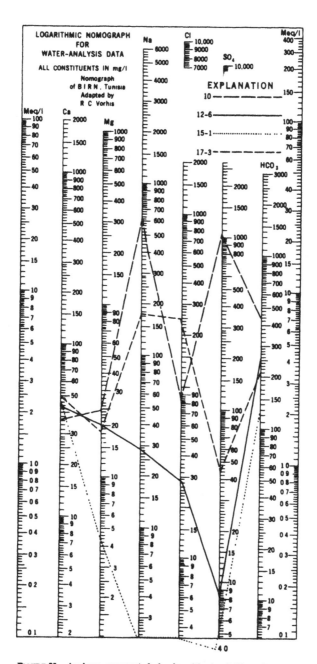

FIGURE 35.—Analyses represented by logarithmic plotting of concentration in milligrams per liter.

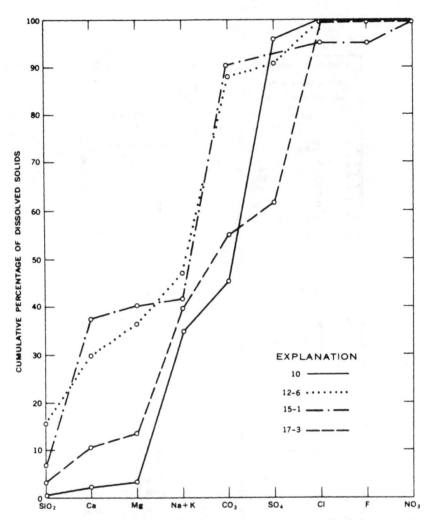

FIGURE 36.—Analyses represented by linear plotting of cumulative percentage composition based on milligrams per liter.

the writer in surveying published literature on water composition, even though Emmons and Harrington do not claim originality for the idea of using trilinear plots for this purpose. In the form used by Emmons and Harrington, the cation triangle lumps calcium with magnesium at one vertex and sodium with potassium at another. This leaves the third vertex for "other metals" which might be present in the mine waters that were of principal interest to these investigators. For most natural water, the content of these other metals is not a significant percentage of the total.

De la O. Careño (1951, p. 87-88) described a method of trilinear plotting which he attributed to Hermion Larios, which combines the plotting with a classification and reference system. The three principal cations are plotted conventionally in one triangle and the three anions in another. Each triangle is divided into 10 approximately equal areas numbered from zero to nine. A two-digit number is then used to characterize the water. The first digit is the number of the area within the cation triangle in which the water plots. The second digit is the number of the area in the anion triangle where the plotted point falls. Similar classification and plotting schemes have appeared in other countries, especially in the U.S.S.R.

A considerable number of authors have described trilinear plots in which the cation and anion triangles are combined or projected in several ways. The expression of analyses by two points in separate graphs is obviously rather inconvenient. By further combining the variables, it is possible to express the water composition in terms of two groups of cationic and two groups of anionic components, with the proviso that because all are percentages, the sums of cations and anions must equal 100. This is equivalent to specifying two variables and permits the analysis to be expressed as a single point in a two-coordinate diagram.

The first trilinear diagram incorporating this combination of the anion and cation fields to be published in the United States was that of Hill (1940). In the Hill diagram, the anion and cation triangles occupy positions at the lower left and lower right and have their bases aligned vertically and their vertices pointing toward each other. The upper central portion of the diagram is diamond shaped. In using this diagram, the proportions of anions and cations are plotted as points in each of the lower triangles. The points are then extended into the central plotting field by projecting them along lines parallel to the upper edges of the central field. The intersection of these projections represents the composition of the water with respect to the combination of ions shown. Hill's original diagram was so arranged that bicarbonate and sulfate were grouped together, and the point in the central field amounted to a plot of sodium percentage versus chloride percentage. Hill also divided the central plotting field into 10 areas and proposed a classification scheme involving 10 types of water, depending on the area in which the analysis is plotted. In later revisions of his procedure, Hill (1941; 1942) combined sulfate with chloride rather than sulfate with bicarbonate.

Langelier and Ludwig (1942) proposed a diagram in which rectangular coordinates were used representing percent sodium plus potassium on the ordinate and percent carbonate plus bicarbonate on the abscissa. This diagram had no cation or anion triangles.

Piper (1944) suggested the form of the trilinear diagram which is reproduced in figure 37. The circles plotted in the central field have areas proportional to dissolved-solids concentrations and are located by extending the points in the lower triangles to points of intersection. The diagram is fundamentally the same as Hill's.

A number of forms of the trilinear diagram have appeared in literature of the U.S.S.R. The procedure of Durov (1948) is very similar to that of Piper. Filatov (1948) proposed a two-point system with cation and anion triangles having a common side.

All trilinear plotting techniques are, in a sense, descendants of the geochemical classification scheme of Palmer (1911). In Palmer's classification, the composition of a water was expressed by percentage of total ions with consideration, in effect, of the proportions of sodium to total cations and of bicarbonate to total anious. Palmer, however, was apparently thinking more in terms of combined salts in solution than more modern chemists do, and his classification scheme is only of historical interest.

The trilinear diagram constitutes a useful tool in water-analysis interpretation. Most of the graphical procedures described here are of value in pointing out features of analyses and arrays of data which need closer study. The graphs themselves do not constitute an adequate means of making such studies, however, unless they can demonstrate that certain relationships exist among individual samples. The trilinear diagrams sometimes can be used for this purpose.

Applications of the diagram pointed out by Piper include testing groups of water analyses to determine whether a particular water may be a simple mixture of others for which analyses are available or whether it is affected by solution or precipitation of a single salt. It can be easily shown that the analysis of any mixture of waters A and B will plot on the straight line AB in the plotting field (where points A and B are for the analyses of the two components) if the ions do not react chemically as a result of mixing. Or, if solutions A and C define a straight line pointing toward the NaCl vertex, the more concentrated solution represents the more dilute one spiked by addition of sodium chloride.

Plotting of analyses for samples from wells successively downslope from each other may show linear trends and other relationships that can be interpreted geochemically. Although it is theoretically possible to make rather rigorous tests of hypotheses about mixing of waters or precipitation of components by the Piper diagram, few examples of its successful use for such purposes can be cited. The usual ground-water system is too complex and sampling is too uncertain to justify an exacting interpretation. Poland, Garrett, and Sinnott (1959) used trilinear diagrams extensively in studying

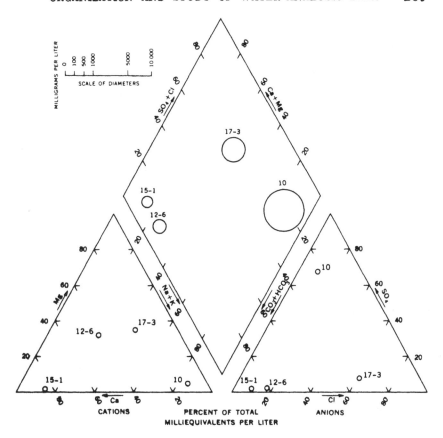

FIGURE 37.—Analyses represented by three points plotted in trilinear diagram (after Piper, (1944)).

contamination of ground water by sea water and other brines along the California coast near Los Angeles. The relationships shown by the diagrams usually constitute supporting evidence for conclusions regarding water sources that also have other bases of support.

Maderak (1966) used trilinear diagrams to show trends in composition as streamflow volume changed at several sampling points on the Heart River in western North Dakota. The diagrams have been extensively used in other U.S. Geological Survey reports as a means of generally indicating similarities and differences in the composition of water from certain geologic and hydrologic units (Hendrickson and Krieger, 1964; Feth, Roberson, and Polzer, 1964).

The value of the diagram for some purposes is decreased because it is difficult to show adequately the differences in total ion concentration among waters and because nonionic solutes, especially silica, are not considered. In many investigations, the diagrams will be found useful, but they cannot be expected to answer all the questions the hydrologist should ask himself when he is studying water-quality data.

METHODS FOR EXTRAPOLATION OF RECORDS

A considerable part of the task of the interpretation of water-quality data can be one of extrapolation or interpolation. For example, a few analyses of river-water samples taken at irregular intervals of time may need to used to estimate a continuous record, or analytical records may need to be extended backward or forward in time by correlating the analyses with some other measured variable. For ground-water studies, the time variability is usually less important than the variations from place to place in the composition of a ground-water body, and procedures are needed for extrapolation of analyses representing individual wells or springs to cover the whole volume of ground water of an area.

The extension of individual observations of river-water composition can be accomplished by several averaging techniques or other statistical treatments already discussed. As the technology of in situ observations of water quality improves, there will be increased need for computing more complete chemical records from a continuously measured property such as specific conductance. This sort of calculation can provide fairly dependable values for major ions, at least for many streams. The accuracy of the calculated value depends on how good a correlation of the measured with the calculated properties can be established from previous records of a more complete nature. Where correlation with conductance cannot be shown, as might be the case for many minor constituents, the continuous observations will need to be supplemented by sampling and analysis.

WATER-QUALITY HYDROGRAPHS

A graph showing the changes over a period of time of some property of water in a stream, lake, or underground reservoir is commonly termed a "hydrograph." Hydrographs showing variability of a property of river water with time are often used as illustrations in reports. Examples given here are figures 4 and 5, showing the change with time of the conductivity of water in the Rio Grande at San Acacia, N. Mex. For a stream, where changes with time can be large and occur rapidly, an accurate extrapolation cannot be made on the basis of time alone. On that basis, it is possible to state only approximately what the composition of water would be at different times of the year. In contrast, streams that are controlled by storage reservoirs, such as the Colorado River below Hoover Dam; streams that derive most or all of their flow from ground-water sources, such as the Niobrara River, which drains the sandhill region of northwestern Nebraska; or streams which have very large discharge rates, such as the Mississippi River at New Orleans, may have relatively minor changes in composition from day to day or even from year to year.

Streams whose flow patterns have been extensively altered by man may show definite long-term trends as the water quality adjusts to the new regime. The Gila River at Gillespie Dam, Ariz., for example, shows a deteriorating quality over the years of record as irrigation depleted the upstream water supply (Hem, 1966).

In ground water, the changes in quality with time are usually comparatively slow. The illustrations already given in this paper, however, show that both long-term and short-term trends can be observed. The slow increase in dissolved solids that occurred in the ground water of the Wellton-Mohawk area of Arizona (fig. 7) represents a condition related to water use and development, but shorter term fluctuations may be related to well construction and operation (fig. 5) or to changing recharge rates, evapotranspiration, or other factors that often influence water near the water table observed in shallow wells or seasonal springs (fig. 6). Deep wells that obtain water from large ground-water bodies which are not too extensively exploited and many thermal springs may yield water of constant composition for many years. Analyses published by George and others (1920) for Poncha Springs near Salida, Colo., represented samples collected in 1911. A sample of a spring in this same group collected in 1958 (White and others, 1963) gave an analysis that did not differ by more than ordinary analytical error from the one made 47 years before, with the exception of the sodium and potassium. The potassium values reported by George and others (1920) appear too high and sodium too low, in comparison with the modern analysis, but the total of the two alkali metals is nearly identical in both analyses.

WATER QUALITY IN RELATION TO STREAM DISCHARGE

The concentration of dissolved solids in the water of a stream is related to many factors, but it seems obvious that one of the most direct and important ones is the volume of liquid water available for dilution. Presumably, therefore, the dissolved-solids concentration should be an inverse function of the rate of discharge of water over all or at least a part of the recorded range. In discussing averaging of river-water chemical analyses, some of the complications that influence discharge versus dissolved-solids concentration have been pointed out. There will no doubt, however, be a continuing demand to know how well the composition of a river water can be computed, if at all, from the water-discharge rate, and some further consideration of the subject is needed. The goal would be to obtain a dissolved-solids rating curve for a stream-measuring station: a graph of dissolved-solids concentration versus water-discharge rate, expressing their correlation.

Streams having the most consistent relationship between water discharge and dissolved-solids concentration ought to be streams that receive a large part of their mineral load at all discharge rates from a relatively constant source upstream from the measuring point, so that the runoff from other sources may be considered essentially a dilution of a component of constant flow rate and dissolved-solids concentration. No stream fits these specifications completely, and few streams which might approach them have been studied closely. In this discussion, some simple mathematical relationships that could be expected to exist between solute concentrations and water flow rate will be presented along with suggestions for their application, and some actual data will be examined for the existence of these or more complicated relationships.

An adaptation of the basic solute balance or inflow-outflow equation, which has already been given, is usually used in this kind of work. The total load of solute carried past a measuring point by a stream during a unit time period can be determined by multiplying the concentration of the solute, C, by the rate of water discharge, Q, and by a conversion factor that will convert the result into the desired units. This factor is a constant and for convenience in this discussion can be considered to be included in Q. The solute outflow of the drainage basin of the river upstream from the sampling point is measured by regular sampling and analysis or by a continuous-sensing unit, and discharge data are obtained by stream gaging at the same site. This solute outflow may be represented by the symbol $C_T Q_T$. It constitutes a summation of upstream solute inflows, including overland drainage, tributary inflows, rainfall on water surfaces, ground-water inflow, discharges from waste and outfalls, atmospheric fallout to the water surface, and other sources. These can be represented by as many terms ($C_1 Q_1$, $C_2 Q_2$, and so forth) as desired. Besides the inflows, it often is necessary to consider other factors that might influence composition at the sampling point, including solution or precipitation reactions involving transported solutes and water, sorption or desorption of ions on sediment, and various other processes. If the system is being evaluated over a specific time interval, these items may be considered as changes in storage of solutes within the basin above the sampling point and included as ΔW terms, representing total solute weight gained or lost. The equation would have the form

$$C_T Q_T = \Sigma C_1 Q_1 + C_2 Q_2 \ldots \pm \Delta W_1 \ldots$$

The simplest model for a particular solute in a reach of river is probably one which assumes that a constant total load of solute is entering upstream and that observed concentration of that solute at the sampling point varies owing to dilution by runoff. If other factors are insignificant, this condition can be evaluated from a simplified form of the solute balance equation:

$$C_1Q_1 + C_2Q_2 = (Q_1 + Q_2)C_3$$

where

Q_1 = volume of flow before dilution,
C_1 = concentration of solute before dilution,
Q_2 = volume of dilution water,
C_2 = concentration of solute in diluting water, and
C_3 = final concentration observed.

The three terms in this equation represent loads of solute, and an inflow-outflow balance is assumed (no change in storage).

If C_1Q_1 is constant, as assumed for this model, and if the concentration of solute in the diluting water is zero, then

$$\frac{C_1Q_1}{(Q_1+Q_2)} = C_3.$$

Q_1C_1 is constant and can also be represented as W_1, the total original solute load. This is the equation of a hyperbola. If expressed in logarithmic form, however,

$$\log C_3 = \log W_1 - \log(Q_1 + Q_2),$$

it has the form of a straight line, with slope -1.0. Thus, the degree to which a particular set of concentration and stream-discharge data fits the simple dilution model can be tested by plotting these variables on log-log paper.

Of course, most natural systems can display a simple dilution mechanism only over a limited range of concentration. Usually, the dilution water contains some of the same solute as the water being diluted, and this will cause the high-discharge end of the plot to approach a minimum near the concentration present in the diluting water. The solubility of the ion being considered, or other factors independent of discharge, also may limit its concentration at low discharge, so the slope of the curve will tend to flatten at very low discharge. It is obvious that complexities in most natural systems will tend to make the equation considerably more complicated. For example, if some of the solute being considered is actually present in the runoff, the value of C_2 will not be zero, and the final equation will have the form

$$C_3 = \frac{W_1 + C_2 Q_2}{Q_1 + Q_2}.$$

If expressed in logarithmic terms, this equation would give a curved line, as $C_2 Q_2$ is unlikely to be constant and probably will vary in response to discharge.

Addition of other possible factors as terms in the foregoing equations can make the mathematics much more intricate, but much can be learned about the characteristics of water-quality variations at any sampling station by studying with care the relationship of these variations to discharge.

A practical difficulty inherent in the interpretation of the log-log type of plot is that solute concentrations often show a rather narrow range of variation, and considerable clustering of points may occur. It is of interest, however, to use this approach to explore the relationships among discharge and solute concentration obtainable from existing records.

In past years, in studies by the U.S. Geological Survey, the only chemical determination made for each daily sample usually was specific conductance, and most of these data have never been published. Discharge information readily available for use with these determinations is the daily mean rate of flow. The instantaneous sample is often not a good representation of the daily mean concentration, and some concentrations thus may be poorly suited for establishing a meaningful discharge-concentration relationship. Hopefully, however, a well-defined relationship will not be completely obscured, although individual points may be scattered. Records collected for the specific purpose of exploring water-quality–water-discharge relationships must be carefully obtained with the aim of getting fully compatible data for both variables. This is perhaps most effectively accomplished by simultaneous continuous measurement of water-discharge rate and solute concentration.

A sampling station whose characteristics should be suitable for establishing a dissolved-solids versus discharge relationship was operated for a time by the U.S. Geological Survey on the San Francisco River at Clifton, in southeastern Arizona. About 60 percent of the annual dissolved-solids load of the river there was contributed by saline inflow from the Clifton Hot Springs, a short distance above the gaging station and about 1½ miles above the sampling section. The varying amounts of discharge from the direct runoff and dilute base flow of the river above the springs might be considered as dilution factors controlling the concentration observed at the sampling point.

In general, a plot of water discharge versus specific conductance for this station should be hyperbolic, with conductance approaching a limiting value at low discharge near the conductance of the spring inflow (near 16,000 micromhos per centimeter) and a limiting value at high discharge fixed by the conductance of rainwater plus the solutes added by the effect of rapid reactions of the runoff with exposed rock and soil minerals. The lower conductance limit cannot be very accurately predicted, but probably would not be much less than 200 micromhos.

Figure 38 is a logarithmic plot of specific conductance of daily samples versus daily mean discharge for the 1944 water year for the San Francisco River at Clifton, Ariz. Some daily values were omitted where the points are closely grouped. The scatter of points is substantial, and it would be difficult to fit them to a straight line. Furthermore, the slope of such a line is not -1.0. Both facts suggest that the simple dilution equation derived in the earlier discussion is not directly applicable.

From other information about the quality of water at this point, however, it is possible to derive a more pertinent relationship. The solute load at the sampling point may be considered to have three components, represented by $C_n Q_n$ terms, in which C represents conductance in micromhos and Q discharge of water in second-foot days. $C_1 Q_1$ represents the solute load brought in by spring inflow. This water has a specific conductance of 16,000 micromhos, and the flow is considered constant at 2 second-feet (Hem, 1950). $C_2 Q_2$ represents base solute load of the river above the springs that dilutes the spring inflow. For simplicity this water is assumed from available analyses of samples taken above the inflow zone to have a specific conductance of 500 micromhos and a maximum flow rate of 100 second-feet. $C_3 Q_3$ represents flood runoff, constituting all flow above 102 second-feet, and is assigned a specific conductance of 200 micromhos. The mixture of all these is represented by concentration C_F. The equation for predicting concentration at this station from water-discharge data becomes

$$C_F = \frac{C_1 Q_1 + C_2 Q_2 + C_3 Q_3}{Q_1 + Q_2 + Q_3}.$$

Substituting appropriate values from quantities postulated above gives the points plotted with the symbol "x" in figure 38. The dashed curve fitted to these points smoothes out some of the irregularities that would result if more calculated points were used, but demonstrates an approximate fit of the model to actually observed conditions. Slightly different postulates could perhaps improve the fit somewhat. The scatter of observed points is substantial, and, therefore, there

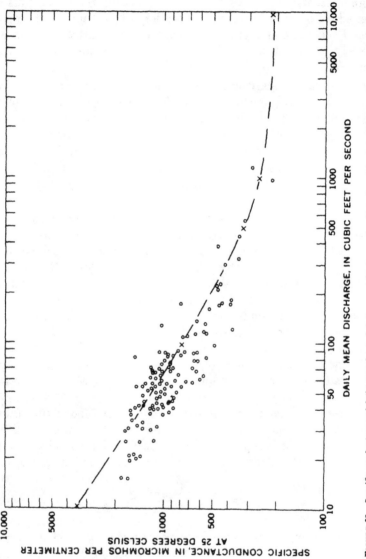

FIGURE 38.—Specific conductance of daily samples and daily-mean discharge, San Francisco River at Clifton, Ariz., October 1, 1943, to September 30, 1944.

would be substantial uncertainty in conductance values calculated from discharge by means of any curve fitted to the points. Some of the spread in points in figure 38 may have resulted from changes in rate of spring inflow, and some might have been caused by incomplete mixing of the inflow with the river water at the sampling point. Most of the scatter, however, is probably related to the failure of individual samples to represent the average composition of the river for a whole day.

The San Francisco River is subject to rapid changes in stage, especially during the summer when heavy local thunderstorms often occur in the afternoon and early evening. The resulting runoff events commonly occur at night, and the river stage was frequently near the low point for the day when samples were collected. Sampling was always done during the daytime, and the probability of a sample representing the day's average composition was not always good during the summer storm period.

Steele (1968a, b) studied the relationships of dissolved-ion concentration to water discharge of Pescadero Creek, a small northern California coastal stream. His data fitted the relation $CQ^N=K$, or $\log C = \log K - N \log Q$. The exponent N can be evaluated from the slope of the regression line and may represent several kinds of hydrologic effects. Ledbetter and Gloyna (1962), Smith, O'Brian, Lefeuvre, and Pogge (1967), and Gunnerson (1967) published papers relating to water-quality modeling. Other investigators who have published dissolved-solids rating curves include Durum (1953), who found a reasonably good correlation between chloride concentration and discharge for the Saline River near Russell, Kans., and Hendrickson and Krieger (1964), who found reasonably consistent relationships between specific conductance and discharge for the Salt River at Shepherdsville, Ky., for low-flow periods.

In streams where runoff patterns and solute-load sources are relatively complicated, a poorly defined discharge–dissolved-solids concentration relationship is to be expected. An interesting feature that has been observed in many streams, although closely studied in only very few places, is a tendency for the water of a rising stage to have a considerably higher dissolved-solids concentration than the water passing the sampling point at an equal flow rate after the peak discharge has passed. Hendrickson and Krieger (1964) presented a graph of conductance versus discharge which shows a counterclockwise loop that ideally would be characteristic of observations during a flood event. At first the conductance is nearly constant as discharge rises; then the conductance begins to decrease as the peak stage is reached, continues to decrease after the flow begins to recede, and then, finally, increases with falling stage until the starting point on the graph is again reached. Flood observations showing several looping cycles were presented by these investigators for the Salt River at Shepherdsville, Ky.

Toler (1965) observed a clockwise loop in his study of the relation of dissolved solids to discharge for Spring Creek in southwestern Georgia. The effect was attributed to an increasing contribution from ground water during falling stages that was made possible by the rapid circulation of water through the limestone of the drainage basin.

The factors that control the concentration of water early in a flood event obviously are different for different streams and sampling points. In general, however, the flood wave moving down the channel tends to push water already in the channel ahead of it, and after a time, a considerable volume of this more highly mineralized water may have accumulated in the wave front.

Daily sampling records of many years' duration are available for some streams. Most of these records have been obtained because of a need by water users for actual measurements of solute concentrations. The routine collection and publication of such records with annual averages of the composition, however, may not provide a very good basis for understanding the processes controlling water composition at the sampling site. Contrary to the intuitive supposition that the longer the record, the better the understanding of the system, it often can be demonstrated that a briefer, more intensive investigation of hydrology and chemistry of a stream can form a more useful model than one which can be derived from many years of routinely collected data.

Figure 39 is a log-log plot of discharge-weighted average dissolved-solids concentrations versus mean water discharge for each year of record for two daily sampling stations in New Mexico. The record for the Rio Grande at San Acacia has also been used for other purposes in this paper and covers a period in all of 17 years beginning with 1940. The record for the Pecos River near Artesia began in 1938, and points in figure 39 cover a period of 25 years.

Although the apparent scatter of the points on the log-log graph is not so great that a regression line cannot be drawn through them, there is no rational basis for deciding whether the line should be straight or curved. Nor would the graph give a very satisfactory way of predicting from the mean discharge what the annual average water composition might be.

The scatter of points for the Rio Grande is probably partly related to some of the characteristics of its drainage basin and flow pattern described earlier. The discharge of the river, however, is also extensively influenced by irrigation diversions, and thus many of the

ORGANIZATION AND STUDY OF WATER-ANALYSIS DATA 279

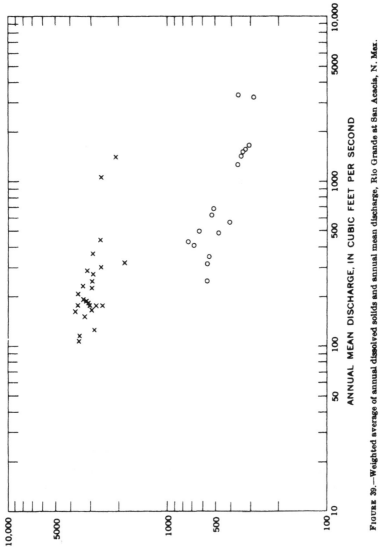

FIGURE 39.—Weighted average of annual dissolved solids and annual mean discharge, Rio Grande at San Acacia, N. Mex. (circles), and Pecos River near Artesia, N. Mex. (crosses).

water-discharge values used in the computation of annual averages are not closely related to water-quality factors. The scatter of points for the Pecos River station probably is also mostly the result of discharge regulation, but the hydrology of the system is further complicated by extensive use of ground water for irrigation upstream. It is interesting to note that the minimum average dissolved-solids concentration, which occurred in 1958, corresponds to a flow less than one-fourth as great as the flow in 1941, the year of highest runoff.

There will be a continuing need for long-term studies of streamwater composition for such purposes as surveillance of pollution, maintaining stream-quality standards, and observing long-term influence of changing land-use or water-use patterns. Some of these studies may require considerable detailed sampling and analysis. This type of work, however, is expensive in money and professional talent, and thus it is essential that the records obtained be significant and interpretable. A program involving the collection of large volumes of data tends to develop a momentum of its own and can degenerate into a repetitive exercise of routine operations. Data must be carefully examined and studied in a logical framework, mathematically designed if possible.

In further emphasis of the need for logical interpretation, it is desirable to use methods of graphing data that can be shown as likely to fit the mathematics of the relationships that are possible. The log-log plot tends to deemphasize the scatter of points that may appear when one or both axes are cartesian, but it fits the mathematcal parameters that can be expected for concentration versus discharge relationships. The use of semilog plots for this relationship should be avoided. The writer was guilty of using semilog plots of concentration versus discharge in the earlier edition of this book, and there are many other examples in the literature.

Natural processes in which time is satisfactorily considered as the independent variable may sometimes be well represented by semilog plotting with time being plotted on the linear axis. Chemical-reaction rates and solute-transport rates may involve this kind of variation.

WATER-QUALITY MAPS

A useful procedure in the study of water-quality data is to enter the information on a map of the area under investigation. A map of this kind is most likely to be useful in the study of underground water in single, widespread aquifers, but mapping also may have some value in surface-water studies. If a systematic areal distribution of water-quality features is observed, correlations with other characteristics of the ground-water system usually can be made. If the map is started early in the investigation of an area, and information added as it is obtained, the areas needing closer field study often can be identified.

Usually, a map of the quality of ground water is prepared by entering numbers or symbols at well and spring locations to represent concentrations of constituents, and the areal distribution of solutes thus can be observed in a general way. If many of the wells are open in more than one aquifer or the water of the aquifers tends to vary in composition with distance below the land surface, these variations may well obscure any lateral changes in water quality in a single aquifer. If the sample sites represent only a few of the existing wells and springs, extrapolation of data between sampling sites may be unwise. If the investigator has many analyses, however, and can reasonably assume they represent a water body that is close to being uniform in composition through a vertical section at all well locations, the results may be best expressed in the form of an isogram map. This type of map extrapolates data between sampling points and often gives useful hydrologic and geologic clues. If may help show areas of recharge and discharge, areas of leakage from other aquifers, and directions of water movement.

MAP SYMBOLS

One type of water-quality map is prepared by entering a symbol at each sampling point to represent the quality observed there. The symbol can be a bar graph, a pattern diagram, or perhaps a distinctive color traced along a stream. Figures 40 and 41 are two types of symbol maps.

Figure 40, published in a report by Sever (1965) on ground water in part of southwestern Georgia, shows iron content of wells in a single aquifer and the tendency for the higher iron concentrations to occur in one area, under the influence of geologic and topographic variations. Figure 42 shows the composition of water from wells at Minot, N. Dak., by "pie" diagrams. According to Pettyjohn (1967) the similarities and differences in composition in areas A-D on the map can be correlated with sources of recharge and the nature of aquifer materials.

Water-quality maps have appeared in the U.S. Geological Survey Hydrologic Atlas series, for example, maps showing stream-water quality in the western United States by Feth (1965), saline ground-water resources of the United States by Feth and others (1965), and river-water quality by Rainwater (1962) for the United States. Some European hydrologists also have published water-quality maps (Langguth, 1966).

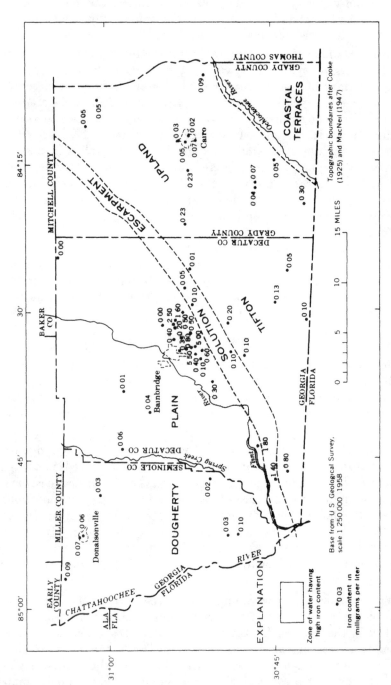

FIGURE 40.—Iron content of ground water from the principal artesian aquifer and topographic regions, southwestern Georgia.

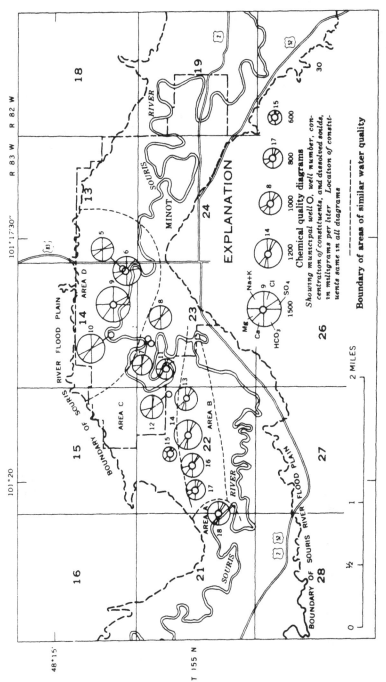

FIGURE 41.—Chemical quality of water and dissolved solids in the Minot aquifer, North Dakota.

ISOGRAM MAPS

The technique of mapping ground-water quality characteristics by drawing lines of equal concentration of dissolved solids or of single ions has been used in reports published by various investigators for more than 50 years. A related procedure identifies concentration ranges by distinctive colors or by shading.

The applicability of this technique depends mainly on two factors: the homogeniety of water composition in the vertical direction at any given point and the spacing between sampling points. An isogram map is particularly useful in studies of ground water in widespread single-aquifer systems. In alluvial fill of the basins within the Basin and Range physiographic province of the western United States, water-bearing zones tend to be lenticular, but in many places are closely enough interconnected to constitute single hydrologic systems over large areas. Isogram maps have been used successfully in studying several areas of this type.

Figure 42 consists of two isogram maps selected from a series of similar maps in a report by Moore (1965). These maps show the iron content and pH of water from the "500-foot" sand of the Claiborne Group in the western part of Tennessee. The outcrop area of the sand, where recharge occurs, is also shown. Iron and pH both increase as water moves away from the area of outcrop. A similar increase in hardness of the water occurs downdip, as the recharge attacks minerals in the aquifer.

In a study of the Phoenix, Ariz., area McDonald, Walcott, and Hem (1947) prepared a map showing ground-water quality over a large area. Subsequent work has shown considerably more variation of composition with depth than had been supposed in 1947, and the importance of this effect seems to have been increased by declining ground-water levels; nevertheless, isogram maps have been useful in water-resources studies of that region. A more recent map of part of this area uses both isograms and symbols to show ground-water composition (Kister and Hardt, 1966). An isogram map of ground-water quality in the Douglas basin, Arizona, was published by Coates and Cushman (1955), and other reports using this device are numerous.

Detailed hydrologic studies in the lower part of Safford Valley, Ariz., were described by Gatewood, Robinson, Colby, Hem, and Halpenny (1950). Quality of ground water in the alluvial fill of Holocene geologic age, which lies adjacent to the Gila River and constitutes a fairly well defined hydrologic unit, is shown by a map in that report, part of which is reproduced here (pl. 2). During the investigation, a large number of shallow observation wells was installed in the river-bottom land, and the concentration of dissolved solids in

ORGANIZATION AND STUDY OF WATER-ANALYSIS DATA

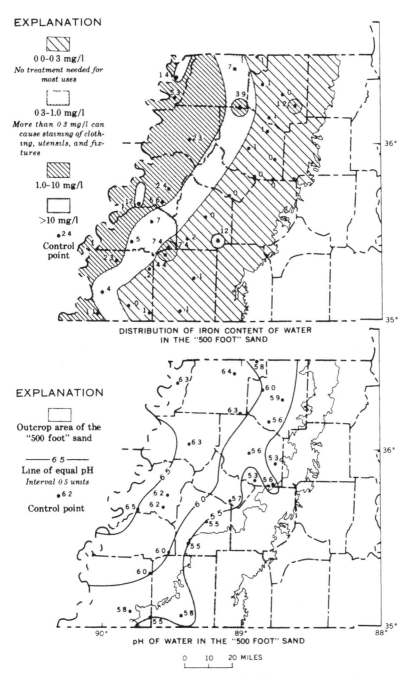

FIGURE 42.—Distribution of iron and pH in water from the "500-foot" sand, western Tennessee.

ground water of the area can be extrapolated with reasonable confidence between wells. Figure 43 shows there is saline ground water throughout the river-bottom area near Fort Thomas. Farther downstream in the vicinity of Geronimo, the saline water appears to have been entirely replaced by much more dilute water. This dilute inflow to the system comes from underflow of Goodwin Wash and other ephemeral tributaries that drain a large area to the south. The effect of this underflow on water levels in the area also is substantial, and maps of the water-table elevation used along with water-quality maps afford better means for gaining a good understanding of hydrologic conditions in the area than either type of map alone.

WATER-QUALITY PROFILES

A diagram showing differences in water quality along a cross section through a stream or through a surface of underground water body may be a useful way of imparting information, and for the purposes of this discussion such diagrams are called water-quality profiles. The differences of water composition across the Susquehanna River at Harrisburg, Pa., are shown in figure 2, and the lack of mixing in the river there has already been noted and explained. Another kind of profile could be a diagram showing river-water composition over a distance along the channel. McCarren (1967) used a diagram covering a long distance on the Allegheny River to show changes in composition from sampling station to sampling station. Profile diagrams may be utilized in studies of estuaries where water composition tends to change frequently, in response to streamflow and tidal effects.

Stratification in lakes and storage reservoirs caused by temperature differences may bring about depletion of dissolved oxygen and can affect many other constituents. Reservoir-profile diagrams are commonly used to show some of these effects. A less common type of reservoir stratification caused by differences in salinity of inflow in Lake Whitney and Possum Kingdom Reservoir on the Brazos River in Texas was shown in a profile diagram in a report by Irelan and Mendieta (1964). Figure 43 shows dissolved-solids concentration and temperature of water observed by Howard (1960) at several depths at two locations in Lake Mead, the reservoir behind Hoover Dam on the Colorado River. A relatively warm layer of dilute water was present near the surface at both sites during the August observation.

Profile diagrams are not commonly used for ground-water reservoirs. Where a considerable difference in water composition occurs with depth, however, some sort of three-dimensional diagram to represent water-quality conditions is useful. The "fence" diagram often used by geologists can be adapted to this purpose. Diagrams of this type were used by Back (1960) to show distribution of anions and cations in

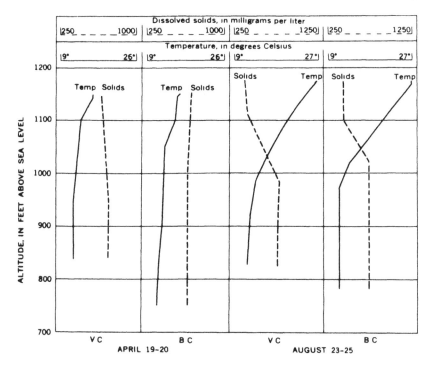

FIGURE 43.—Temperature and dissolved solids of water in Lake Mead in Virgin and Boulder Canyons, 1948. Virgin Canyon=V C , Boulder Canyon=B.C.

ground water of the Atlantic Coastal Plain in the eastern United States.

RELATIONSHIP OF WATER QUALITY TO LITHOLOGY

The extent to which the composition of natural water can be related to the lithology of the area in which it occurs has been studied by many investigators. The analyses selected for use in this book give some indications of what relationships can be expected under the best conditions.

From consideration of the processes described in this paper, it seems reasonable to conclude that the constituents of natural water most likely to be derived directly from solution of minerals in rocks and in soil are dissolved silica and the cations. Although cation-anion balance must be maintained in all solutions, the anions may be derived in large part from nonlithologic sources. For example, the bicarbonate present in most water is derived mostly from carbon dioxide that has been extracted from the air and liberated in the soil through biochemical activity. Some rocks serve as sources of chloride and sulfate through direct solution. The circulation of sulfur, however, may be greatly influenced by biologically mediated oxidation and reduction, and atmospheric chloride circulation may be a significant factor influencing the anion content in many natural waters. The anions in rainfall are balanced by cations, partly H^+ but also including other cations. Consequently, a part of the cation content of natural water may be derived from nonlithologic sources. The importance of this effect on cation concentrations is usually rather small.

Many rock minerals may be viewed as hydrogen-ion acceptors, as indicated in the reactions shown previously. Hydrogen ions that participate in weathering reactions are produced in respiration and decay of organic matter and in the oxidation of sulfides. Ideally, half the bicarbonate dissolved in water associated with carbonate rock came from the rock, and the remainder from soil air, or other similar carbon dioxide sources. In noncarbonate rock environments, the hydrogen ions introduced through solution of carbon dioxide react with rock minerals, but are mostly retained in solid residual species formed in the reaction, such as silica or clay minerals. In a sense, the hydrogen ions replace cations in the rock minerals, and the cations released into weathering solutions are balanced in solution by equivalent quantities of bicarbonate. Some of the H^+ and OH^- in the reaction is also derived from hydrolysis. An excellent summary of weathering reactions has been published by Keller (1957), who pointed out the significance of hydrolysis in these reactions.

A general relationship between mineral composition of a natural water and that of the solid minerals with which the water has been in contact is certainly to be expected. This relationship may be comparatively simple and uncomplicated, as in the case of an aquifer receiving direct recharge by rainfall and from which water is discharged without contacting any other aquifer or other water. Or the situation may be rendered very complex by influence of one or more interconnected aquifers of different composition, mixing of unlike waters, chemical reactions such as cation exchange, adsorption of dissolved ions, and biological influences. Processes involved in soil formation and the soil composition of the area may have a considerable influence on composition of both surface and underground water.

Most rocks are complex mixtures of minerals which differ widely in their stability toward, or solubility in, water. The bulk of all rocks except the evaporites, however, is made up of minerals that are not readily soluble. The classification to which a particular rock is assigned is decided on the basis of the bulk composition. This may not adequately show the presence of the more soluble components that may exert an influence on the composition of water circulating through the rock. In some kinds of rock, relatively minor components may control major features of the composition of circulating ground water. A carbonate-cemented sandstone, for example, might be largely composed of silica in the form of quartz, but yield a water containing mostly calcium and bicarbonate ions. Resistate and hydrolyzate rocks may contain remnants of highly mineralized connate water, which may strongly affect the quality of ground water or of surface water associated with such rocks.

The rock and water and air system obviously imposes difficulties in applying a strict chemical-equilibrium model. Some rock species, such as the carbonates, do lend themselves to this approach. Others are better considered as influenced by reaction rates and irreversible mechanisms. Garrels (1967) described some of the effects that might be anticipated from solution of igneous-rock minerals. If viewed as irreversible reactions, one might expect similar stoichiometric relationships among silica and cations in solution for certain solution processes for considerable periods of time. Upper limits on most of the solute species would represent saturation with respect to some new solid phase, and the stoichiometry thus might be altered after long contact time of the solution with the rock minerals.

In spite of the acknowledged difficulties, many investigators have wished for a classification system for waters based on the chemical composition of the rocks from which the waters have dissolved their load of mineral matter. Regarding such a classification, Clarke (1924b, p. 8), in a paper on the composition of river and lake water, stated

"A classification of waters according to their origin is something quite different * * * its purpose is geologic and although no complete scheme for it has yet been developed, the analyses as arranged in this memoir give some suggestions as to what may be possible."

Clarke goes on to mention limestone and dolomite waters and siliceous water which he states "might be termed granitic or feldspathic, at least until a better name can be found."

In his "Data of Geochemistry," Clarke cited a considerable number of studies, made mostly in central and western Europe, in which the writers had studied river-water composition and correlated it with drainage-basin geology. Some of these studies were made as early as the middle of the 19th century; hence interest in this subject cannot be said to be a recent development.

Investigators since Clarke's time have not been able to develop any simple classification system to relate water to rock types on the basis of the dissolved matter in the water. Indeed, analyses that seem much alike are sometimes ascribed in the literature to entirely different kinds of rock, and obviously the water obtained from a granite exposed in a tropical rain forest can be expected to be very different from the water obtained from a similar rock in Scandinavia or in northern

Canada. If extreme conditions are left out, however, some systematic evaluation is possible. The analyses presented here to show general trends represent conditions occurring within the United States. The analyses were selected from tables given earlier in this report, and part of the basis for their selection was that their composition clearly showed the properties one might expect on a theoretical basis. A more exacting type of study, in which the water and associated rock species would both be analyzed, has been made for granitic terranes in the California Sierra Nevada by Feth, Roberson, and Polzer (1964) and for granitic and other terranes in a mountainous area near Santa Fe, N. Mex., by Miller (1961). The general findings of these investigators and of Garrels (1967), who studied published data from several sources, are in accord with the statements made about water from igneous terranes in the first edition of this paper, and the discussion is, therefore, repeated here with only minor changes. The analyses shown probably represent somewhat idealized situations, but they may aid the reader to cultivate an understanding of the composition of water that might be expected from different rock types.

IGNEOUS ROCKS

Rocks of igneous origin may be classified as extrusive or intrusive. The extrusive rocks include those ejected at the land surface as lava flows and the pyroclastic deposits resulting from explosive volcanic activity. The latter include such deposits as volcanic ash and cinders, tuff, and breccia. Some of the extrusive deposits are permeable enough to permit rapid and extensive circulation of water and can be good aquifers. The intrusive deposits include batholiths, stocks, dikes, sills, and other forms that are intruded below the surface of the earth and usually are dense and nearly impermeable to water except along fractures. Both the extrusive and intrusive rocks are further classified by geologists on the basis of chemical and mineral composition, texture, and other characteristics. The chemical and mineral composition is of principal importance in the relationships to be discussed here, and classifications based on other characteristics will not be considered. Extrusive rocks and intrusive rocks of the same chemical and mineral composition have different names, but tend to yield similar weathering products to water. As a class, the extrusive rocks react more readily than the intrusive rocks for two reasons: (1) the extrusive rocks generally expose more surface area for reaction and (2) many are partly glassy—that is, not crystalline.

Igneous rocks consist predominantly of silicates, although in glassy species the crystal organization may be poorly developed. The classification scheme earlier presented for silicate minerals aids somewhat in understanding the bases of the classification of igneous rocks. If oxygen is relatively abundant compared to silicon, minerals having inosilicate structures can be formed as the molten mass crystallizes. If oxygen is relatively depleted, the structures of minerals will more likely have the tectosilicate pattern. The intermediate classes of silicate structures will be formed in systems lying between the extremes. The crystallization process is complex, however, and products formed depend on availability of other elements and on temperature and pressure. The subject cannot be considered here, but is discussed more extensively in geochemical textbooks. (See Krauskopf, 1967, p. 355–452.)

Generally, rock analyses are expressed in terms of oxides. These can be recalculated to a statement that shows the proportions of the principal mineral species in the rock. Analyses in this form are the basis for the classification used here.

Rocks of igneous origin compose a wide range of mineral composition, for rock masses are generally mixtures of many different mineral species. Although it is not possible to design a simple classification scheme for these rocks, a general basis for the commonly used names for different rock types can be shown readily in terms of the major mineral components. Readers who are unfamiliar with petrology may thus gain some understanding of the meaning of such terms as granite or basalt and of the possible significance of the rock composition thus indicated in predicting the chemical composition of solutions of weathering products. The classification scheme used here is in general that of Peterson (1961), who considered it to represent a general concensus of authorities on the subject. All rocks form a continuous series having a very wide range of composition, and any assignment of classification and name must be entirely arbitrary.

The proportion of pure silica, generally in the form of quartz, which is present forms a convenient base from which to begin a rock classification scheme: (1) rocks rich in quartz ($SiO_2 > 10\%$), (2) rocks impoverished in quartz and in feldspathoids (both $<10\%$), and (3) rocks rich in feldspathoids and impoverished in quartz. Feldspathoids are a class of minerals chemically similar to feldspar, but somewhat impoverished in silicon. Rock masses which belong to the third class above are not common and will not be considered further here.

The second property used here in classifying igneous rocks is derived from the amount and type of feldspar present. Thus, rocks rich in quartz and having a large proportion of feldspar of which more than two-thirds is of the potassium or sodium type would be called granite if intrusive in origin and rhyolite if extrusive. A rock rich in quartz and feldspar in which plagioclase predominates and the proportion of albite to anorthite is greater than 1:1, would be called quartz diorite if intrusive and dacite if extrusive. Both these two kinds of rock should contain less calcium than sodium and potassium.

Rocks relatively impoverished in quartz but rich in feldspar, more than two-thirds of which is of the sodium or potassium type, are called syenite if intrusive and trachyte if extrusive. Rocks impoverished in quartz but rich in feldspar in which plagioclase predomintaes and the proportion of albite to anorthite is greater than 1:1, are called diorite if intrusive and andesite if extrusive. Rocks relatively impoverished in quartz and containing more than 10 percent feldspar, mostly plagioclase in which anorthite predominates, are called gabbro or diabase if intrusive and basalt if extrusive.

Rocks in the latter class generally contain considerable amounts of ferromagnesian minerals such as hornblende, amphibole, and olivine. In the peridotite rocks, quartz and feldspar are essentially absent, and the ferromagnesian species of minerals predominate. Rocks of this type are commonly called ultrabasic.

Further information concerning classification of rocks can be found in standard texts on petrology and geochemistry. In practice, the application of any system requires careful examination of the rocks, and the classification given to a particular rock based only on examination in the field may not always be as indicative of its composition as the foregoing scheme might suggest.

In considering igneous rocks in their relation to water composition, the texture and structure of the rocks is significant as it relates to the surface area of solid rock that may be exposed to attack. Ground water may be recovered in large amounts from some of the extrusive igneous rocks where there are shrinkage cracks and other joints, interflow zones, or other openings through which water may move. Most igneous rocks, however, are relatively impermeable. Surface water originating in areas where igneous rocks are exposed is low in dissolved solids because in general the weathering attack on igneous rocks is slow. Concentrations reached are likely to be a function of contact time and area of solid surface exposed to water. Where vigorous soil-forming processes and plant growth are occurring, an enhanced supply of carbon dioxide and hence hydrogen ions becomes available to circulating water. The amount of attack that occurs can be considered a function of the availability of H^+, as well as reaction time.

Surface water in areas of igneous rocks may display the effects of rock solution less distinctly than underground waters that have better opportunity to participate in reactions with the rock minerals. A considerable and usually uncertain fraction of the solute load in a very dilute river or lake water may be related to solutes in rainfall or to dust and other atmospheric fallout transported from other localities. The examples later cited in this section are all underground waters.

Detritus may be derived from igneous rocks by erosion processes that are largely mechanical, and subsequent circulation of water through detritus of this type can give rise to an assemblage of solutes very similar to that to be expected from the unaltered rock. The detrital material, however, has greatly increased surface area.

From the above generalizations, a very much simplified view of the process of attack by water may be expressed in terms of chemical equations. The reaction of carbon dioxide and water supplies hydrogen ions;

$$CO_2(aq) + H_2O = HCO_3^- + H^+.$$

Hydrogen ions aid in the attack on feldspars that causes them to be changed to clay minerals, here represented by kaolinite, and silica, and cations are released;

$$2NaAlSi_3O_8 + 9H_2O + 2H^+ = Al_2Si_2O_5(OH)_4 + 2Na^+ + 4H_4SiO_4$$

and

$$CaAl_2Si_2O_8 + H_2O + 2H^+ = Al_2Si_2O_5(OH)_4 + Ca^{+2}.$$

Attack on ferromagnesian species may be represented by the decomposition of forsterite to form antigorite, silica, and magnesium ions;

$$5Mg_2SiO_4 + 8H^+ + 2H_2O = Mg_6(OH)_8Si_4O_{10} + 4Mg^{+2} + H_4SiO_4.$$

A good many other similar reactions have been written for other mineral species, but for this discussion it is not necessary to go into further detail. The direct solution of quartz to give H_4SiO_4 is a possible source of silica in water also, but this is a slow reaction and is generally a less significant source than silicate decomposition.

All these reactions involving dissolution of silicates probably are more complicated than the equations imply, but they can serve as a basis for some useful generalizations. The reactions as written are not reversible, but the rate at which the reaction proceeds is speeded if H^+ activity increases and will tend to be retarded by the attainment of high activities of the dissolved products. The rate of reaction also is a function of the area of surface of solids exposed per unit volume of solution and rates of ion transport away from the solid surfaces by water movement through the reaction sites. The reaction rate also is increased by increased temperature.

Very little is specifically known about rates or mechanisms of any of these reactions. The stoichiometric relationships, however, and other considerations displayed in the simple equations can be used to draw certain inferences. For example, it would appear that sodium feldspar produces more silica than the other reactants considered, and water influenced by disintegration of this mineral to form kaolinite should have a high silica concentration. As a result of this reaction, the molarity of silica released to the water would be about double that of the sodium.

In figure 44, analyses of several waters known to be associated with igneous rocks of different kinds are shown graphically. Analysis 12-2 represents water from a spring issuing from rhyolitic terrane; the effect of dissolution of sodium-rich silicate minerals typical of this rock type is demonstrated both by the high silica and by the high molar ratios of silica to sodium and to bicarbonate, although the ratio calculated for albite dissolution is not reached. The water represented by this analysis issues from the ground at a temperature of 38°C. The elevated temperature increases the solubility of silica and also probably causes a more rapid attack on the rhyolitic glass and feldspars and a higher concentration of sodium than might otherwise be observed.

Analysis 16-3 represents water from a spring issuing from an ultrabasic rock. The molar ratio of magnesium to silica in this water is 3.3, which is a fairly close approach to the theoretical value of 4 suggested by the equation for dissolution of olivine.

In general, the production of silica per hydrogen ion used up is higher for sodium-rich feldspar than for the other rock minerals considered. Thus one might expect higher silica concentrations in relation to total ion content for water from granite or rhyolite than for water associated with basalt or ultrabasic rock where sodium-rich feldspar would be rare or absent. Ideally, the ratio of calcium to sodium in water should be related to the composition of plagioclase feldspar present in associated rocks. Also, the proportion of magnesium to the other cations in the water could be an index to the relative abundance of ferromagnesian rock minerals in the environment.

Garrels (1967) made a number of generalizations, like the ones above, that he believed useful in considering broadly the geochemistry of natural water. It is evident, of course, that this is only a starting point in considering the highly complex systems that usually are found in nature.

Analyses 12-4 and 12-5 represented in figure 44 are for water associated with basalt transitional between the rhyolite of 12-1 and the ultrabasic rock of 16-3. The decreasing relative importance of sodium and the decrease in $SiO_2:HCO_3$ ratio is evident in the change in the rock type to a form less likely to contain alkali feldspar; however, the relationship is obscured to some extent by the differences in total ion content among the different waters.

SEDIMENTARY ROCKS

Several systems exist for classification of sedimentary rocks. These systems are based on chemical composition, grain size, and general origin. For the purposes of this discussion, the classification already

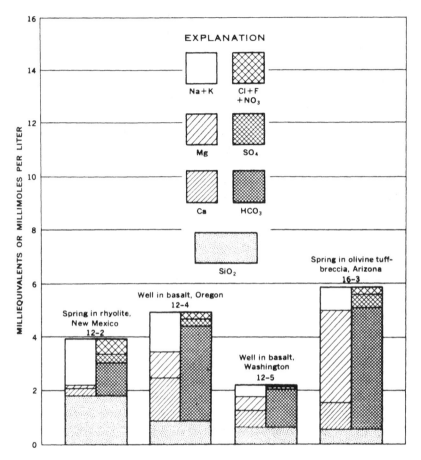

FIGURE 44.—Composition of ground water obtained from igneous rocks.

introduced for sedimentary rocks in the section "Composition of the earth's crust" is continued. Resistates include sandstone and other species made up of relatively unaltered fragments of other rocks. Hydrolyzates are fine-grained species made up, in large part, of clay minerals. Precipitates are, for the most part, carbonate rocks. Evaporites consist of relatively soluble minerals, such as gypsum and halite, that were deposited from water that became very concentrated owing to evaporation. This classification scheme relates mainly to chemical composition.

RESISTATES

The sedimentary rocks classed as resistates have a considerable range of chemical composition. An ideal example is a sandstone composed of quartz grains and other highly resistant mineral fragments such as garnet, tourmaline, and zircon. The conditions under which the parent rock was weathered and the detritus transported, however, may be almost as important in determining the composition of the detrital particles as the resistance of the minerals themselves. In many resistate sediments, therefore, chemical attack may not have been strong enough to break down feldspars or even some of the less resistant minerals. Some sands and gravels are made up principally of comminuted fragments of the original rock that have undergone little chemical change. These fragments will be subject to later attack and solution by water in the same way as the original rock, except perhaps at an accelerated rate owing to the increased surface area.

The consolidated resistate sediments, such as sandstone, contain cementing material deposited on the grain surfaces and within the openings between grains. This cementing material is usually deposited from water that has passed through the rock at some past time and can be redissolved. The most common cementing materials include calcium carbonate, silica, and ferric hydroxide or ferrous carbonate, with admixtures of other materials such as clay minerals.

Many resistate sedimentary rocks are relatively permeable to water and may, therefore, easily receive and transmit solutes acquired by water from some other type of rock. In the course of moving through the sedimentary formations, several kinds of alteration processes may occur that may influence the composition of the transmitted water. One of these processes has already been mentioned—the precipitation of cementing material. This process is commonly a reversible one, but it may be more complicated than simple bulk precipitation and re-solution of a solid. For example, the electrostatic field in a solution very near a mineral surface probably is different from that prevailing farther away from the surface, and precipitation of solids may be catalyzed by this effect. The precipitation of a coating on the grain surfaces of a detrital rock may alter the composition of water which has reached this rock from some other formation. As this "foreign" water passes through its new environment, the coating of grain surfaces will progress until all the active surfaces in the formation are covered, a process that may require a long time and a large volume of water.

Processes that might be involved in precipitation on active surfaces are not well known and will not be further discussed here. The inclusion of minor metals in such precipitates is an interesting possibility. The coprecipitation of certain metals with calcite has been studied by Alekin and Moricheva (1960). The possibility that coprecipitation might occur with some other kinds of cementing material seems at least as great. Thus, the concentrations of minor elements in water from resistate rocks may be rather closely related to processes of deposition or re-solution of cementing materials.

Other factors that may control concentrations of major, as well as minor, constituents of water circulating through resistate sediments are adsorption and ion exchange. Most mineral surfaces have some cation-exchange capacity, and some minerals have anion-exchange capacity. The processes that are involved have been described earlier. As some ion always will be present at an exchange site, the reactions are replacements whereby ions from solution are adsorbed and sorbed ions are released, in a manner generally predictable from equilibrium theory. The manner in which exchange reactions occur in aquifers, however, is in need of much more study than it has yet received.

A marine sandstone would be expected to have sodium ions adsorbed on exchange sites while the rock is saturated with sea water. Fresh water entering the sandstone at some later period can be expected to have its cation composition altered by ion-exchange reactions. Commonly, this takes the form of removal of most of the divalent cations from the entering water and their replacement by sodium present in exchange positions on the mineral surfaces. This natural softening effect has been observed in many aquifers. One would expect that an aquifer having this property would show general behavior like that of an ion-exchange column or water softener and have a characteristic break-through point at which the exchange capacity became exhausted. There is little published information on the exchange behavior of natural systems which is complete enough to demonstrate this effect.

The adsorptive capacity of aquifers for minor elements and the effect of changing composition of the water moving through the system on possible movement or release of the minor elements has been studied to some extent (Robinson, 1962) in connection with proposals for disposal of radioactive waste; however, the bulk ion-exchange capacity of aquifer material is seldom ascertained and most features of the subject are still in need of further study.

The most likely anion-exchange process in natural systems might seem to be replacement of adsorbed OH^-, and there are sites on kaolinite particles where OH^- is available (Halevy, 1964). These hydroxide ions could perhaps be replaced by fluoride ions because these have the same charge and approximate shape, and Bower and Hatcher (1967) studied this process in soils. Some ground waters that are high in pH show high concentrations of fluoride, possibly owing to this effect (analysis 1, table 18). Other anions generally are larger than fluoride and seem less likely to be involved in ion-exchange equilibria, although Yamabe (1958) has considered the reaction for carbonate ions.

The participation of clay minerals in ion-exchange reactions has been indicated in some of the foregoing discussions. These minerals have high exchange capacity and may be present in considerable amounts in some resistate sediments, although the clay minerals are more characteristically found in hydrolyzates. Exchange reactions also are important, of course, in water in hydrolyzate sediments.

Many resistate sediments may remain under relatively reducing conditions for long periods of time and may contain some reduced mineral species, such as pyrite. The solutions moving through the sedimentary rock at later periods in its history may participate in oxidation or reduction reactions. Many such reactions may be biologically mediated—that is, they may be intimately involved with life processes of microorganisms. Oxidation reactions can be expected in the part of the rock lying above the water table and to some extent at greater depth, depending on availability of oxygen. Species stable in reduced environments, such as pyrite, will be altered by oxidation, with a release of energy. The sulfur of pyrite, for example, will be oxidized to sulfate, and ferrous iron released. Reduction processes that can occur where oxygen is depleted may convert sulfate to sulfur or sulfide, and both oxidation and reduction reactions involving nitrogen may be significant. Any of these reactions for which an outside source of energy is needed either require some other material that bacteria can use for food or must occur where biota can obtain radiant energy from sunlight.

Figure 45 describes graphically the analyses of four waters from resistate sediments. Analysis 12–7 represents water from the Santa Fe Formation that underlies the Rio Grande valley at Albuquerque, N. Mex. The high silica content and other properties suggest the water has attacked particles of igneous rock included in this formation. The ground water associated with a pure quartz sand can be among the lowest in dissolved-solids content of any ground water.

Analysis 14–8 is for a much more highly mineralized water from the Dakota Sandstone in North Dakota, where the concentration of sodium, chloride, and sulfate is high. These ions can best be explained as having migrated into the sandstone from other rock formations associated with the sandstone. The sluggish circulation of water in the sandstone retards flushing of solutes.

Analysis 15–6 represents water from valley fill at Phoenix, Ariz., that reflects the influence of irrigation-drainage water. The proportions of cations in this water probably have been altered by ion-exchange reactions in the overlying soil.

Analysis 15–9 represents water from a well in the Dakota Sandstone in Kansas and is characterized by predominance of calcium and bicarbonate. The pH of the water, along with the concentrations of

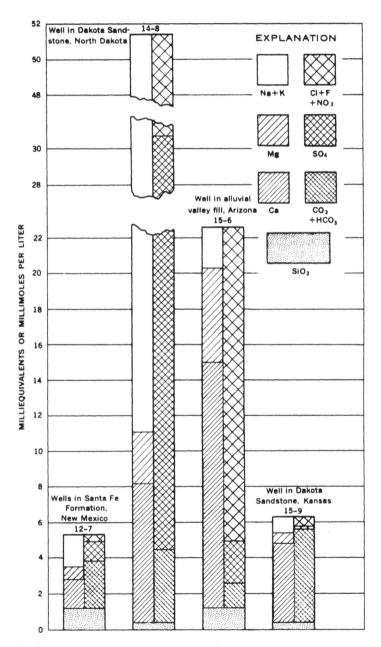

FIGURE 45.—Composition of ground water obtained from resistate sedimentary rock types.

calcium and bicarbonate, indicates that the water is a little above saturation with respect to calcite. Very likely, the calcium and bicarbonate in the water are related to cementing material in the sandstone.

Some rock formations may yield notable concentrations of unusual constituents to ground water and thus provide a natural tracer by which the lithologic source of the water can be recognized. One of the more outstanding examples of such an effect is the occurrence of strontium in water from certain sandstones and other rock formations in Wisconsin; as noted by Nichols and McNall (1957). An increased amount of information has been made available in recent years on minor-element contents of water. Whether other minor-element contents will be useful in correlating water with source rock is not yet known.

Because of the many ways in which their composition may be affected, the water from resistate rocks has a wide range of chemical quality. Although water from a particular aquifer may have distinctive properties, these must generally be determined by actual sampling and analysis and can seldom be definitely predicted in advance.

HYDROLYZATES

Shale and other fine-grained sedimentary rocks are composed, in large part, of clay minerals and other particulate matter that has been formed by chemical reactions between water and silicates. Almost all such rocks also contain finely divided quartz and other minerals characteristic of resistates, but such minerals are present as smaller particles. Shale and similar rocks are porous but do not transmit water readily because openings are very small and poorly connected. In many instances the original deposition of the rock occurred in salt water, and some of the solutes may remain in the pore spaces and attached to the particles for very long periods after the rock has been formed. As a result, the water that can be obtained from a hydrolyzate rock may contain rather high concentrations of dissolved solids. Because environmental conditions commonly change from time to time during deposition, many hydrolyzate sediments contain layers of coarser material, and water may be transmitted in those layers at a rate sufficient to supply a well or spring. Even where the material is too fine grained to transmit water to wells in useful amounts, the very wide lateral exposure of such formations to more porous overlying, and underlying, layers may result in a significant migration of water and solutes from the hydrolyzates into the aquifers with which they are interbedded. These effects can be noted in many places and can become important when hydrostatic pressures in the zone of saturation are extensively changed by withdrawals of ground water. The effect of shale beds containing solutes on quality of surface runoff is worth noting also.

Davis (1961), in a study of surface-water composition, noted that the runoff from areas underlain by certain geologic formations in California was characterized by low ratios of bicarbonate to sulfate. He correlated this fact with the presence of sulfide minerals in the sedimentary formations that had been laid down in reducing environments. Weathering caused the oxidation of the sulfide to sulfate. Reduced iron minerals, notable pyrite, frequently may be associated with hydrolyzates.

Analyses in figure 46 represent four waters associated with hydrolyzate sediments. They share one dominant characteristic in that sodium is their principal cation. This is not necessarily true of all water associated with hydrolyzates, however, as some may contain large amounts of calcium and magnesium. Clay minerals can have high cation-exchange capacities and may exert a considerable influence on the proportionate contents of the different cations in water associated with them. Direct solution of the hydrolyzate minerals themselves, however, is a less significant factor except in environments where water circulation is rapid (high rainfall) or where a low pH in the circulating water is maintained.

Analysis 14-3 represents water from the Chattanooga Shale and has a low dissolved-solids content because of the abundant supply of water from rainfall. The rather high silica concentration in this water is somewhat unusual for water from hydrolyzates and may be the result of solution of unaltered silicate minerals. Analysis 17-2 represents water from sandstone and shale in the Fort Union Formation. This analysis shows the effect of both cation-exchange softening and sulfate reduction. The former process has increased the sodium concentration at the expense of the calcium and magnesium. The latter process has caused the bicarbonate to increase partly at the expense of the sulfate.

The composition of water at very low flow in the Moreau River at Bixby, S. Dak., is shown by analysis 17-4. The drainage basin of this stream contains a high proportion of hydrolyzate rocks. Analysis 17-9 represents water from the Chinle Formation, which is sandy enough in some places to yield small amounts of water.

The effect of hydrolyzates containing readily soluble material can be observed in floodwaters of some streams, for example, in the basin of the Rio Puerco, which is tributary to the Rio Grande above San Acacia, N. Mex. Floodwaters from this tributary are heavily laden with suspended sediment and dissolved solids. Figure 47 shows the average analyses for normal flow and for summer-flood periods for the

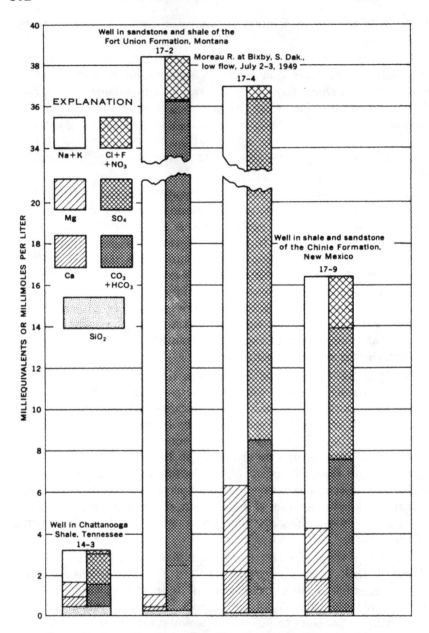

FIGURE 46.—Composition of water obtained from hydrolyzate sedimentary rock types.

Rio Grande at San Acacia that were presented in table 23. The difference between the averages is in large part the effect of inflow from the Rio Puerco. Somewhat similar effects can be demonstrated for other streams draining hydrolyzate sediments, for example, the chemical

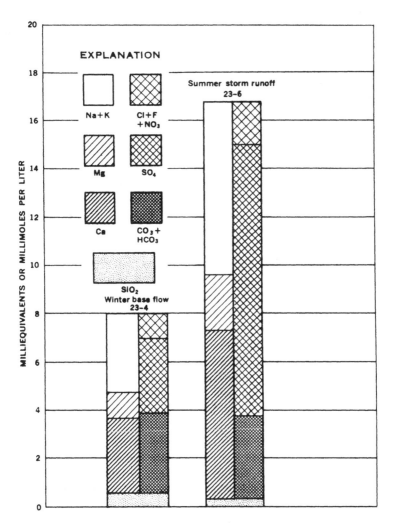

FIGURE 47.—Average composition of water from Rio Grande during two periods in 1945 and 1946.

relationships in the Powder River Basin of Montana. The chemical relationships of this basin were described by Swenson (1953).

PRECIPITATES

The sedimentary rocks termed "precipitates," for the purposes of this discussion, are the carbonates: limestone, which is mainly calcium carbonate, and dolomite, which is principally the compound $CaMg(CO_3)_2$. The impure and magnesian limestones have a wide range of Ca:Mg ratios, up to equimolar. There are no massive carbonate-rock formations in which magnesium is more abundant than calcium, although several kinds of magnesium carbonate minerals with and without added hydroxide do exist. Carbonate minerals other than those of calcium and magnesium occur as impurities in many limestones and dolomites. As a class, the carbonate rocks can be viewed as fairly simple chemical substances, as compared to the silicate rocks.

The solution of limestone is principally the process of solution of calcium carbonate, which already has been discussed at length. This reaction is fairly rapid (Weyl, 1958) and is reversible, although the reverse reaction does not always appear to take place rapidly. In any event, although some degree of supersaturation with respect to calcite is not uncommon in water associated with limestone, the general processes of solution and deposition can be readily explained by application of chemical-equilibrium calculations. The proportion of magnesium to calcium in such water reflects to some extent the composition of the limestone. This interpretation must be made cautiously, however, because there are many complications that can influence the calcium-to-magnesium ratio of water from limestone. Silica is normally a minor constituent of water from limestone.

Analysis 15-1 in figure 48 represents water from a relatively pure limestone. According to the calcite-solubility graph (fig. 18), this water is somewhat below saturation with respect to calcite. The source is a spring of large discharge in a region where water is plentiful and probably the circulation is too rapid to permit equilibrium to be reached. The predominance of calcium and bicarbonate in the analysis, however, is obvious. This analysis also is shown graphically in figures 30-37.

Dolomite does not dissolve reversibly—that is, the conditions required for direct precipitation of dolomite are not generally reached, at least in the kind of environment where ground water of potable quality occurs. As noted for limestone, however, the ratio of calcium to magnesium in water from dolomite tends to reflect the 1:1 composition of those ions in the rock, so long as the solution is not subjected to too many influences that might cause some calcium carbonate precipitation. Calcite precipitation from such a water can decrease the Ca:Mg ratio to values below 1. Analysis 16-1 in figure 48 represents water from the Knox Dolomite. It has a Ca:Mg ratio that is near 1.0 and from its pH seems to be virtually at saturation with respect to calcite. The influence of impurities in the rocks on the first two analyses in figure 48 obviously is minor.

Analysis 16-9 is for a water from impure limestone that contains both gypsum and dolomite, and the water also has been influenced by solution of sodium chloride. The hydrology of the aquifer from which this water came, the San Andres Limestone of the Roswell basin, New Mexico, was first described in detail by Fiedler and Nye (1933).

RELATIONSHIP OF WATER QUALITY TO LITHOLOGY 305

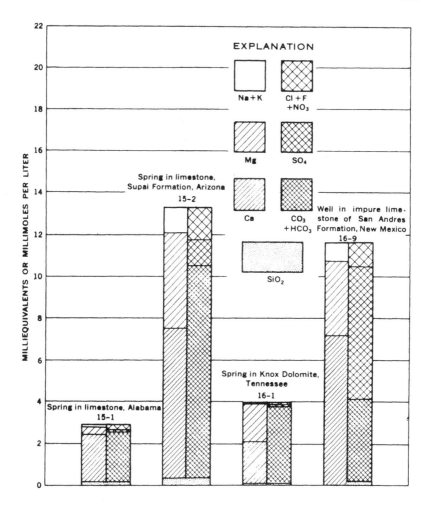

FIGURE 48 —Composition of ground water obtained from precipitate sedimentary rock types

Analysis 15-2 represents a water from limestone in the deeply buried Supai Formation in the Grand Canyon region, Arizona. The constituents other than calcium, magnesium, and bicarbonate in this water may originate from other formations, as little is known of the circulation of the water in this aquifer. A point of interest, however, is the high degree of supersaturation with respect to calcite. The water deposits calcite on exposure to the air. Obviously, at some point in its circulation path, this water achieved a considerably greater capacity than most waters for solution of limestone. Some writers, notably in the U.S.S.R., have ascribed this kind of effect to carbon dioxide released by metamorphic alteration of carbonate rock at depth (Orfanidi, 1957). Other processes, however, seem to offer possible alternative explanations.

The apparent enrichment in magnesium of water from carbonate rocks that results when calcite is precipitated can be observed in analysis 5, table 16. This analysis is not shown graphically.

Readers who are interested in pursuing further the subject of relationships between carbonate rocks and water composition will find more details in discussions in this book under "Calcium," "Magnesium," "Alkalinity," and "Hydrogen-ion activity," and in the large literature on this subject that is available elsewhere (for example, Garrels and Christ, 1964, p. 74–92).

EVAPORITES

The highly soluble nature of evaporite sediments causes water that may be associated with them to have high dissolved-solids concentrations. Gypsum and anhydrite are the least soluble of the rocks considered here as evaporites, and gypsum can transmit water through solution channels as limestone does. A water moving through gypsum eventually becomes saturated with respect to that solid and has a composition like the one shown in analysis 15–3, figure 49. The distribution of solutes in two nearly saturated brines from southeastern New Mexico is shown graphically by two analyses in figure 49. One of these, analysis 17–8, is essentially a saturated $Na^+ + Cl^-$ solution. The other, analysis 16–6, has a high proportion of Mg^{+2} and SO_4^{-2}. Analysis 18–2, figure 49, is for a water associated with a Wyoming Na_2CO_3 deposit. The vertical scale in figure 49 is condensed by a factor of 50 as compared with figures 44–48; hence the graphs are not directly comparable. The scale used in figure 49 causes the graph for analysis 15–3, a water which has more than 30 meq/l of cations, to appear small even though the water is above the usual concentration limit for potability.

The composition of natural brines is usually closely related to the composition of the evaporite deposits from which they are derived and may include minerals of economic value. Some of the elements commercially recovered from such brines are magnesium, potassium, boron, lithium, and bromine. Many analyses of natural brines have been published by White, Hem, and Waring (1963).

METAMORPHIC ROCKS

Rocks of any kind may be metamorphosed. The process as considered here consists of the alteration of rock by heat and pressure to change the physical properties and sometimes the mineral composition. All degrees of alteration exist, up to complete reassembly of rock

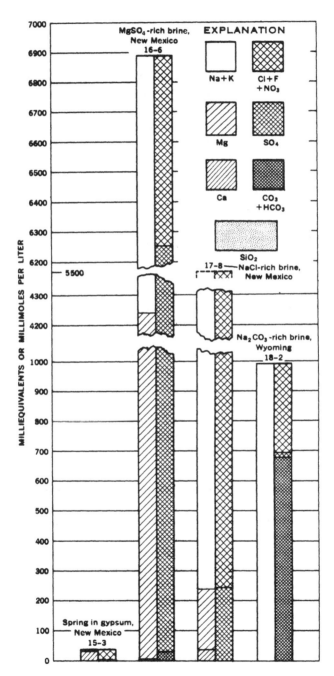

FIGURE 49.—Composition of ground water obtained from evaporite sedimentary rock types.

components into new minerals, and considering the wide possible variety of original rocks, it is evident that few simple generalization about the composition of water to be expected in association with metamorphic rocks can be made.

If sediments are subjected to sufficiently intense heat and pressure, they can be completely melted into a new igneous mass. Gneiss and schist result from heat and pressure that do not completely reorganize the initial rock. Although these rock types may be dense and nonwater bearing, there are some places where ground water can be recovered from them. Water from such formations can generally be expected to be low in solute concentrations and to resemble more closely the water from igneous terranes than that from the sediments as they existed before being metamorphosed. Analysis 12-9, represented graphically in figure 50, is for water from schist. Analysis 13-5 represents water from a granitic gneiss. The rather high proportion of silica in both these waters shows the influence of silicate minerals.

The dense structures of slate, a metamorphosed form of shale, and of quartzite, a metamorphosed form of quartzose sandstone, tend to restrict water movement to fracture zones and thereby prevent contact of ground water with as large surface areas of minerals as is normally expected in shale or in sandstone. The opportunity for water to pick up solutes from these metamorphosed rocks is, therefore, relatively small, and water from slate or quartzite also is likely to contain low concentrations of solutes. Analysis 16-8, figure 50, represents water from a quartzite in Alabama. Limestone may be converted to marble without much chemical change, and the processes of solution in such a marble are very much the same as those in limestone. If the original limestone was impure, however, it could yield a marble containing some minerals less soluble than calcite.

Processes of metamorphism generally involve aqueous solutions, and the alteration products that result may include hydrolyzate minerals. The release of water from some types of rocks during metamorphic processes can be expected. White (1957b) examined natural waters for possible features of their composition that could be attributed to metamorphism of rocks. Among properties that were cited by White as possible indications of metamorphic influence on water composition are high concentrations of sodium, bicarbonate, and boron and relatively low concentrations of chloride.

Some writers (for example, Chebotarev, 1955) speak of metamorphism with respect to changes in composition of natural water which occurs as it moves about the hydrologic circulation process. This use of the term differs from the use made of it here. In general, it would seem desirable to reserve the term "metamorphism" for rock alteration.

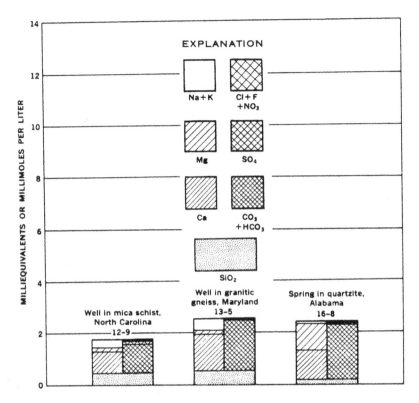

FIGURE 50.—Composition of ground water obtained from metamorphic rocks.

INFLUENCE OF ACTIVITIES OF MAN

The power of the human species to alter natural environments is extensive and far reaching. Sometimes these alterations are byproducts of actions which had other purposes in view. It is mainly this kind of activity that has caused changes in water quality that are both obvious and undesirable. The effects, commonly termed "pollution," have been developing for centuries wherever the population density was high enough to produce the effects and have long been a matter of concern to thoughtful observers. The term "pollution" is not exactly definable, but it seems desirable for this discussion to restrict it to adverse effects on natural-water quality that are definitely man produced.

Pollution may take many forms. Some are obvious and direct, such as release of untreated sanitary sewage and industrial organic or inorganic chemicals to streams or lakes. Other forms of pollution may be more subtle and indirect, for example, the excessive growth of algae due to the addition of nutrients to streams and lakes, the intrusion of sea water into wells due to the lowering of water levels in coastal aquifers, or the increase in sédiment loads of streams due to deforestation and overgrazing.

High densities of population of human or other species inevitably encounter waste-disposal problems, and these problems may be greatly magnified in industrial civilizations that produce large volumes of waste products. Waste accumulations in excessive amounts may occur in colonies of some of the lower forms of life and may cause the colonies to die out. The intelligence of men as individuals is certainly sufficient to recognize, understand, and cope with waste problems. Unfortunately, as in many other aspects of human behavior, the intelligence and ethical standards of social groups seem to lag far behind the levels attainable by individuals. Thus, many aspects of both cause and cure of water pollution lie in the area of the social rather than the physical or biological sciences.

It is not the purpose here to discuss pollution problems or remedies at length or in detail, but it is necessary to consider certain ways in which they may be related to natural processes controlling water quality. Better understanding of the natural processes is essential in order to cope with the large and rapidly growing problems of water pollution.

Waste-disposal practices may be classified into three major categories on the basis of the general aim of the process: (1) processes that convert the waste to innocuous or to reusable material, (2) processes that disperse the waste into a diluting medium, and (3) processes in which wastes are stored in a place where it is supposed they cannot later become a nuisance or hazard. Failure of any of these processes to work satisfactorily can cause pollution problems.

Techniques of waste disposal in which organic waste is converted to innocuous forms are commonly utilized in sewage-disposal plants. Natural processes that produce similar results occur in oxygenated river water: however, heavy loading of streams with organic waste produces many undesirable effects, and the natural process of purification may easily become overwhelmed. Many inorganic wastes must either be dispersed into the environment in such a way as to avoid objectionable concentrations in any locality or be stored where they can do no harm.

Some common ionic species may be dispersed into the environment with no serious problems. Chloride, for example, may not be objectionable if maintained at a low enough concentration, and it can be conveyed to the ocean, where it will cause no significant effects. Some other solutes, however, may tend to accumulate and become concentrated in such places as stream sediment or biota and can be released from such accumulations in unexpected ways to cause troublesome local concentrations.

A method of disposal used increasingly for obnoxious wastes is to inject them into deep aquifers that contain salty water. This generally amounts to storing the waste, but it is a method of storage whereby control cannot be maintained over the material. More positive controls generally are required for the highly dangerous wastes produced in some nuclear processes. Highly radioactive material is extensively stored in tanks or in solid form where the possibility of escape is minimized.

Whatever the technique of waste disposal used, the increasing density of population in many areas makes it more and more likely that any water sampled will be affected in some way by the activities of man. The added material may not fit the term "pollution" in the sense that the water becomes unusable, but it may have measurable effect on major or minor dissolved-ion concentrations. Addition of organic matter to ground water may lower the redox potential and cause changes in the geochemical balance of the system, and changes in pH may cause rock minerals to dissolve.

Although some polluted surface waters can be restored to reasonable quality levels fairly rapidly by decreasing waste loads or concentrations, the process can be costly. A polluted ground water, on the other hand, may be so slow to recover from that condition that it becomes necessary to think of the pollution of aquifers as almost irreversible once it has occurred. For this reason great care is needed to protect ground-water aquifers. Incidents of contamination of ground water from septic tanks, sewage and industrial waste-disposal systems, solid-waste disposal practices, and natural-gas and petroleum-storage leaks as well as other topics related to the general subject of ground-water pollution were described in a publication of the U.S. Public Health Service (1961). Control and abatement of the pollution which has resulted from faulty waste-disposal practices are obviously essential.

One aspect of a form of indirect pollution of water mentioned above is the intrusion of sea water. This subject has been selected for further discussion here because it illustrates some of the subtleties of pollution effects and is an area in which both hydrologic and chemical knowledge is essential in achieving control.

SALT-WATER INTRUSION

Both surface and underground waters may be contaminated by saline water where opportunity for contact occurs. The activities of man may aid in bringing about contamination, but are by no means always required. Along sea coasts there is a salt-water–fresh-water contact zone both in streams and in aquifers that extend under the sea.

The zone in which fresh water mixes with sea water in an estuary commonly extends upstream, as well as seaward from the river mouth, and fluctuates in position depending on river flow and ocean tides. The actual boundary between fresh and salty water is, therefore, dynamic and shifting. The zone of mixing effects extends farthest upstream in regions where the coastal plain is nearly flat and the lower reaches of rivers are tidal estuaries or submerged valleys whose bottoms are below sea level for some distance inland.

The energy of fresh water flowing down the river tends to push a fresh-water front out into the estuary, and this force is continuous, but fluctuates in response to the volume of flow. The cyclic rise and fall of the ocean in response to tides is opposed to the flushing action of the stream, and as a result salt water periodically is pushed back upstream as far as the tidal force can move it in opposition to the fresh-water movement.

Because of its higher density, sea water tends to move along the bottom of the channel, whereas the fresh water is pushed upward nearer the surface. There is considerable turbulence, however, along the fresh-water–salt-water boundary that results in mixing. The effects of turbulence and other physical factors promoting mixing, such as channel obstructions, are probably more important than ionic diffusion in bringing about mixing in tidal streams.

As a result of these effects, the water of a tidal stream varies in composition vertically and horizontally in the cross section at any point, and the water changes in quality with passage of time in response to changes in streamflow and to tidal fluctuations. The general conditions that can be expected in the tidal zone can be predicted when the effects of the various factors in the particular area of interest have been evaluated satisfactorily. Study of tidal streams requires extensive sampling on horizontal and vertical cross sections and integration of the results with data on streamflow and tides. A number of such studies have been made. The subject was discussed by Keighton (1966) with respect to the special problems of the Delaware River estuary.

Because of the obvious hydraulic complexity of estuaries, they pose special problems in the behavior of pollutants, some of which have been described by Pyatt (1964).

The relation between sea water and fresh water in aquifers along the seacoast is generally describable by hydraulic relationships, although the effect of man's withdrawal of water from the landward parts of these aquifers may have far-reaching effects on the position of the salt-water–fresh-water interface. The hydraulics of coastal aquifers have been studied extensively in the field and by means of laboratory models. It is generally agreed by hydrologists that the boundaries between fresh water and salt water in coastal aquifers depend on the balance of forces in a dynamic system in which fresh water continuously moves seaward at a rate sufficient to balance the head in the fresh-water aquifer. As Hubbert (1940) pointed out, the result of this movement is that fresh water discharges into the sea through the saturated zone from the high-tide line for some distance offshore. Cooper (1959) described how the movements of fresh and salt water along a contact zone tend to produce a diffuse zone of mixing rather than a sharp interface.

In a confined or semiconfined aquifer that is open to the sea at some distance offshore, a similar dynamic circulation pattern can be expected. So long as a high head of fresh water inland is maintained in the aquifer, fresh water will discharge into the ocean at the outcrop of the aquifer and maintain the zone of contact with salty water in the aquifer a considerable distance offshore. Pumping inland will reduce the head of the fresh water, and because head changes are transmitted rapidly through the system, the flow of fresh water seaward will be decreased. The head may decline enough to stop entirely the seaward flow of fresh water past the interface. With the decreased fresh-water flow, the system will be unstable, and salt water will invade the aquifer The salt-water front will move inland to the point where the reduced fresh-water head is again sufficient to produce a balancing seaward movement of fresh water past the interface.

Overdevelopment of coastal aquifers can greatly decrease the fresh-water head and bring about conditions that are favorable for migration of salt water inland. The actual migration of the salt-water front, however, is relatively slow, as it represents actual movement of water in the system under low energy gradients with high resistance. The appearance of salty water in a well may not occur until some years after the head decline has reached serious proportions. The rate of movement of some of the ions in the salt-water front will be influenced by ion exchange, and diffusion and head fluctuation will cause the interface to become a transitional zone rather than a sharp front.

Salt-water intrusion into highly developed aquifers is a serious problem in many places along continental margins and can occur in other places where nonoceanic salt water can be drawn into an aquifer when hydraulic heads are altered. Hydrologists are frequently confronted with the need for recognizing incipient stages of salt-water intrusion so that steps can be taken to correct the situation.

The composition of average sea water is given in table 2. Chloride is the major anion and moves through aquifers at nearly the same rate as the intruding water. Increasing chloride concentrations may well be the first indication of the approach of a sea-water contamination front. In an area where no other source of saline contamination exists, high chloride concentrations in ground water can be considered rather definite proof of sea-water contamination. If important amounts of chloride could come from other sources, however, the establishment of definite proof of sea-water intrusion may be difficult.

Components of sea water other than chloride may be used to identify contamination, but difficulties often are found in using them. Magnesium is present in sea water in much greater concentration than calcium. A low calcium: magnesium ratio may sometimes be indicative of sea-water contamination. The presence of sulfate in anionic proportion similar to that of sea water also might be indicative. Because of possible cation-exchange reactions and sulfate reduction in the aquifers, however, where sea water is introduced suddenly, the proportions of anions and cations in the first contaminated water to reach the sampling point cannot be expected to be exactly the same as those of a simple mixture of sea water and fresh water. It is indeed very likely that the water in the advancing salt-water front even after moving only a short distance through an aquifer will have little superficial resemblance to a simple mixture. After the exchange capacity of the aquifer has been satisfied and equilibrium reestablished, the water transmitted inland will be virtually unaltered sea water; however, a considerable volume of water may need to pass before this stage is reached. Because of the cation-exchange effects, the removal of the introduced ions from a contaminated aquifer by restoring seaward movement of fresh water also will tend to be slow and difficult.

Minor constituents of sea water may aid in some instances in determining whether a particular aquifer has been contaminated by sea water or by some other saline source. Incipient stages of contamination cannot generally be detected by these constituents. Piper, Garrett, and others (1953) were confronted with the need to differentiate sea-water contamination from contamination by connate brine and obtained some success in doing this by comparing iodide, boron, and barium concentrations in the suspected water The constituents that might be expected to be useful in identifying sources should be selected by using knowledge of composition of contaminating solutions and considering the chemical and exchange behavior of the solutes

In table 24 are given analyses of water from a well in the Los Angeles, Calif., area before and after contamination with sea water The calcium concentration in the contaminated water is higher than that which would be expected from a simple admixture of sea water sufficient to produce the observed chloride concentration in the

TABLE 24.—*Analyses showing effects of sea-water contamination in Gaspur water-bearing zone, Dominguez Gap, Los Angeles County, Calif.*

[Date below sample number is date of collection Source of data Piper, Garrett, and others (1953, p. 227)]

Constituent	(1) Jan. 8, 1923		(2) Apr. 4, 1928	
	mg/l	meq/l	mg/l	meq/l
Silica (SiO₂)			20	
Iron (Fe)			.97	
Calcium (Ca)	27	1 35	438	21.86
Magnesium (Ng)	11	.90	418	34.38
Sodium (Na)	}82	3.57	{1,865	81.10
Potassium (K)			56	1.43
Bicarbonate (HCO₃)	235	3.85	193	3.16
Sulfate (SO₄)	40	.83	565	11.76
Chloride (Cl)	40	1 13	4,410	124.38
Fluoride (F)			0.0	.00
Nitrate (NO₃)			1.8	.03
Dissolved solids				
Calculated	318		8,200	
Hardness as CaCO₃	113		2,810	
Noncarbonate	0		2,650	

1. Well 4/13-35 M3, So Calif Edison Co., West Gaspur Well, Los Angeles, Calif, before contamination by sea water
2. Same well, water contaminated by intrusion of sea water.

contaminated water, probably owing to the exchange of calcium held on exchange sites on aquifer minerals for sodium in solution in the advancing salt-water front.

Detection and tracing of sea-water contamination using stable isotopes of hydrogen and oxygen, or long-lived radioactive isotopes such as carbon-14, should have a considerable potential. Hanshaw, Back, Rubin, and Wait (1965) described this technique.

Large withdrawals of ground water from wells near the shoreline of Los Angeles County, Calif., has caused extensive inland contamination of ground water by sea-water intrusion. Remedial measures undertaken there since the 1940's to control the situation have been described in many published reports, for example, that of Banks, Richter, and Harder (1957). Sea-water intrusion was controlled by injecting fresh water through wells into the contaminated aquifer close to the shoreline so as to build up a fresh-water barrier.

APPLICATION OF WATER-QUALITY MEASUREMENTS TO QUANTITATIVE HYDROLOGY

Certain types of water-quality determinations can be used to measure or estimate water-discharge rates or quantities of stored water. Two kinds of "chemical" measuring techniques suitable for surface streams have been in use for many years (Corbett and others, 1945, p. 88-90). One of these has been extensively applied to determine the rate of solute movement, or time of travel of water through a reach of a river, a value that cannot be accurately estimated with usual flow-rate measurements. In this procedure, a readily detectible solute is added to the stream in the form of a concentrated slug, and the length of time required for the material to appear at a downstream point is measured. In recent years, many measurements of this kind have been made with dyes such as rhodamine B (U.S. Geological Survey, 1963) in response to the need for time-of-travel data in pollution studies and other applications. Inorganic materials also can be used as tracers, and Nelson, Perkins, and Haushild (1966) used radioactive materials introduced into the Columbia River by nuclear facilities at Hanford, Wash., to trace the flow times in that stream.

The salt-dilution method of measuring flow rate of a stream consists of adding at a constant rate a known quantity of tracer, usually sodium chloride, and measuring the concentration of chloride in the water upstream from the point of addition and far enough downstream that mixing is complete. This provides enough information for calculaing the discharge rate of the stream. The method can be used in systems where other procedures are impossible because of inaccessibility or extremes of turbulence or velocity. Amounts of salt added are not large enough to affect water quality adversely.

The mathematical basis for calculating discharge by the tracer method is the ion-balance equation cited previously in connection with the relation of stream-water quality to discharge. The water discharges in a system of this type are related by the equation

$$q_1 + q_2 = q_3,$$

where q_1 is the discharge upstream, q_2 the discharge of tagging solution, and q_3 the discharge downstream from the point of addition of q_2. To express the discharge rate for the tracer ion, one may write the equation

$$q_1 c_1 + q_2 c_2 = q_3 c_3,$$

where the c terms are the concentrations known or observed at the points where the q values are taken. When the technique is used as described avove, all the c terms and q_2 are known: hence the two equations provide values for q_1 and q_3.

This procedure for evaluating discharges has potential hydrologic uses which seem not to have been widely realized. Water-quality changes occur in river systems owing to inflows of water of a different composition, and a combination of sampling and flow measurement might be used to help determine quantities that cannot be directly measured.

A simple example of a system that can be evaluated this way is stream low in dissolved salts that receives inflow from a salt-water spring. If the spring inflow is accessible for measurement, the streamflow can be measured by analyzing samples of water collected from the spring and from the stream above and below the inflow point and then determining the spring flow. If the spring flow cannot be measured directly, as is more commonly the case, a single measurement of river flow and analysis of samples at the three necessary points can provide a basis for calculating spring inflow.

It is true that the usual procedure for the stream gager would be to measure flow above and below the inflow zone and calculate inflow by difference. If the river flows are fairly large, however, the probable magnitude of the measuring error may easily exceed the quantity of inflow.

At Clifton, Ariz., the San Francisco River receives enough inflow from hot saline springs that issue from gravel in the streambed to alter the composition of the river water considerably at normal stages of flow. The composition of the inflowing water was determined by sampling one point of spring discharge that was above the river level. And samples of river water were obtained upstream and downstream from the inflow zone. The discharge was measured by means of a current meter at the downstream site. Table 25 contains the analyses and the calculated and measured discharges. Although the calculated inflow is a little more than 2 cfs, almost all this amount occurs below the water surface in the stream, where it is not noticeable to the casual observer

TABLE 25.—*Analyses of water from Clifton Hot Springs and from San Francisco River above and below Clifton, Ariz.*

[Date below sample number is date of collection Source of data Hem (1950)]

Constituent	(1) Jan. 10, 1944		(2) Jan. 10 1944		(3) Jan. 10, 1944	
	mg/l	meq/l	mg/l	meq/l	mg/l	meq/l
Silica (SiO_2)					58	
Iron (Fe)					.19	
Calcium (Ca)	44	2.20	72	3 59	860	42.91
Magnesium (Mg)	12	.99	13	1.07	41	3.37
Sodium (Na)	58	2.52	156	6.78	2,670	116.10
Potassium (K)					142	3.63
Bicarbonate (HCO_3)	204	3.34	206	3.38	109	1.79
Sulfate (SO_4)	17	.35	19	.40	153	3.19
Chloride (Cl)	70	1.97	270	7.62	5,800	163.58
Fluoride (F)	.8	.04	.8	04	3 0	.16
Nitrate (NO_3)	2	00	5	01	7 5	.12
Dissolved solids Calculated	302		633		9,790	
Hardness as $CaCO_3$	160		233		2,310	
Noncarbonate	0		64		2,220	
Specific conductance (micromhos per cm at 25°C)	547		1,160		16,500	
Discharge (cfs)	58 (calc)		60		2.1 (calc)	

1. San Francisco River, 2 miles above Clifton, Ariz.
2. San Francisco River at gaging station at Clifton, Ariz
3. Clifton Hot Springs seep opposite Southern Pacific Lines depot Temp, 43.4°C.

and is not measurable by direct means. The correlation of dissolved-solids and discharge data for the San Francisco River at Clifton was discussed earlier in this paper.

In areas where interconnections between ground-water and surface-water systems are of interest, detailed studies often include seepage measurements. These consist of a series of measurements of river flow and tributary inflow taken in a downstream direction, with unmeasured gains and losses between measuring points being ascribed to ground-water inflows or losses of streamflow to the ground-water reservoir. Such measurements can provide considerably more information if water samples are taken at all measuring sites. The analyses of the samples help show where both inflow and outflow may be occurring between measuring points and help refine the investigators' understanding of the hydrologic system. Data of this kind were used to help evaluate details of the hydrology of the Safford Valley, Ariz., (Gatewood and others, 1950).

In studies of surface streams and related inflows and outflows, enough measurements and observations usually can be made to provide for a detailed evaluation. Although it would be helpful to be able to estimate ground-water flow rates and other characteristics with similar techniques, the available technology usually is not adequate. For example, there are serious problems inherent in deciding whether ground-water samples from wells adequately represent the entire thickness of aquifer penetrated and whether the tracer ions used move with the same speed as the water itself. There are, however, a number of examples that can be cited in which chemical techniques of measurement have been useful.

Many investigations of direction and rates of ground-water movement have been made by studies that involved injection of slugs of salt, dye, or radioactive material. The idea is by no means new. Slichter (1902) conducted a number of studies in which salt solution was added to one well and its appearance monitored in adjacent wells; he referred to similar work done in Europe at earlier times by Adolph Thiem. Objections can be made to the use of salt solutions, as they are heavier than naturally occurring water and may move differently if there is no density difference; also large amounts might constitute objectionable pollution.

A common tracing technique that can be used where the ground water moves through large openings such as rock fissures or solution channels is to inject dye at a convenient entry point and watch for it to appear elsewhere. The application of tritium or carbon-14 measurements to ground-water flow tracing has veen discussed earlier under the section "Other radioactive nuclides." Any material that is added must pass through the ground-water system from point of application to point of recovery without being held back significantly by adsorption or being altered by chemical reactions. Most dyes are likely to be adsorbed by mineral surfaces and usually can be used successfully only in systems where surface effects are minimal. In a limestone aquifer there may be large channels through which water moves, and dyes often are used to trace water movement through them.

It is characteristic of ground-water circulation patterns that some components of flow traverse an indirect route from points of recharge to points of discharge. A tracer may thus be diluted with a very large volume of water and be difficult to recover in measurable amount. Sometimes the volume of diluting water can be approximated, however, by chemical considerations.

Water stored in McMillan Reservoir in the Pecos River north of Carlsbad, N. Mex., escapes through solution openings in gypsum, but the outflow apparently is all or nearly all returned to the river through springs located a few miles downstream from the reservoir. Theis and Sayre (1942, p. 54–58) used the observed behavior of chloride concentrations in the reservoir and spring water to develop an empirical equation relating water emerging at the springs to water that had been stored in the reservoir at earlier times. From this, the volume of storage in the ground-water system feeding the springs was estimated. Kaufman and Orlob (1956) observed that chloride ions seemed to move at essentially the same rate as water through porous materials. In fact, the behavior of chloride in their experiment was a little better than that of tritium, which was partly retarded by exchange reactions, even though the tritium actually was incorporated in water molecules.

Some aquifers can be considered as conduits with fairly well defined boundaries. Water moving down such a channel may show quality changes related to side inflows or other factors. Plate 2, which is a water-quality map of part of Safford Valley, Ariz., shows how the composition of ground water in the alluvial fill, which was considered by Gatewood and others (1950) to be a hydrologic unit, changes downstream. The inflow of relatively fresh water from the south side of this section of the valley is evidently large compared to the amount of rather saline water moving downstream in the fill above Fort Thomas. The relative magnitudes of flow can be estimated by the composition of the influent water and the mixture.

Admittedly, the effects of other factors, such as river stage and pumping for irrigation, may influence the results of calculations. Where the data can be obtained, however, water-quality maps offer a method of extending estimates of both ground-water flows and the relative amounts of water contributed from different sources.

In a study of ground-water and surface-water relationships along the Illinois River at Peoria, Ill., Larson (1949) used water-quality data to estimate the proportions of river water and ground water in a pumped well near the stream. Chemical studies to evaluate sources of water in wells along the Platte River in Nebraska were described by Barnes and Bentall (1968).

Where changes in the chemical composition of water can be reasonably ascribed to mixing of water from different sources and a satisfactorily accurate measurement of the composition of water from the sources can be made, the hydrologist should be able to utilize methods like those given here for estimating quantities. Although only a rather small number of examples of such calculations can be cited from published reports, the use of water quality for estimating quantities of water probably has a considerable potential value. As the processes by which water quality is controlled in nature become better known, more applications of this kind will become possible.

RELATIONSHIP OF QUALITY OF WATER TO USE

An immediate purpose of the usual quality-of-water study is to determine if the water is satisfactory for a proposed use. Accordingly, the subject of water-analysis interpretation must often include some consideration of standards and tolerances that have been established for water that is to be used for various purposes. Standards for water to be used for drinking and other domestic uses have been established in many countries. Published literature contains tolerance levels and related data for constituents of water to be used in agriculture, in industry, for propagation of fish, and for a number of other specific purposes.

Water that is to be used as a public supply may be employed for many purposes. Therefore, the standards used to evaluate the suitability of water for public supply are generally more restrictive than those which would be applied to water for a small domestic or farm supply, although not necessarily more rigorous for individual components than the limits that apply for many industrial uses.

Water from mineral and hot springs is used medicinally in some places, although probably much less extensively today than in the past. This application has been discussed by Licht (1963) and will not be considered here.

A very extensive survey of water-quality standards and review of the literature on effects of solutes on uses of water has been prepared by McKee and Wolf (1963). A more recent but less detailed survey was published by the U.S. Federal Water Pollution Control Administration (1968).

DOMESTIC USES AND PUBLIC SUPPLIES

Besides being chemically safe for human consumption, water to be used in the home should be free from undesirable physical properties such as color or turbidity and have no unpleasant taste or odor. Harmful microorganisms should be virtually absent; however, they are not usually considered in ordinary chemical analyses, although some are directly related to solutes that may be present. The presence of harmful microorganisms is considerably more difficult to ascertain than most other properties of water, but it is a very important consideration. Over the years, great progress has been made in decreasing the incidence of water-borne diseases, and in the United States such diseases have largely disappeared.

The standard test for bacteriologic quality is the determination of coliform bacteria concentration. The common species *Escherichia coli* occurs in great numbers in the intestinal tracts of warm-blooded animals, and the presence of these and related bacteria in water is generally considered an index of fecal pollution. The coliform bacilli are not themselves directly harmful, but their presence in excessive numbers is considered suggestive of the possible presence of other species that are pathogenic. Although direct determination of pathogens is often advocated, the problems involved are large, and the substitution of such determinations for the simple coliform count is not likely to occur. As noted above, the sanitary condition of a water is not indicated by chemical testing alone. Thus, a water that is safe for drinking on the basis of its chemical composition may not be safe bacteriologically. This subject is not considered in detail here. Standards for public water supplies were recommended by the U.S. Public Health Service (1962).

Because of differences in individual tolerances, different amounts of water used, and other factors, it is difficult to establish exact limits for any of the mineral components that are commonly found in natural water to be used for drinking. The limits usually quoted in the United States for drinking water are based on the U.S. Public Health Service drinking-water standards. These standards were established first in 1914 to control the quality of water supplied by interstate common carriers to passengers for drinking and to chefs for cooking or preparing food. The standards have been revised several times and those in current use date from 1962 (U.S. Public Health Service, 1962). The concentrations were arrived at from considering industrial-exposure tolerances and probable amounts of water intake, and they contain a considerable safety factor.

CHEMICAL CHARACTERISTICS OF NATURAL WATER

According to the 1962 standards, the following substances should not be present in amounts greater than those shown:

Substance	Concentration (mg/l)
Arsenic (As)	0.05
Barium (Ba)	1.0
Cadmium (Cd)	.01
Chromium (hexavalent, as Cr)	.05
Cyanide (CN)	.2
Fluoride (F)	(1)
Lead (Pb)	.05
Selenium (Se)	.01
Silver (Ag)	.05

[1] Fluoride limits depend on average daily maximum air temperature. (See table 26.)

An excessive concentration of any of these constituents constitutes a basis for rejection of the supply. If the water is acceptable on the basis of this set of standards, a second set is considered. These are to be complied with unless no better supply is available.

Substance	Concentration (mg/l)
Alkyl benzene sulfonate (ABS)	0.5
Arsenic (As)	.01
Chloride (Cl)	250
Copper (Cu)	1
Carbon chloroform extract (CCE)	.2
Cyanide (CN)	.01
Fluoride (F)	(1)
Iron (Fe)	.3
Manganese (Mn)	.05
Nitrate (NO_3)	45
Phenols	.001
Sulfate (SO_4)	250
Total dissolved solids	500
Zinc (Zn)	5

[1] Fluoride limits depend on average daily maximum air temperature. (See table 26.)

Alkyl benzene sulfonate is a synthetic detergent formerly in common use. The carbon chloroform extract is a measure of the total concentration of organic solutes.

TABLE 26.—*Recommended limits of fluoride concentration in drinking water (U.S. Public Health Service (1962))*

Annual average of maximum daily air temperatures (°F)[1]	Recommended control limits (mg/l)		
	Lower	Optimum	Upper
50.0–53.7	0.9	1.2	1.7
53.8–58.3	.8	1.1	1.5
58.4–63.8	.8	1.0	1.3
63.9–70.6	.7	.9	1.2
70.7–79.2	.7	.8	1.0
79.3–90.5	.6	.7	.8

[1] Based on temperature data obtained for a minimum of 5 years.

Regarding fluoride concentrations, the limits state:

"When fluoride is naturally present in drinking water the concentration should not average more than the appropriate upper limit * * *. Presence of fluoride in average concentrations greater than two times the optimum values * * * shall constitute grounds for rejection of the supply."

Although these standards have been widely quoted for drinking water, they are not completely practicable for all areas. In some sections of the United States the water available from all or nearly all individual domestic supplies and public supplies does not meet the standards in the second list above in one or more respects. Thus residents of some areas have used water containing more than 1,000 mg/l of dissolved solids all their lives. Although detailed medical data are not available, there do not seem to be any obvious detrimental effects on public health that can be attributed to such water supplies. Some of the undesirable impurities in the tables may be ingested in larger quantities from sources in the diet other than drinking water. The limits given in the first list, however, refer to substances known or thought to be toxic and should be followed as closely as possible.

The limits for radioactive substances in drinking water are viewed somewhat differently from those for nonradioactive solutes. It is generally agreed the effects of radioactivity are harmful, and unnecessary exposure should be avoided. If the concentration of raduim-226 does not exceed 3 picocuries per liter, and strontium-90 does not exceed 10 picocuries per liter, the supplies are considered satisfactory by the Public Health Service standards, without considering the total exposure of consumers to radiation. Above those levels, approval is given only when it can be shown that the total radiation exposure of people using the water will not exceed recommended levels.

Strontium-90 is a fission product, but radium-226 occurs naturally and in some waters may exceed the limit shown above. Both nuclides are preferentially absorbed in bone structure and are, therefore, especially undesirable in drinking water. When both nuclides are well below the specific limits given here, a total content of beta emitters amounting to no more than 1,000 picocuries per liter is permitted.

The lower limit of detection of solutes in water by taste is, of course, a function of individual sensitivity. Some substances can be detected in very low concentrations, however. Chlorophenols may be present in industrial wastes and sometimes are produced by chlorination of a water containing some phenolic impurities. Chlorophenols can impart a noticeable taste when only a few micrograms per liter are present (Burttschell and others, 1959). Free chlorine concentrations of a few tenths milligram per liter are usually noticeable, also.

On the other hand, many common solute ions cannot be detected by taste until concentrations of tens or even hundreds of milligrams per liter are attained. Chloride concentrations of 400 mg/l impart a noticeable salty taste for most people. Users may become accustomed to waters containing high concentrations of major ions, however, and prefer the taste of such water to the "tastelessness" of more dilute solutions.

A considerable literature exists on the subject of taste effects in drinking water. Taste thresholds for zinc, copper, iron, and manganese were explored by Cohen, Kamphake, Harris, and Woodward (1960). Bruvold, Ongerth, and Dillehay (1967) made a statistical study of consumer attitudes toward tastes imparted by major ions.

"Ideal" standards of water quality are set up from time to time within the water-supply industry. A paper on this subject by Bean (1962) gave a set of upper limits that are more stringent than the U.S. Public Health Service standards for most constituents.

AGRICULTURAL USE

Water required for nondomestic purposes on farms and ranches includes that consumed by livestock and that used for irrigation.

Water to be used by stock is subject to quality limitations of the same type as those relating to quality of drinking water for human consumption. Most animals, however, are able to use water that is considerably higher in dissolved-solids concentration than that which is considered satisfactory for humans. There are few references in published literature that give maximum concentrations for water to be used for livestock. Range cattle in the western United States may get accustomed to highly mineralized water and can be seen in some places drinking water which contains nearly 10,000 mg/l of dissolved solids. To be used at such a high concentration, however, these waters must contain mostly sodium and chloride. Water containing high concentrations of sulfate are much less desirable. An upper limit of dissolved solids near 5,000 mg/l is recommended by some investigators for water to be used by livestock, and it would seem obvious that for best growth and development of the animals their water supply should have concentrations considerably below the upper limit.

In the report by McKee and Wolf (1963), the upper limits of concentration for stock water include the following:

Stock	Concentration (mg/l)
Poultry	2,860
Pigs	4,290
Horses	6,435
Cattle (dairy)	7,150
Cattle (beef)	10,100
Sheep (adult)	12,900

These, in turn, were quoted from a publication of the Department of Agriculture of Western Australia.

Selenium is known to be highly toxic to animals, but generally, poisoning results from eating selenium-concentrating plants. A concentration in water of 0.4–0.5 mg/l was reported by McKee and Wolf (1963) to be nontoxic to cattle. Water that has selenium concentrations higher than 0.5 mg/l but is satisfactory in other respects may exist, but would certainly be unusual. Toxic substances of other kinds may be present in polluted water, and incidents of poisoning of animals by drinking polluted water are not uncommon.

The chemical quality of water is an important factor to be considered in evaluating its usefulness for irrigation. Features of the chemical composition that need to be considered include the total concentration of dissolved matter in the water, the concentrations of certain potentially toxic constituents, and the relative proportions of some of the constituents present. Whether a particular water can be used successfully for irrigation, however, depends on many factors not directly associated with water composition. A brief discussion of some of these factors is included here to show the complexity of the problem of deciding whether or not a given water is suitable. Readers interested in the subject can find more information in the other publications to which reference is made.

The portion of the irrigation water that is actually consumed by plants or evaporated is essentially free from dissolved material. The growing plants do selectively retain some nutrients and a part of the mineral matter originally dissolved in the water, but the amount of major cations and anions so retained is not a large part of their total content in the irrigation water. Eaton (1954, p. 12) determined the quantity of mineral matter retained by crop plants and showed that it consists mostly of calcium and magnesium salts. The bulk of the soluble material originally present in the water that was consumed remains behind in the soil, normally in solution in residual water. The concentration of solutes in soil moisture cannot be allowed to rise too high because excessive concentrations interfere with the osmotic process by which plant root membranes are able to assimilate water and nutrients. Some substances of low solubility, especially calcium carbonate, may precipitate harmlessly in the soil as solute concentrations are increased, but the bulk of the residual solutes present a disposal problem that must be solved effectively to maintain the productivity of irrigated soil. The osmotic processes by which plants are able to absorb water through their roots are discussed under the section "Membrane effects."

The extent and severity of salt-disposal problems in irrigated areas depend on several factors. Among these factors are the chemical composition of the water supply, the nature and composition of the soil and subsoil, the topography of the land, the amounts of water used and the methods of applying it, the kinds of crops grown, the climate of the region, especially the amount and distribution of rainfall, and the nature of the ground-water and surface-water drainage system.

In most areas, the excess soluble material left in the soil from previous irrigations is removed by leaching the topsoil and permitting part of the resulting solution to percolate below the root zone and downward to the ground-water reservoir. In those areas where the water table beneath the irrigated land can be kept far enough below the surface, this process of drainage is reasonably effective. The necessary leaching may be accomplished by rainfall in those areas where the precipitation is sufficient to saturate the soil deeply. The leaching process also occurs during the irrigation season either with this purpose specifically in mind or when extra amounts of water are added in an effort to store an extra supply of water in the soil or to utilize unusually large supplies of water when they happen to be available. The need for leaching the soil is generally recognized by farmers who use highly mineralized irrigation water.

In areas where natural drainage is inadequate, the irrigation water that infiltrates below the root zone will eventually cause the water table to rise excessively, a process resulting in serious problems. "Seeped" (waterlogged) land in which the water table is at or very near the land surface has become common in many irrigated areas of the United States and elsewhere. Such land has little value for agriculture, but instead often provides a site for water-loving vegetation. Transpiration by such vegetation and direct evaporation from wet soil and open water surfaces can result in waste of large and economically significant quantities of water; also, the soil in such areas soon becomes highly charged with residual salts.

Although waterlogged areas generally have extensive saline accumulations, the quality of the water available was not always the primary cause of abandonment of the land. The reasons for failure of ancient irrigation enterprises include some things unrelated to water-management practices. According to Eaton (1950), however, failure to provide proper drainage to remove excess water and solutes must have played a large part in the decline of areas along the Tigris and Euphrates Rivers in what is now Iraq and may have been important elsewhere also.

In most large irrigated areas, it has become necessary to provide some means of facilitating the drainage of ground water so that the water table is held well below the land surface. In some areas this is accomplished by systems of open drainage ditches or by buried tile systems. In many places, the excess ground water is pumped out by means of wells. The extracted drainage water is higher in dissolved-solids concentration than the original irrigation water, owing to the depletion of the water itself and the leaching of soil and subsoil of saline material, fertilizers, and soil amendments. The amount of solutes in the drainage water, however, is commonly small enough that the water can be reused for irrigation. In areas of intensive development, several such cycles can occur before the water is finally released.

For long-term successful operation of an irrigation project, all the ions present in the irrigation water that were not extracted by plants must be disposed of either by flowing away from the area in drainage or by storage in an innocuous form within the area. The relationship among total ion loads in an irrigated area can be expressed in terms of the ion inflow-outflow or salt-balance equation:

$$W_I - W_o \pm \Delta W_S = 0.$$

In this equation, the W_I and W_o terms represent total ion loads into and out of the area over a finite time period; the ΔW_S term is the change in storage within the area during that period. Although this equation tends to obscure the importance of the time factor and is certainly an oversimplification, it has frequently been used to evaluate the performance of irrigation developments.

Scofield (1940) termed this general relationship the "salt balance" and made calculations of solute inflow and outflow on the bases of streamflow and water analyses. These calculations did not attempt to evaluate changes in ion storage except to term the salt balance unfavorable when storage was increasing and favorable when it was decreasing.

In the more highly developed irrigated areas where most of the water comes from a surface supply, as along the Rio Grande from the San Luis Valley in Colorado to Fort Quitman, Tex., and along the Gila and its tributaries in Arizona, the drainage water returned to the stream by the upper irrigated areas is used again for irrigation in the next area downstream. The cycle of use and reuse may be repeated six times or more until the residual drainage is too small in quantity and too saline to have any further value.

Deterioration of ground-water quality associated with irrigation development is commonly observed. Moreover, the rate of deterioration is increased when water pumped from wells for drainage is reused in the vicinity for irrigation, for this practice increases the intensity of recycling and converts part of what would have been ion outflow into stored ions. If very large amounts of water are pumped, the ground-water circulation pattern in the affected area may come to resemble a closed basin with no outflow. Obviously, the end result of such an overdevelopment will be an excessive dissolved-solids concentration in the ground water. The rates of movement of solutes through systems of this kind, however, are not well enough understood to permit a prediction of how long a period of time might be required to attain that end result. In some areas the decline of the water table from pumping may be more rapid than the rate at which residual solutes move downward from the soil zone, at least for a considerable length of time.

The need for research on the processes and concepts involved in the salt-balance equation is obvious. The storage term in the equation includes material held in the soil and subsoil, that in the saturated zone, and that in the material between. It includes dissolved ions and material that has precipitated as sparingly soluble minerals as well as ions held by adsorption on mineral surfaces. The length of time required for a given addition to the storage item to be felt as a change in salinity of some specific fraction of the water supply can be estimated only when more is known about the processes involved. Some fairly sophisticated attempts to model solute circulation through irrigated areas have appeared in recent papers. A study of this type by Orlob and Woods (1967) pointed up the need for more information to improve ability to predict the effect of changes in water use on the quality of residual water.

In addition to problems caused by excessive total amounts of dissolved solids, certain specific constituents in irrigation water are especially undesirable, and some may be damaging even when present only in small quantities. One of these constituents that has received considerable attention is boron. This element is essential in plant nutrition and is sometimes added to fertilizer in small amounts because some soils in humid regions are deficient in boron. A small excess over the needed amount, however, is toxic to some types of plants.

Work done about 1930 by the U.S. Department of Agriculture showed that the plants most sensitive to excess boron included citrus fruit trees and walnut trees. Later work summarized by the U.S. Salinity Laboratory staff (1954) is shown in part in table 27. A rating table was developed by the Salinity Laboratory that indicates the permissible boron concentrations in irrigation water for the three classes of plants (table 28).

The toxicities of other constituents of irrigation water generally are less striking. The U.S. Salinity Laboratory (1954), however, has ascertained that some plants are specifically affected by sulfate in excessive amounts, and some are adversely affected by magnesium. Consideration of details of plant response to water composition is not appropriate here.

The term "alkali" is commonly used throughout the western United States to refer to efflorescent deposits of white material or salt crusts that appear where water evaporates from soil surfaces. The word does not necessarily imply anything about the composition of the material. In some irrigated soils, patches of alkali with a dark color occur, and this material is commonly called black alkali. The dark color is caused by organic material leached from the soil. Black alkali is mostly sodium carbonate, and soil where it occurs has a very high pH along with other undesirable properties. White alkali deposits commonly are predominantly sodium sulfate and chloride, but may also contain calcium and magnesium. Soils of high salinity interfere with crop growth, and a high pH may decrease the solubility of some essential elements.

TABLE 27.—*Relative tolerance of crop plants to boron*

[In each group, the plants first named are considered as being more sensitive and the last named more tolerant. After U.S. Dept. of Agriculture Handbook 60 (1954)]

Sensitive	Semitolerant	Tolerant
Lemon	Lima bean	Carrot
Grapefruit	Sweetpotato	Lettuce
Avocado	Bell pepper	Cabbage
Orange	Pumpkin	Turnip
Thornless blackberry	Zinnia	Onion
Apricot	Oat	Broadbean
Peach	Milo	Gladiolus
Cherry	Corn	Alfalfa
Persimmon	Wheat	Garden beet
Kadota fig	Barley	Mangel
Grape (Sultanina and Malaga).	Olive	Sugar Beet
Apple	Ragged Robin rose	
Pear	Field pea	Palm (*Phoenix canariensis*)
Plum	Radish	Date palm (*P. dactylifera*)
	Sweet pea	Athel (*Tamarix aphylla*)
American elm		Asparagus
Navy bean	Tomato	
Jerusalem-Artichoke	Pima cotton	
Persian (English) walnut.	Acala cotton	
Black walnut	Potato	
Pecan	Sunflower (native)	

TABLE 28.—*Rating of irrigation water for various crops on the basis of boron concentration in the water*

	Classes of water	Sensitive crops (mg/l)	Semitolerant crops (mg/l)	Tolerant crops (mg/l)
Rating	Grade			
1	Excellent	<0.33	<0.67	<1.00
2	Good	.33–.67	.67–1.33	1.00–2.00
3	Permissible	.67–1.00	1.33–2.00	2.00–3.00
4	Doubtful	1.00–1.25	2.00–2.50	3.00–3.75
5	Unsuitable	>1.25	>2.50	>3.75

The process of cation exchange already has been discussed. It occurs in irrigated soil and may influence soil properties especially when concentrations of solutes are relatively high. An irrigation water having a high proportion of sodium to total cations tends to place sodium ions in the exchange positions on the soil-mineral particles, and water having mostly divalent cations reverses this process. The particles in the soil having the highest exchange capacity are the clay minerals. These minerals preferentially adsorb divalent ions, and when their exchange sites are occupied by calcium and magnesium, the physical properties of the soil are optimal for plant growth and cultivation. If the exchange positions become saturated with sodium, however, the soil tends to become deflocculated and relatively impermeable to water. A soil of this type can be very difficult to cultivate and may not support plant growth.

The cation-exchange process is reversible and, therefore, can be controlled either by adjusting the composition of the water, where this is possible, or by using soil amendments. The condition of a sodium-saturated soil can be improved by applying gypsum that releases calcium to occupy exchange positions. The soil also may be treated with sulfur, sulfuric acid, ferrous sulfate, or other chemicals that tend to lower the pH of the soil solution. The lowered pH brings calcium into solution by dissolving carbonates or other calcium minerals.

The tendency of a water to replace adsorbed calcium and magnesium with sodium can be expressed by the sodium-adsorption-ratio, which has been discussed under the section "Sodium-adsorption-ratio (SAR)."

All investigators agree that deposition of some of the dissolved ions in irrigation water can occur as a result of concentration effects and related changes and that some of the calcium and bicarbonate can be expected to be precipitated as calcium carbonate. Such a reaction might be considered a form of ion storage in the salt-balance equation. Doneen (1954) pointed out that gypsum also could be deposited from irrigation water without doing any harm to soils. Neither of these precipitation reactions can be expected to remove all the ionic species involved, and rather than make arbitrary subtractions of these ions from inflow or outflow terms in the salt-balance equation as most writers have done, it might be better to incorporate them into the storage term. A working salt-balance equation that would be complete enough to have practical usefulness should contain a storage term that includes a number of different components, including the precipitates.

In considering the effects that might follow from precipitation of carbonates from irrigation water, Eaton (1950) suggested that if much of the calcium and magnesium originally present were precipitated, the residual water would be considerably higher in proportion of sodium to the other cations than it had been originally. Some waters in which the bicarbonate content is greater than an amount equivalent to the total of the calcium and magnesium could thus evolve into solutions containing mostly sodium and bicarbonate and would have a high pH and a potential for deposition of black alkali.

Although the relationships suggested above are somewhat oversimplified, the U.S. Salinity Laboratory (1954, p. 75) made studies that showed some additional hazard does exist when waters high in bicarbonate and relatively low in calcium are used for irrigation. It was proposed, therefore (U.S. Salinity Laboratory, 1954, p. 81), that waters containing more than 2.5 meq/l of residual sodium carbonate are not suited for irrigation, that those containing 1.25-2.5 meq/l are marginal, and that those containing less than 1.25 meq/l are probably safe. Residual sodium carbonate is defined as twice the amount of carbonate or bicarbonate a water would contain after subtracting and amount equivalent to the calcium plus the magnesium.

The residual sodium carbonate concept appears to ignore two major factors in the chemical behavior of carbonate species that must have some influence. As pointed out in earlier sections of this paper, calcium carbonate tends to precipitate independently and thus leaves magnesium in solution. Although the conditions in irrigated soils might favor deposition of mixed carbonates, their existence does not seem to have been domonstrated. A more recent paper by Eaton, McLean, Bredell, and Doner (1968) suggested the precipitation of magnesium might occur by combination with silica. The other factor is the production of carbon dioxide in the soil by plants. This is a major source of bicarbonate ions and in some irrigated soils may provide much more of this material than the irrigation water originally contained.

From this brief discussion, it should be evident that relationships between water quality and the feasibility of the use of water for irrigation are not simple. Increasing difficulties that can be expected as salinity of the water supply increases could be translated into economic effects. A decrease in crop yield accompanied by increased costs in water and land management occurs as the quality of the water becomes less suitable.

Generalizations regarding sensitivity of crops to salinity of water supply was made by the U.S. Salinity Laboratory staff (1954). A list of crops arranged in three groups with respect to their tolerance toward salinity shows that generally crops tolerant to boron also are tolerant to salinity. According to this list, the more sensitive species include fruit trees and beans and the most tolerant include the date palm, asparagus, beets, bermudagrass, cotton, sugar beets, and barley.

Most vegetables and field and forage crops are included in the moderately tolerant group.

A diagram widely used for evaluating waters for irrigation, published by the U.S. Salinity Laboratory (1954), is reproduced here (fig. 51).

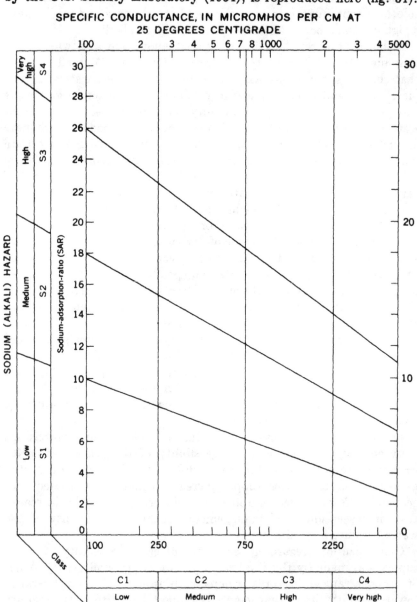

FIGURE 51.—Diagram for use in interpreting the analysis of irrigation water Adapted from U S. Salinity Laboratory Staff (1954).

The specific conductance, as an index of dissolved-solids concentration, is plotted on one axis and sodium-adsorption-ratio on the other. The diagram is divided into 16 areas that are used to rate the degree to which a particular water may give rise to salinity problems and undesirable ion-exchange effects.

Water having a specific conductance greater than 5,000 micromhos per centimeter is used with some success in certain areas, where soil conditions, crops grown, and irrigation techniques are suitable. The hydrologist needs to consider local experience before arbitrarily stating whether a given water is usable or not. Salinity problems, however, may be slow to develop and may be observable only in lowered yields or other factors not easy to evaluate. A water of high salinity must always be viewed with suspicion. A summary of water quality as it relates to agriculture is contained in a recent report by the U.S. Federal Water Pollution Control Administration (1968, p. 111–184).

INDUSTRIAL USE

The quality requirements for industrial water supplies range widely, and almost every industrial application has its own standards. For some uses such as single-pass condensing or cooling or the concentrating of ores, chemical quality is not particularly critical, and almost any water may be used. At the opposite extreme, water approaching or equaling the quality of distilled water is required for processes such as the manufacture of high-grade paper or pharmaceuticals, where impurities in the water would seriously affect the quality of the product. Modern maximum-pressure steam boilers may require makeup water less concentrated than the average distilled water of commerce, and very pure water is desirable in nuclear reactors to minimize induced radioactivity by neutron activation of the dissolved species.

It is not the purpose of this paper to review the subject of industrial water-quality standards in any detail. Some idea as to the varied nature of the requirements for certain industries can be obtained from table 29. This table is based on data presented by the U.S. Federal Water Pollution Control Administration (1968). Additional consideration of industrial water-quality requirements is given by McKee and Wolf (1963).

The standards given in the table represent maximum values to be present in the water at the point of use, after any necessary treatment but before adding any internal conditioners that may be needed during the process. The increasing stringency of requirements for boiler-feed water as the steam pressure increases is particularly noticeable. Constituents for which no entry appears in the table represent either

TABLE 29.—*Water-quality requirements*

[Concentrations represent upper limits for water at point of use before addition of internal conditioners

Constituent	Boiler feedwater pressure (pounds per square inch gauge)				Textiles (scouring, bleaching, and dying)
	0-150	150-700	700-1,500	1,500-5,000	
Silica (SiO_2)	30	10	0.7	0.01	
Aluminum (Al)	5	.1	.01	.01	
Iron (Fe)	1	.3	.05	.01	0.1
Manganese (Mn)	.3	.1	.01		.01
Calcium (Ca)		0	0		
Magnesium (Mg)		0	0		
Ammonium (NH_4)	.1	.1	.1	.7	
Copper (Cu)	.5	.05	.05	.01	.01
Zinc (Zn)		0	0		
Bicarbonate (HCO_3)	170	120	48		
Sulfate (SO_4)					
Chloride (Cl)					
Fluoride (F)					
Nitrate (NO_3)					
Hardness (as $CaCO_3$)	20	0	0	0	25
Alkalinity (as CO_3)	140	100	40	0	
Acidity (as CO_2)	0	0	0	0	
pH	8.0–10.0	8.2–10.0	8.2–9.0	8.8–9.2	(³)
Dissolved solids	700	500	200	.5	100
Color (units)					5
Organics:					
CCl_4 extract	1	1	.5	0	
Methylene-blue active substances	1	1	.5	0	
Chemical oxygen demand	5	5	.5	0	
Dissolved oxygen	2.5	.007	.007	.007	
Temperature (°F)					
Suspended solids	10	5	0	0	5

¹ Not to exceed U.S. Public Health Service drinking-water standards.
² Limit for noncarbonate hardness, 70 mg/l as $CaCO_3$.
³ Ranges from 2.5 to 10.5 depending on process and product.

things for which no limit is given or which cannot attain objectionable levels if the water meets the other specifications. Water used for processing food or beverages must also meet the U.S. Public Health Service (1962) drinking-water standards.

It is technically possible to treat any water to give it a composition suitable for special uses. If the water requires extensive treatment, however, especially if large amounts of water are involved, it may not be feasible economically to use some supply sources. Industrial plants having large water requirements are commonly located with reference to avaliability of water.

Although water temperature is not a chemical property and has not received much consideration here, the temperature of a supply and the seasonal fluctuation of that temperature are major considerations in use of water for cooling by industry. In some areas, ground water is extensively used for this purpose because of its low and uniform temperature. Some industries have recharged ground-water aquifers with cold water from surface streams each winter and have withdrawn the cool stored water in the summer when the regularly used surface supply is too warm for efficient cooling. In some areas industrial plants

RELATIONSHIP OF QUALITY OF WATER TO USE 335

for selected industries and processes

and are in milligrams per liter except as indicated (U.S. Federal Water Pollution Control Adm., 1968)]

Chemical pulp and paper		Wood chemicals	Synthetic rubber	Petroleum products	Canned, dried, and frozen fruits and vegetables	Soft drinks bottling	Leather tanning (general finishing processes)	Hydraulic cement manufacture
Unbleached	Bleached							
50	50	50			50			35
1.0	.1	.3	0.1	1	.2	0.3	0.3	25
.5	.05	.2	.1		.2	.05	.2	.5
20	20	100	80	75	100			
12	12	50	36	30				
		250						
		100			250	500	250	250
200	200	500		300	250	500	250	250
					1	(¹)		
		5			10			
100	100	900	350	²350	250		soft	
		200	150		250	85		400
					0			0
6-10	6-10	6.5-8.0	6.2-8.3	6.0-9.0	6.5-8.5		6.0-8.0	6.5-8.5
		1,000		1,000	500			600
30	10	20	20		5	10	5	
					.2	(⁴)	(⁵)	1
	95							
10	10	30	5	10	10			500

⁴ Carbon chloroform extract limit 0.2 mg/l; also specified to be free from taste and odor.
⁵ Carbon chloroform extract limit 0.2 mg/l.

located along the same stream use and reuse the water for cooling until the temperature of the water is far above normal levels for many miles of river. Excessively high temperatures deplete dissolved oxygen and interfere with the normal stream ecology. This effect is sometimes called thermal pollution.

Industrial expansion has contributed greatly to the increasing per capita utilization of water in the United States. Much of the industrial use, however, is nonconsumptive. That is, the water is not evaporated or incorporated into the finished product, but is released after use by the industry without significant change in quantity, possibly with an increased load of dissolved or suspended material or perhaps with very little change from its original composition. As water supplies have become more fully utilized, however, many industries have found it necessary to conserve and reuse water that in former years would have been allowed to flow down a sewer or back to a surface stream. In some places, reclaimed sewage has been used for certain noncritical industrial purposes.

Recirculation of water that is depleted by evaporation, as in a cooling tower, introduces concentration factors, and intensified reuse can result in effluents of high dissolved-solids concentration. Most industrial effluents that have high solute concentrations, however, are the result of other effects.

RECREATIONAL AND ESTHETIC USES

Considerable attention is now being paid to recreational uses of rivers and lakes for such purposes as swimming, fishing, boating, and for plain esthetic enjoyment. The cost of restoring water bodies that have lost their value for such purposes because of pollution may be substantial, but there is strong public support in many places for the the aim of creating or protecting waters for these purposes.

Water for swimming and other sports in which water is in contact with the skin must obviously conform to sanitary requirements. Fish that are sought by anglers require relatively clean water with a good supply of dissolved oxygen. Certain metal ions may be lethal to fish at levels well below those permissible in the U.S. Public Health Service (1962) drinking-water standards. Most species of fish, for example, are sensitive to zinc. Some species of fish are more sensitive than others to ions and organic solutes, and certain combinations of ions may exert synergistic effects. McKee and Wolf (1963) compiled many references on the effects of dissolved material on fish. Water-quality requirements for fish have been summarized by the U.S. Federal Water Pollution Control Administration (1968, p. 28–110).

Although highly impure water is attractive in appearance when viewed from a distance, it is obvious that even the lowest standards of pollution control must aim to produce a product reasonably pleasing to the senses of the viewer from close at hand, when he walks along the shore or rides over the water in a boat. The surroundings of the water body are an important part of this esthetic impression.

WATER-MANAGEMENT CONCEPTS AND PROBLEMS

The term "water-quality management" is frequently used in recent literature. Sometimes it is used as a synonym for "pollution control." Most of the time, however, it implies the utilization and development of water resources in a way that maintains the quality at the optimum level. This may involve many administrative and engineering factors concerned with decreasing the pollution loads contributed to streams by better and more complete sewage treatment, cleaning up existing pollution, and designing and building storage facilities to increase low flows of streams and thus to decrease quality fluctuations, or any of a number of related activities. The use of the term also implies that enough is known about the natural-water circulation systems so that quality indeed can be effectively managed.

The concept of water-quality management is related in a general way to broader concepts of management of water resources for full and efficient utilization by man. This kind of water utilization may involve extensive storage and transport facilities to make water available when and where it is needed. Sometimes a considerable degree of chemical control is required to make the quality of the water satisfactory, as in the treatment of public water supplies.

As the intensiveness of development of water supplies increases, the interweaving of chemical effects with the various physical processes in the circulation of water becomes more and more evident and of greater and greater practical importance. Some of the chemical effects of water storage in reservoirs have already been mentioned. There are some undesirable chemical effects in many impoundments; perhaps most significant is the tendency for water to become stratified at times and for previously accumulated sediments to contribute undesirable impurities to the water near the reservoir bottom. In arid climates, however, a more visible and generally objectionable feature of large open-storage reservoirs is the loss of water by evaporation. Evaporation losses from Lake Mead on the Colorado River average 849,000 acre-feet per year (Meyers, 1966, p. 94), equivalent to a depth of more than 6 feet over the surface of the lake.

A means of storage that avoids some of the disadvantages and ineffectiveness of surface reservoirs would have considerable appeal. The method most frequently suggested, and one that is generally believed to have the greatest promise, is the injection of surface water into the ground-water body for later removal by pumping. The integration of surface-water and ground-water systems that would result if this technique is adopted on a large scale has a very logical appeal to the water-resources planner.

It should be evident from the descriptions of chemical systems and influences in surface streams and in ground-water bodies that the two environments are very different. Conversion of a surface water to a ground water in large quantity at a rapid rate and with a minimum of effort and expense is necessary in successful artificial recharge of ground-water reservoirs. Chemical factors that appear subtle and unimportant to the casual observer may create great difficulties in the recharging process.

Recharging techniques most commonly used include spreading water on the land surface and allowing it to percolate to the water table and direct injection of water through wells, as a reversal of the usual pumping process. Some success has, of course, been attained in both ways, partly inadvertently. For example, most irrigation projects have brought about extensive ground-water recharge by infiltration from irrigated fields as well as from unlined ditches and canals. "Water flooding" as a means of recovering petroleum is extensively practiced, and highly unfavorable chemical conditions have been overcome as brines are pumped back down wells in these operations.

Nevertheless, it is evident from literature on the subject of artificial recharge that the techniques for accomplishing recharge are still in a rather early stage of development, and most attempts at injection through wells have encountered difficulties. It is obvious that often insufficient thought has been given to the effect on water compositions that would be brought about by injecting them into a new environment or to reactions that might occur between native and introduced solutes. Changes in Eh and in pH may occur altering solubility of solutes such as calcium or iron and causing precipitates that clog openings around the injection well. Or the new solution may dissolve objectionable amounts of impurities from solids that were stable in the previously existing environment.

Research on artificial recharge is in progress, and a better understanding of problems and means of coping with them should result in time. There is an extensive literature on the nonchemical aspects of artificial recharge. An introduction to earlier literature can be obtained from the bibliography by Todd (1959). More recent papers describing recharging experiments in which some consideration was given to chemistry include one by Reed, Deutsch, and Wiitala (1966) on work done at Kalamazoo, Mich., and two reports on experiences in Oregon and Washington by Price, Hart, and Foxworthy (1965) and by Foxworthy and Bryant (1967). Artificial-recharge experiments in the Grand Prairie region of Arkansas included some chemical considerations described by Sñiegocki (1963).

Wise management of water resources includes reasonable controls of present utilization practices so that quality is maintained for all present and potential future users of the resource. The discussions of principles of natural-water chemistry and their application in this volume have often referred to uncertainties and need for further study. Nevertheless, the general level of knowledge of the principles has greatly improved over the past few decades, and at least in the field of water supply and utilization there is every reason for optimism that man will be able to learn how to develop and use the resource wisely. Continuing effort both in research and in action programs will accomplish this goal.

SELECTED REFERENCES

Akabane, Nasasake, and Kurosawa, Akira, 1958, Silicic acid solutions, 1, Systems of SiO_2, NaOH, $CaCl_2$, and H_2O: Chem. Soc. [Japan] Jour., v. 61, p. 303.

SELECTED REFERENCES

Alekin, O. A., and Moricheva, N. P., 1960, Investigation of the sorption of microelements by the carbonate system of natural water: Akad. Nauk SSSR Doklady, v. 133, nos. 1-6, p. 644-647 [English translation by Am. Geol. Inst.].

American Public Health Association, 1965, Standard methods for the examination of water and wastewater [12th ed.]: New York, 769 p.

American Society for Testing and Materials, 1964, Manual on industrial water and industrial waste water [2d ed.]: Philadelphia, 856 p.

Anderson, M. S., Lakin, H. W., Beeson, K. C., Smith, F. F., and Thacker, Edward, 1961, Selenium in agriculture: Agricultural Research Service, U.S. Dept. Agriculture Handb. 200, 65 p.

Anderson, P. W., 1963, Variations in the chemical character of the Susquehanna River at Harrisburg, Pennsylvania: U.S. Geol. Survey Water-Supply Paper 1779-B, 17 p.

Anderson, V. S., 1945, Some effects of atmospheric evaporation and transpiration on the composition of natural waters in Australia: Jour. Australian Chem. Inst. Proc., v. 12, p. 41-68, 83-98.

Askew, H. O., 1923, Solubility and hydrolysis of calcium carbonate: New Zealand Inst. Trans. and Proc., v. 54, p. 791-796.

Association of Official Agricultural Chemists, 1965, Official methods of analysis of the Association of Official Agricultural Chemists [10th ed.]: 957 p.

Baas-Becking, L. G. M., Kaplan, I. R., and Moore, Derek, 1960, Limits of the natural environment in terms of pH and oxidation-reduction potentials: Jour. Geology, v. 68, p. 243-284.

Babcock, H. M., Brown, S. C., and Hem, J. D., 1947, Geology and ground-water resources of the Wellton-Mohawk area, Yuma County, Ariz.: U.S. Geol. Survey open-file report, 22 p.

Back, William, 1960, Origin of hydrochemical facies of ground water in the Atlantic Coastal plain: Internat. Geol. Cong. 21st, Copenhagen 1960, Rept., pt. 1, p. 87-95.

—— 1963, Preliminary result of a study of the calcium carbonate saturation of ground water in central Florida. Internat. Assoc. Sci. Hydrology, v. 8, no. 3, p. 43-51.

Back, William, and Barnes, Ivan, 1961, Equipment for field measurement of electrochemical potentials, in Short papers in the geologic and hydrologic sciences: U.S. Geol. Survey Prof. Paper 424-C, p. C366-C368.

—— 1965, Relation of electrochemical potentials and iron content to groundwater flow patterns U S. Geol. Survey Prof. Paper 498-C, 16 p.

Back, William, Cherry, R. N, and Hanshaw, B. B., 1966, Chemical equilibrium between the water and minerals of a carbonate aquifer Natl Speleol. Soc. Bull., v. 28, no 3, p 119-126.

Baker, M. N, 1949, The quest for pure water. New York, Am. Water Works Assoc., 527 p.

Banks, H. O, Richter, R. C., and Harder, James, 1957, Sea-water intrusion in California Am. Water Works Assoc. Jour, v 49, p 71-88.

Barker, F. B., and Johnson, J. O, 1964, Determination of radium in water: U.S. Geol. Survey Water-Supply Paper 1696-B, 29 p.

Barker, F. B., Johnson, J. O, Edwards, K. W., and Robinson, B. P., 1965, Determination of uranium in natural waters: U.S. Geol. Survey Water-Supply Paper 1696-C, 25 p.

Barker, F. B., and Robinson, B. P., 1963, Determination of beta activity in water. U.S. Geol. Survey Water-Supply Paper 1696-A, 32 p.

Barnes, Ivan, 1964, Field measurement of alkalinity and pH: U.S. Geol. Survey Water-Supply Paper 1535-H, 17 p.

——— 1965, Geochemistry of Birch Creek, Inyo County, Calif., a travertine depositing creek in an arid climate: Geochim. et Cosmochim. Acta, v. 29, p. 85-112.

Barnes, Ivan, and Back, William, 1964a, Dolomite solubility in ground water *in* Short papers in geology and hydrology: U.S. Geol. Survey Prof. Paper 475-D, p. D179-D180.

——— 1964b, Geochemistry of iron-rich ground water of southern Maryland: Jour. Geology, v. 72, p. 435-447.

Barnes, Ivan, and Bentall, Ray, 1968, Water-mineral relations of Quaternary deposits in the lower Platte River drainage area in eastern Nebraska: U.S. Geol. Survey Water-Supply Paper 1859-D, 39 p.

Barnes, Ivan, and Clarke, F. E., 1969, Chemical properties of ground water and their corrosion and encrustation effects on wells: U.S. Geol. Survey Prof. Paper 498-D, 58 p.

Barnes, Ivan, LaMarche, V. C., Jr., and Himmelberg, Glen, 1967, Geochemical evidence of present-day serpentinization: Science, v. 156, p. 830-832.

Barnes, Ivan, Stuart, W. T., and Fisher, D. W., 1964, Field investigation of mine waters in the northern anthracite field, Pennsylvania: U.S. Geol. Survey Prof. Paper 473-B, 8 p.

Barnett, P. R., Skougstad, M. W., and Miller, K. J., 1969, Chemical characterization of a public water supply: Am. Water Works Assoc. Jour. v. 61, p. 61-67.

Barth, T. W. F., 1952, Theoretical petrology, a text book on the origin and the evolution of rocks: New York, John Wiley and Sons, 416 p.

——— 1961, Abundance of the elements, areal averages, and geochemical cycles: Geochim. et Cosmochim. Acta, v. 23, p. 1-8.

Bates, R. G., 1964, Determination of pH—theory and practice: New York, John Wiley and Sons, 435 p.

Bean, E. H., 1962, Progress report on water quality criteria. Am. Water Works Assoc. Jour., v. 54, p. 1313-1331.

Begemann, Friedrich, and Libby, W. F., 1957, Continental water balance, groundwater inventory and storage times, surface ocean mixing rates and worldwide water circulation patterns from cosmic-ray and bomb tritium· Geochim. et Cosmochim. Acta, v. 12, p. 277-296.

Benson, S. W., 1960, The foundations of chemical kinetics: New York, McGraw-Hill Book Co., 703 p.

Berner, R. A., 1967, Comparative dissolution characteristics of carbonate minerals in the presence and absence of aqueous magnesium ion Am. Jour. Sci., v. 265, p. 45-70.

Berry, D. W., 1952, Geology and ground-water resources of Lincoln County, Kansas: Kansas Geol. Survey Bull. 95, 96 p.

Billings, G. K., and Williams, H. H., 1967, Distribution of cholrine in terrestrial rocks—a discussion. Geochim. et Cosmochim. Acta, v. 31, p. 2247.

Bingham, F. T., Page, A. L., and Bradford, G. R., 1964, Tolerance of plants to lithium: Soil Sci., v. 98, p. 4-8.

Black, A. P., and Christman, R. F., 1963a, Chemical characteristics of fulvic acids: Am. Water Works Assoc. Jour., v. 55, p. 897-912.

——— 1963b, Characteristics of colored surface waters: Am. Water Works Assoc. Jour., v. 55, p. 753-770.

Black, C. A., Evans, D. D., White, J. L., Ensmigner, L. E., and Clark, F. E., 1965, Methods of soil analysis; Part 2, Chemical and microbiological properties: Madison, Wis., Am. Soc. Agronomy, 1572p.

Bloss, F. D., and Steiner, R. L., 1960, Biogeochemical prospecting for manganese in northeast Tennessee: Geol. Soc. America Bull., v. 71, p. 1053-1066.

Bond, G. W., 1946, A geochemical survey of the underground water supplies of the Union of South Africa: South Africa Dept. Mines Geol. Survey Memoir 41, 208 p.

Bower, C. A., and Hatcher, J. T., 1967, Adsorption of fluoride by soils and minerals: Soil Sci., v. 103, p. 141-154.

Bredehoeft, J. D., Blyth, C. R., White, W. A., and Maxey, G. B., 1963, Possible mechanism for concentration of brines in subsurface formations: Am. Assoc. Petroleum Geologists Bull., v. 47, p. 257-269.

Brocas, J., and Picciotto. E., 1967, Nickel content of Antarctic snow—Implications of the influx rate of extraterrestrial dust: Jour. Geophys. Research, v. 72, p. 2229-2236.

Broom, M. E., 1966, "Iron water" from wells—causes and prevention: Ground Water, v. 4, no. 1, p. 18-21.

Bruvold, W. H., Ongerth, H. J., and Dillehay, R. C., 1967, Consumer attitudes towards mineral taste in domestic water: Am. Water Works Assoc. Jour., v. 59, p. 547-556.

Buckman, H. O., and Brady, N. C., 1960, The nature and properties of soils [6th ed.]: New York, MacMillan Co., 567 p.

Burttschell, R. H., Rosen, A. R., Middleton, F. M., and Ettinger, M. B., 1959, Chlorine derivatives of phenol causing taste and odor: Am. Water Works Assoc. Jour., v. 51, p. 205-214.

Butler, J. N., 1964, Ionic equilibrium—a mathematical approach: Reading, Mass., Addison-Wesley Publishing Co., 547 p.

Canney, F. C., and Post, E. V., 1964, Suggested exploration target in west-central Maine *in* Geological Survey research: U.S. Geol. Survey Prof. Paper 501-D, p. D61-D64.

Carlston, C. W., 1964, Use of tritium in hydrologic research problems and limitations: Internat. Assoc. Scientific Hydrology Bull., v. 9, no. 3, p. 38-42.

Carlston, C. W., and Thatcher, L. L., 1962, Tritium studies in the United States Geological Survey *in* Tritium in the physical and biological sciences—a symposium on the detection and use of tritium: Internat. Atomic Energy Agency, Symposium Proc., v. 1, Vienna, May 1961, p. 75-81.

Carozzi, A. V., 1960, Microscopic sedimentary petrography: New York, John Wiley and Sons, 485 p.

Chebotarev, I. I., 1955, Metamorphism of natural waters on the crust of weathering: Geochim. et Cosmochim. Acta, v. 8, p. 22-48; 137-170, 198-212.

Cheney, E. S., and Jensen, M. L., 1965, Stable carbon isotopic composition of biogenic carbonates: Geochim. et Cosmochim. Acta, v. 29, p. 1331-1346.

Clark, F. M., Scott, R. M., and Bone, Ester, 1967, Heterotrophic iron-precipitating bacteria: Am. Water Works Assoc. Jour., v. 59, p. 1036-1042.

Clarke, F. E., 1966, Significance of chemistry on water well development· Central Treaty Organization Symposium on hydrology and water resources development, 1st, Ankara, Turkey, 1966, p. 367-390.

Clarke, F. W., 1924a, The data of geochemistry [5th ed.]: U.S. Geol. Survey Bull. 770, 841 p.

——— 1924b, The composition of river and lake waters of the United States: U.S. Geol. Survey Prof. Paper 135, 199 p.

Clarke, F. W., and Washington, H. S., 1924, The composition of the earth's crust: U.S. Geol. Survey Prof. Paper 127, 117 p.

Cloke, P. L., 1963, The geologic role of polysulfides; Part 1, The distribution of ionic species in aqueous sodium polysulfide solutions: Geochim. et Cosmochim. Acta, v. 27, p. 1265–1298.

Coates, D. R., and Cushman, R. L., 1955, Geology and ground-water resources of the Douglas Basin, Ariz., *with a section on* Chemical quality of ground water by J. L. Hatchett: U.S. Geol. Survey Water-Supply Paper 1354, 56 p.

Cohen, J. M., Kamphake, L. J., Harris, E. K., and Woodward, R. L., 1960, Taste threshold concentrations of metals in drinking water: Am. Water Works Assoc. Jour., v. 52, p. 660–670.

Colby, B. R., Hembree, C. H., and Jochens, E. R., 1953, Chemical quality of water and sedimentation in the Moreau River Drainage basin, South Dakota: U.S. Geol. Survey Circ. 270, 53 p.

Colby, B. R., Hembree, C. H., and Rainwater, F. H., 1956, Sedimentation and chemical quality of surface waters in the Wind River Basin, Wyo.: U.S. Geol. Survey Water-Supply Paper 1373, 336 p.

Collins, W. D., 1923, Graphic representation of analyses: Indus. and Eng. Chemistry, v. 15, p. 394.

Conway, E. J., 1943, Mean geochemical data in relation to oceanic evolution: Royal Irish Acad. Proc., v. 48, sec. B, no. 8, p. 119.

Cooke, C. W., 1925, The Coastal Plain *in* Physical geography of Georgia: Georgia Geol. Survey Bull. 42, p. 19–54.

Cooper, H. H., Jr., 1959, A hypothesis concerning the dynamic balance of fresh water and salt water in a coastal aquifer: Jour. Geophys. Research, v. 64, p. 461–467.

Corbett, D. M., and others 1945, Stream gaging procedure. U.S. Geol. Survey Water-Supply Paper 888, 245 p.

Correns, C. W., 1956, The geochemistry of the halogens, *in* Ahrens, L. H., Rankama, K., and Runcorn, S. K., eds., Physics and Chemistry of the Earth, v. 1: New York, McGraw Hill Book Co., p. 181–233.

Covington, A. K., Robinson, R. A., and Bates, R. C., 1966, The ionization constant of deuterium oxide from 5° to 50°: Jour. Phys. Chemistry, v. 70, p. 3820–3824.

Craig, Harmon, 1963, The isotope geochemistry of water and carbon in geothermal areas: [Italian] Consiglio Nazionale delle Richerche, Laboratorio di Geologia Nucleare, Univ. Pisa, [Internat. Conf.] Nuclear geology on geothermal areas, Spoleta, Italy, 1963, p. 17–51.

Davis, G. H., 1961, Geologic control of mineral composition of stream waters of the eastern slope of the southern Coast Ranges, Calif.: U.S. Geol. Survey Water-Supply Paper 1535-B, 30 p.

Davis, S. N., 1964., Silica in streams and ground water: Am. Jour. Sci., v. 262, p. 870–891.

Dawdy, D. R., and Feth, J. H., 1967, Applications of factor analysis in study of chemistry of ground water quality, Mojave Valley, Calif.: Water Resources Research, v. 3, p. 505–510.

De la O. Careño, Alfonso, 1951, Las provincias geohidrológicas de México (primera parte): Univ. Nac. México, Inst. geología, Bol. 56, 137 p.

Dole, R. B., 1909, The quality of surface waters of the United States; Part 1, Analyses of waters east of the one hundredth meridian: U.S. Geol. Survey Water-Supply Paper 236, 123 p.

Doneen, L. D., 1954, Salination of soil by salts in the irrigation water: Am. Geophys. Union Trans., v. 35, p. 943–950.

Drost-Hansen, W., 1967, The structure of water and water-solute interactions *in* Equilibrium concepts in natural water systems: Am. Chem. Soc. Advances in Chemistry Ser., v. 67, p. 70–120.

Durfor, C. N., and Becker, Edith, 1964, Public water supplies of the 100 largest cities in the United States, 1962: U.S. Geol. Survey Water-Supply Paper 1812, 364 p.

Durov, S. A., 1948, Natural waters and graphic representation of their composition: Akad. Nauk SSSR Doklady, v. 59, p. 87–90.

Durum, W. H., 1953, Relationship of the mineral constituents in solution to streamflow, Saline River near Russell, Kans.: Am. Geophys. Union Trans., v. 34, p. 435–442.

Durum, W. H., and Haffty, Joseph, 1961, Occurrence of minor elements in water: U.S. Geol. Survey Circ. 445, 11 p.

───── 1963, Implications of the minor element content of some major streams of the world: Geochim. et Cosmochim. Acta, v. 27, p. 1–11.

Durum, W. H., Heidel, S. G., and Tison, L. J., 1960, World-wide runoff of dissolved solids: Internat. Assoc. Scientific Hydrology Pub. 51, p. 618–628.

Eaton, F. M., 1950, Significance of carbonates in irrigation waters: Soil Sci., v. 69, p. 127–128.

───── 1954, Formulas for estimating leaching and gypsum requirements of irrigation waters: Texas Agriculture Expt. Sta. Misc. Pub. 111, 18 p.

Eaton, F. M., McLean, G. W., Bredell, G. S., and Doner, H. E., 1968, Significance of silica in the loss of magnesium from irrigation waters: Soil Sci., v. 105, p. 260–280.

Egner, H., and Eriksson, Erik, 1955, Current data on the chemical composition of air and precipitation: Tellus, v. 7, p. 134–139.

Ehrlich, G. G., and Schoen, Robert, 1967, Possible role of sulfur-oxidizing bacteria in surficial acid alteration near hot springs, *in* Geological Survey Research 1967: U.S. Geol. Survey Prof. Paper 575–C, p. C110–C112.

Eichholz, G. G., Galli, A. N., and Elston, L. W., 1966, Problems in trace element analysis in water: Water Resources Research, v. 2, p. 561–566.

Emmons, W. H., 1917, The enrichment of ore deposits: U.S. Geol. Survey Bull. 625, 530 p.

Emmons, W. H., and Harrington, G. L., 1913, A comparison of waters of mines and of hot springs: Econ. Geology, v. 8, p. 653–669.

Epstein, Samuel, and Mayeda, T., 1953, Variation of O^8 content of waters from natural sources: Geochim. et Cosmochim. Acta, v. 4, p. 213–224.

Erickson, J. E., 1957, Geochemical prospecting abstracts July 1952–December 1954: U.S. Geol. Survey Bull. 1000–G, p. 357–395.

Eriksson, Erik, 1955, Airborne salts and the chemical composition of river waters: Tellus, v. 7, p. 243–250.

───── 1960, The yearly circulation of chloride and sulfur in nature; meteorological, geochemical and pedological implications—Parts 1 and 2: Tellus, v. 11, p. 375–403, and Tellus, v. 12, p. 63–109.

Evans, H. T., Jr., and Garrels, R. M., 1958, Thermodynamic equilibria of vanadium in aqueous systems as applied to the interpretation of the Colorado Plateau ore deposits: Geochim. et Cosmochim. Acta, v. 15, p. 131–149.

Fersman, A. E., Borovik, S. A., Gorshkov, S. D., Popov, Sosedko, 1952, Geochemical and mineralogical methods of prospecting for mineral depostis: U.S. Geol. Survey Circ. 127, 37 p. [Translated from Leningrad Akad. Nauk. Moskva· by Hartsock and Pierce.]

Feth, J. H., 1965, Calcium, sodium, sulfate, and chloride in stream water of the western conterminous United States to 1957: U.S. Geol. Survey Hydrol. Inv. Atlas HA-189.

—— 1966, Nitrogen compounds in natural water—a review: Water Resources Research, v. 2, p. 41–58.

Feth, J. H., and Hem, J. D., 1963, Reconnaissance of headwater springs in the Gila River drainage basin, Ariz.: U.S. Geol. Survey Water-Supply Paper 1619-H, 54 p.

Feth, J. H. and others, 1965, Preliminary map of the conterminous United States showing depth to and quality of shallowest ground water containing more than 1,000 ppm dissolved solids: U.S. Geol. Survey Hydrol. Inv. Atlas HA-199.

Feth, J. H., Roberson, C. E., and Polzer, W. L., 1964, Sources of mineral constituents in water from granitic rocks in Sierra Nevada, California and Nevada: U.S. Geol. Survey Water-Supply Paper 1535-I, 70 p.

Feth, J. H., Rogers, S. M., and Roberson, C. E., 1961, Aqua de Ney, California, a spring of unique chemical character: Geochim. et Cosmochim Acta, v. 22, p. 75–86.

—— 1964, Chemical composition of snow in the northern Sierra Nevada and other areas: U.S. Geol. Survey Water-Supply Paper 1535-J, 39 p.

Feulner, A. J., and Hubble, J. H., 1960, Occurrence of strontium in the surface and ground waters of Champaign County, Ohio: Econ. Geology, v. 55, p. 176–186.

Fiedler, A. G., and Nye, S. S., 1933, Geology and ground water resources of the Roswell artesian basin, New Mexico: U.S. Geol. Survey Water-Supply Paper 639, 372 p.

Filatov, K. V., 1948, Graphic representation of chemical analyses of water: Akad. Nauk SSSR Doklady, v. 59, p. 91–94.

Fisher, D. W., 1968, Annual variations in chemical composition of atmospheric precipitation in eastern North Carolina and southeastern Virginia: U.S. Geol. Survey Water-Supply Paper 1535-M, 21 p.

Fishman, M. J., Robinson, B. P., and Midgett, M. R., 1967, Water analysis: Anal. Chemistry, v. 39, p. 261R–294R.

Fleischer, Michael, 1953, Recent estimates of the abundance of elements in the earth's crust: U.S. Geol. Survey Circ. 285, 7 p.

—— 1954, The abundance and distribution of the chemical elements in the earth's crust: Jour. Chem. Education, v. 31, p. 446–455.

Ford, M. E., Jr., 1963, Air injection for control of reservoir limnology: Am. Water Works Assoc. Jour., v. 55, p. 267–274.

Fournier, R. O., and Rowe, J. J., 1962, The solubility of cristobalite along the three-phase curve, gas plus liquid plus cristobalite: Am. Mineralogist, v. 47, p. 897–902.

Foxworthy, B. L., and Bryant, C. T., 1967, Artificial recharge through a well tapping basalt aquifers at The Dalles, Oreg.: U.S. Geol. Survey Water-Supply Paper 1594-E, 55 p.

Frear, G. L., and Johnston, John, 1929, The solubility of calcium carbonate (calcite) in certain aqueous solutions at 25°C: Am. Chem. Soc. Jour., v. 51, p. 2082–2093.

Friedlander, Gerhart, and Kennedy, J. W., 1955, Nuclear and radiochemistry: New York, John Wiley and Sons, 468 p.

Friedman, Irving, 1953, Deuterium content of natural waters and other substances Geochim. et Cosmochim. Acta, v. 4, p. 89–103.

Gambell, A. W , and Fisher, D. W., 1966, Chemical composition of rainfall, eastern North Carolina and southeastern Virginia: U.S. Geol. Survey Water-Supply Paper 1535-K, 41 p.

Garrels, R. M., 1960, Mineral equilibria at low temperature and pressure: New York, Harper and Bros., 254 p.

―― 1965, Silica, role in the buffering of natural waters: Science, v. 148, p. 69.

―― 1967, Genesis of some ground waters from igneous rocks *in* Abelson, P. H., Researches in Geochemistry, v. 2: New York, John Wiley and Sons, p. 405-420.

Garrels, R. M., and Christ, C. L., 1964, Solutions, minerals, and equilibria· New York, Harper and Row, 450 p.

Garrels, R. M., and MacKenzie, F. T., 1967, Origin of the chemical compositions of some springs and in lakes *in* Equilibrium concepts in natural water chemistry: Am. Chem. Soc. Advances in Chemistry Ser., v. 67, p. 222-242.

Garrels, R. M., and Naeser, C. R., 1958, Equilibrium distribution of dissolved sulfur species in water at 25°C and one atmosphere total pressure: Geochim. et Cosmochim. Acta, v. 15, p. 113-130.

Garrels, R. M., Thompson, M. E., and Siever, R., 1960, Stability of some carbonates at 25°C and one atmosphere total pressure· Am. Jour. Sci., v. 258, p. 402-418.

Gast, J. A., and Thompson, T. G., 1959, Evaporation of boric acid from sea water: Tellus, v. 11, p. 344-347.

Gatewood, J. S., Robinson, T. W., Colby, B. R., Hem, J. D., and Halpenny, L. C., 1960, Use of water by bottom-land vegetation in lower Safford Valley, Ariz.: U.S. Geol. Survey Water-Supply Paper 1103, 210 p.

Gatewood, J. S., Wilson, Alfonso, Thomas, H. E., and Kister, L. R., 1964, General effects of drought on water resources of the Southwest: U.S. Geol. Survey Prof. Paper 372-B, 53 p.

George, R. D., and others, 1920, Mineral waters of Colorado: Colorado Geol. Survey Bull. 11, 474 p.

Germanov, A. L., Volkov, G. A., Lisitsin, A. K., and Serebrennikov, V. A., 1959, Investigation of the oixidation-reduction potential of ground waters: Geokhimiya, v. 1959, p. 322-329.

Gidley, H. K., 1952, Installation and performance of radial collector wells in Ohio River gravels: Am. Water Works Assoc. Jour., v. 44, p. 1117-1126.

Ginzburg, I. I., and Kabanova, E. S., 1960, Silica and its form in natural waters: Kora Vyvetrivaniya, vyp. 3, Akad. Nauk SSSR, Inst. Geologii Rudnykh Mestorozhdenii Petrografii., Mineralogii i Geokhimii, p. 313-341. [In Russian.]

Glagoleva, M. A., 1959, The forms of migration of elements in river waters: K poznaniyu diageneza osadkov., Akad. Nauk SSSR, Sbornik Statei 1959, p. 5-28.

Glasstone, Samuel, and Lewis, David, 1960, Elements of physical chemistry: Princeton, N. J., D. Van Nostrand Co., 758 p.

Goldberg, E. D., 1954, Marine geochemistry; Part 1, Chemical scavengers of the sea: Jour. Geology, v. 62, p. 249-265.

―― 1961, Chemistry in the oceans *in* Sears, Mary, ed., Oceanography: Internat. Oceanographic Cong., Am. Assoc, Advancement Sci. Pub. 67, p. 583-597.

―― 1963, Chemistry—the oceans as a chemical system *in* Hill, M. N., Composition of sea water, comparative and descriptive oceanography, Volume 2 *of* The Sea New York, Interscience Publishers, p. 3-25.

Goldberg, E. D., and Arhennius, G. O. S, 1958 Chemistry of Pacific pelagic sediments: Geochim. et Cosmochim. Acta, v. 13, p. 153–212.

Goldschmidt, V. M, 1933, Grundlagen der quantitativen chemie· Fortschr. Mineralog. Krist. Petrog., v. 17, p. 112.

—— 1937, The principles of distribution of chemical elements in minerals and rocks: Chem. Soc. [London] Jour , v. 1937, p. 635.

—— 1954, Geochemistry (Alex Muir, ed.). Oxford, England, Clarendon Press, 730 p.

Gorham, Eville, 1955, On the acidity and salinity of rain: Geochim. et Cosmochim. Acta, v. 7, p. 231–239.

—— 1961, Factors influencing supply of major ions to inland waters with special reference to the atmosphere: Geol. Soc. America Bull., v. 72, p. 795–840.

Graf, D. L., 1962, Minor element distribution in sedimentary carbonate rocks: Geochim. et Cosmochim. Acta, v. 26, p. 849–856.

Greenberg, S. A., and Price, E. W., 1957, The solubility of silica in solutions of electrolytes Jour. Phys. Chemistry, v. 61, p. 1539–1541.

Greenwald, Isidor, 1941, Dissociation of calcium and magnesium carbonates and bicarbonates· Jour. Biol. Chemistry, v. 141, p. 789–796.

Griggs, R. L., and Hendrickson, G. E., 1951, Geology and ground-water resources of San Miguel County, N. Mex.: New Mexico Bur. Mines and Mineral Resources, Ground-Water Report 2, 121 p.

Gunnerson, C. G., 1967, Streamflow and quality in the Columbia River basin: Am. Soc. Civil Engineers Proc., Sanitary Eng. Div. Jour , v. 39, p. 1–16.

Haffty, Joseph, 1960, Residue method for common minor elements. U S. Geol. Survey Water-Supply Paper 1540–A, 9 p.

Halevy, E., 1964, The exchangeability of hydroxyl groups in kaolinite: Geochim. et Cosmochim. Acta, v. 28, p. 1139–1145.

Hanshaw, B. B., 1964, Cation-exchange constants for clay from electrochemical measurements: Natl. Conf. on Clays and Clay minerals, 12th, Atlanta 1963, Proc., p. 397–421.

Hanshaw, B. B., Back, William, and Rubin, Meyer, 1965, Radiocarbon determinations for estimating ground water flow velocities in central Florida: Science, v. 148, no. 3669, p. 494–495.

Hanshaw, B. B., Back, William, Rubin, Meyer, and Wait, R. L., 1965, Relation of carbon 14 concentration to saline water contamination of coastal aquifers· Water Resources Research, v. 1, p 109–114.

Hanshaw, B. B., and Zen, E-An, 1965, Osmotic equilibrium and overthrust faulting· Geol. Soc. America Bull , v. 76, p. 1379–1386.

Harbaugh, J. W., 1953, Geochemical prospecting abstracts through June 1952. U.S. Geol. Survey Bull. 1000–A, p. 1–50.

Hawkes, H. E , and Webb, J. S , 1962, Geochemistry in mineral explorations New York, Harper and Row, 415 p.

Heidel, S. G., 1965, Dissolved oxygen and iron in shallow wells at Salisbury, Md Am. Water Works Assoc. Jour., v 57, p. 239–244.

Helfferich, Friederich, 1962, Ion Exchange New York, McGraw-Hill Book Co., 624 p.

Hely, A. G., Hughes, G. H., and Irelan, Burdge, 1966, Hydrologic regimen of Salton Sea, Calif.: U.S. Geol Survey Prof. Paper 486–C, 32 p

Hem, J. D., 1950, Quality of water in Gila River basin above Coolidge Dam, Ariz.: U.S. Geol. Survey Water-Supply Paper 1104, 230 p.

―――― 1960, Some chemical relationships among sulfur species and dissolved ferrous iron. U.S. Geol. Survey Water-Supply Paper 1459-C, p. 57-73.

―――― 1961a, Calculation and use of ion activity. U.S. Geol. Survey Water-Supply Paper 1535-C, 17 p.

―――― 1961b, Stability field diagrams as aids in iron chemistry studies: Am. Water Works Assoc. Jour., v. 53, p. 211-228.

―――― 1963a, Chemical equilibria and rates of manganese oxidation: U.S. Geol. Survey Water-Supply Paper 1667-A, 74 p.

―――― 1963b, Manganese complexes with bicarbonate and sulfate in natural water: Jour. Chem. and Eng. Data, v. 8, p. 99-101.

―――― 1964, Deposition and solution of manganese oxides: U.S. Geol. Survey Water-Supply Paper 1667-B, 42 p.

―――― 1966, Chemical controls of irrigation drainage water composition: Am. Water Resources Conf., 2d, Chicago 1966, Proc., p. 64-77.

―――― 1968, Graphical methods for studies of aqueous aluminum hydroxide, fluoride, and sulfate complexes: U.S. Geol. Survey Water-Supply Paper 1827-B, 33 p.

―――― 1968a, Aluminum species in water *in* Trace inorganics in water, by R. Baker, no. 73 *of* Advances in chemistry series, edited by R. F. Gould: Am. Chem. Soc., p. 98-114.

Hem, J. D., and Cropper, W. H., 1959, Survey of ferrous-ferric equilibria and redox potentials· U.S. Geol. Survey Water-Supply Paper 1459-A, 31 p.

Hem, J D., and Roberson, C. E., 1967, Form and stability of aluminum hydroxide complexes: U.S. Geol. Survey Water-Supply Paper 1827-A, 55 p.

Hem, J. D., and Skougstad, M. W., 1960, Coprecipitation effects in solutions containing ferrous, ferric, and cupric ions: U.S. Geol. Survey Water-Supply Paper 1459-E, p. 95-110.

Hendrickson, G. E., and Jones, R. S., 1952, Geology and ground-water resources of Eddy County, N. Mex.: New Mexico Bur. Mines and Mineral Resources, Ground-Water Report 3, 169 p.

Hendrickson, G. E., and Krieger, R. A., 1964, Geochemistry of natural waters of the Blue Grass Region, Kentucky: U.S. Geol. Survey Water-Supply Paper 1700, 135 p.

Hicks, W. B., 1921, Potash resources of Nebraska: U.S. Geol. Survey Bull. 715, p. 125-139.

Hill, R. A., 1940, Geochemical patterns in Coachella Valley, Calif.: Am. Geophys. Union Trans., v. 21, p. 46-49.

―――― 1941, Salts in irrigation water: Am. Soc. Civil Engineers Proc., v. 67, p. 975-990.

―――― 1942, Salts in irrigation water: Am. Soc. Civil Engineers Trans., v. 107, p. 1478-1493.

Hoffman, C. M., and Stewart, G. L., 1966, Quantitative determination of tritium in natural waters: U.S. Geol. Survey Water-Supply Paper 1696-D, 18 p.

Holland, H.D., Kirsipu, T. V., Huebner, J. S., and Oxbough, V. M., 1964, On some aspects of the chemical evolution of cave waters: Jour. Geology, V. 72, p. 36-67.

Horn, M. K., and Adams, J. A. S., 1966, Computer-derived geochemical balances and element abundances: Geochim. et Cosmochim. Acta, v. 30, p. 279-297.

Hostetler, P. B., 1964, The degree of saturation of magnesium and calcium carbonate minerals in natural waters: Internat. Assoc. Sci. Hydrology Commission of subterranean waters Pub. 64, p. 34-49.

Howard, C. S., 1960, Chemistry of the water *in* Comprehensive survey of sedimentation in Lake Mead 1948–49: U.S. Geol. Survey Prof. Paper 295, p. 115–124.

Hsu, K. J., 1963, Solubility of dolomite and composition of Florida ground waters: Jour. Hydrology, v. 1, p. 288–310.

Hsu, Pa Ho, and Bates, T.F., 1964, Formation of X-ray amorphous and crystalline alluminum hydroxides: Mineralog. Mag., v. 33, p. 749–768.

Hubbert, M. K., 1940, The theory of ground-water motion: Jour. Geology, v. 48, p. 785–944.

Hückel, Walter, 1950, Structural chemistry of inorganic compounds: New York, Elsevier Publishing Co., 1094 p.

Hutchinson, G. E., 1957, A Treatise on Limnology; v. 1, Geography, Physics, and Chemistry: New York, John Wiley and Sons, 1010 p.

Iler, R. K., 1955, The colloid chemistry of silica and silicates: Ithaca, N.Y., The Cornell Univ. Press, 324 p.

Ingols, R. S., and Wilroy, R. D., 1963, Mechanism of manganese solution in lake waters: Am. Water Works Assoc. Jour., v. 55, p. 282–290.

Ingram, M. I., Mackenthun, K. M., and Bartsch, A. F., 1966, Biological field investigative data for water pollution surveys: U.S. Dept. Interior, Fed. Water Pollution Control Admin., WP-13, 139 p.

Irelan, Burdge, and Mendieta, H. B., 1964, Chemical quality of surface waters in the Brazos River basin in Texas: U.S. Geol. Survey Water-Supply Paper 1779–K, 70 p.

Jenne, E. A., 1968, Controls on Mn, Fe, Co, Ni, Cu, and Zn concentrations in soils and water, the significant role of hydrous Mn and Fe oxides *in* Trace Inorganics in Water: Am. Chem. Soc. Advances in Chemistry Ser., v. 73, p. 337–387.

Johns, W. D., and Huang, W. H., 1967, Distribution of chlorine in terrestrial rocks: Geochim. et Cosmochim. Acta, v. 31, p. 35–49.

Johnson, J. O., and Edwards, K. W., 1967, Determination of strontium 90 in water: U.S. Geol. Survey Water-Supply Paper 1696–E, 10 p.

Jones, P. H., and Buford, T. B., 1951, Electric logging applied to ground-water exploration: Geophysics, v. 16, p. 115–139.

Junge, C. E., 1958, The distribution of ammonia and nitrate in rain water over the United States: Am. Geophys. Union Trans., v. 39, p. 241–248.

―――― 1960, Sulfur in the atmosphere: Jour. Geophys. Research, v. 65, p. 227–237.

―――― 1963, Air chemistry and radioactivity: New York, Academic Press, 382 p.

Junge, C. E., and Gustafson, P. E., 1957, On the distribution of sea salt over the United States and its removal by precipitation: Tellus, v. 9, p. 164–173.

Junge, C. E., and Werby, R. T., 1958, The concentration of chloride, sodium, potassium, calcium and sulfate in rain water over the United States: Jour. Meteorology, v. 15, p. 417–425.

Kaplan, I. R., Rafter, T. A., and Hulston, J. R., 1960, Sulfur isotopic variations in nature; Part 8, Application to some biogeochemical problems: New Zealand Jour. Sci., v. 3, p. 338–361.

Kaufman, W. J., and Orlob, G. T., 1956, Measuring ground water movement with radioactive and chemical tracers: Am. Water Works Assoc. Jour., v. 48, p. 559–572.

Kavanaugh, J. L., 1964, Water and solute-water interactions: San Francisco, Holden-Day, Inc., 101 p.

Kay, R. L., 1968, The effect of water structure on the transport properties of electrolytes *in* Trace Inorganics in Water: Am. Chem. Soc. Advances in Chemistry Ser., v. 73, p. 1-17.

Keighton, W. B., 1966, Fresh-water discharge—salinity relations in the tidal Delaware River: U.S. Geol. Survey Water-supply Paper 1586-G, 16 p.

Keller, W. D., 1957, The principles of chemical weathering [revised ed.]: Columbia, Mo., Lucas Bros. Publishers, 111 p.

Kelley, W. P., 1948, Cation exchange in soils. New York, Reinhold Publishing Corp., 144 p.

Kennedy, V. C., 1956, Geochemical studies in the southwestern Wisconsin zinc-lead area: U. S. Geol. Survey Bull. 1000-E, p. 187-223.

────── 1965, Mineralogy and cation exchange capacity of sediments from selected streams: U.S. Geol. Survey Prof. Paper 433-D, 28 p.

Kennedy, V. C., and Brown, T. E., 1966, Experiments with a sodium-ion electrode as a means of studying cation-exchange rates; Natl. Conf. on Clays and Clay Minerals, 13th, Madison 1964, Proc., p. 351-352.

Kielland, Jacob, 1937, Individual activity coefficients of ions in aqueous solutions: Am. Chem. Soc. Jour., v. 59, p. 1675-1678.

Kister, L. R., and Hardt, W. F., 1961, Correlation of ground-water quality with different sediment types, lower Santa Cruz basin, Ariz.: Ariz. Geol. Soc. Digest, v. 4, p. 79-85.

Kister, L. R., and Hardt, W. F., 1966, Salinity of the ground water in western Pinal County, Ariz.: U.S. Geol. Survey Water-Supply Paper 1819-E, 21 p.

Kobayashi, Jun, 1957, On geographical relationship between the chemical nature of river water and death-rate from apoplexy: Ber. Ohara Inst. fur landwirtschaftliche Biologie Bd. 11, Heft 1, p. 12-21.

Koga, Akito, 1957, Chemical studies on hot springs of Beppu: Nippon Kagaku Zasshi, v. 78, p. 1713-1725.

Konovalov, G. S., 1959, Removal of microelements by the principal rivers of the USSR: Akad. Nauk SSSR Doklady, v. 129, p. 912-915. [English translation by AGI from Doklady Akad. Nauk, Dec. 1960, p. 1034-1037].

Konzewitsch, Nicolás, 1967, Study of the proposed classifications for natural waters according to their chemical composition: Secretaria de estado de energia y mineria, Agua y energia electrica, div. recursos hidricos, Buenos Aires, Argentina, 108 p.

Krauskopf, K. B., 1956a, Dissolution and precipitation of silica at low temperatures: Geochim. et Cosmochim. Acta, v. 10, p. 1-26.

────── 1956b, Factors controlling the concentrations of thirteen rare metals in sea water: Geochim. et Cosmochim. Acta, v. 9, p. 1-32B.

────── 1967, Introduction to geochemistry: New York, McGraw-Hill Book Co., 721 p.

Kroner, R. C., and Kopp, J. F., 1965, Trace elements in six water systems of the United States: Am. Water Works Assoc. Jour., v. 57, p. 150-156.

Kulp, J. L., Turekian, K. K., and Boyd, D. W., 1952, Strontium content of limestone and fossils: Geol. Soc. America Bull., v. 63, p. 701-716.

Kuroda, P. K., and Sandell, E. B., 1953, Chlorine in igneous rocks. Geol. Soc. America Bull., v. 64, p. 879-896.

Kuroda, P. K., and Yokoyama, Yuji, 1954, Determination of short-lived decay products of radon in natural waters: Anal. Chemistry, v. 26, p. 1509-1511.

Lamar, W. L., and Goerlitz, D. F., 1966, Organic acids in natural colored surface waters: U.S. Geol. Survey Water-Supply Paper 1817-A, 17 p.

Lamb, A. B., and Jacques, A. O., 1938, The slow hydrolysis of ferric chloride in dilute solution; 2, The change in hydrogen-ion concentration: Am. Chem. Soc. Jour., v. 60, p. 1215.

Langbein, W. B., and Durum, W. H., 1967, The aeration capacity of streams: U.S. Geol. Survey Circ. 542, 6 p.

Langelier, W. F., 1936, The analytical control of anticorrosion water treatment: Am. Water Works Assoc. Jour., v. 28, p. 1500–1521.

Langelier, W. F., and Ludwig, H. F., 1942, Graphical methods for indicating the mineral character of natural waters: Am. Water Works Assoc. Jour., v. 34, p. 335–352.

Langguth, H. R., 1966, Die grundwasserverhältnisse im Bereich des Velberter Sattebs (Rheinisches Schiefergebirge): Der Minester fur Ernahrung, Landwirtschaft u. Forsten des Landes Nordrhein-Westfalen, Dusseldorf (Germany) 127 p.

Larson, T. E., 1949, Geologic correlations and hydrologic interpretation of water analyses: Illinois State Water Survey Div., Circ. 27, 8 p.

Latimer, W. M., 1952, Oxidation potentials [2d ed.]: New York, Prentice-Hall, 392 p.

Ledbetter, J. O. and Gloyna, E. F., 1962, Predictive techniques for water quality-inorganics: Univ. Texas Dept. Civil Eng. Environmental Health Eng. Laboratories, 72 p.

Lee, G. F., and Hoadley, A. W., 1967, Biological activity in relation to the chemical equilibrium composition of natural waters, *in* Equilibrium concepts in natural water systems: Am. Chem. Soc. Advances in Chemistry Ser., v. 67, p. 319–338.

Licht, Sidney, ed., 1963, Medical Hydrology: New Haven, Conn., E. Licht Publisher, 714 p.

Lieth, Helmut, 1963, The role of vegetation in the carbon dioxide content of the atmosphere: Jour. Geophys. Research, v. 68, p. 3887–3898.

Lindeman, H. B., 1954, Sodium carbonate brine and trona deposits in Sweetwater County, Wyoming: U.S. Geol. Survey Circ. 235, 10 p.

Lishka, R. J., Kelso, F. S., and Kramer, H. P., 1963, Evaluation of methods for determination of minerals in water: Am. Water Works Assoc. Jour., v. 55, p. 647–656.

Livingstone, D. A., 1963, Chemical composition of rivers and lakes *in* Data of geochemistry [6th ed.]: U.S. Geol. Survey Prof. Paper 440–G, G1–G64

Ljunggren, Pontus, 1951, The biogeochemistry of manganese: Geol. Förem. i Stockholm, Förh., v. 73, p. 639–652.

―――― 1953, Some data concerning the formation of manganiferous and ferriferous bog ores: Geol. Fören. i Stockholm, Förh., v. 75, p. 277–297.

―――― 1955, Geochemistry and radioactivity of some Mn and Fe bog ores: Geol. Fören. i Stockholm, Förh., v. 77, p. 33–44.

Lohr, E. W., and Love, S. K., 1954a, The industrial utility of public water supplies of the United States; Part 1, States east of the Mississippi River: U.S. Geol. Survey Water-Supply Paper 1299, 639 p.

―――― 1954b, The industrial utility of public water supplies of the United States, Part 2, States west of the Mississippi River: U.S. Geol. Survey Water-Supply Paper 1300, 462 p.

McCarren, E. F., 1967, Chemical quality of surface water in the Allegheny River basin, Pennsylvania and New York: U.S. Geol. Survey Water-Supply Paper 1835, 74 p.

McCarty, P. L., Hem, J. D., Jenkins, David, Lee, G. F., Morgan, J. J., Robertson, R. S., Schmidt, R. W., Symons, J. M., and Trexler, M. V., 1967, Sources of nitrogen and phosphorus in water supplies [Part 2]: Am. Water Works Assoc. Jour., v. 59, p. 344-366.

McDonald, H. R., Walcott, H. N., and Hem, J. D., 1947, Geology and groundwater resources of the Salt River Valley area, Maricopa and Pinal Counties, Ariz.: U. S. Geol. Survey open-file report, 45 p.

McKee, J. E., and Wolf, H. W., 1963, Water quality criteria: California State Water Quality Control Board Publ. 3-A, 548 p.

MacKenzie, F. T., and Garrels, R. M., 1966, Chemical mass balance between rivers and oceans: Am. Jour. Sci., v. 264, p. 507-525.

MacNeil, F. S., 1947, Geologic map of the Tertiary and Quaternary formations of Georgia: U.S. Geol. Survey Oil and Gas Inv. (Prelim.) Map 72.

Maderak, M. L., 1966, Sedimentation and chemical quality of surface water in the Heart River drainage basin, N. Dak.: U.S. Geol. Survey Water-Supply Paper 1823, 42 p.

Maliuga, D. P., 1950, Concerning biogeochemical provinces in the southern Urals: (Moscow) Akad. Nauk SSSR Doklady, v. 70, p. 257-259.

Mannheim, F. T., 1965, Manganese-iron accumulations in the shallow marine environment—symposium on marine geochemistry: Rhode Island, Narragansett Marine Laboratory, Univ. Rhode Island, Occ. Pub. 13, p. 217-276.

Markward, E. L., 1961, Geochemical prospecting abstracts January 1955–June 1957: U.S. Geol. Survey Bull. 1098-B, p. 57-160.

Maucha, Rezso, 1949, The graphical symbolization of the chemical composition of natural waters: Hidrol. Közölny, v. 13, p. 117-118.

Meisler, Harold, and Becker, A. E., 1967, Hydrogeologic significance of calcium-magnesium ratios in ground water from carbonate rocks in the Lancaster quadrangle southeastern Pennsylvania in Geological Survey research 1967: U.S. Geol. Survey Prof. Paper 575-C, p. C232-C235.

Meyers, J. S., 1962, Evaporation from the 17 Western States: U.S. Geol. Survey Prof. Paper 272-D, p. 71-100.

Middleton, F. M., Greenberg, A. E., and Lee, G. F., 1962, Tentative method for carbon chloroform extract (CCE) in water: Am. Water Works Assoc. Jour., v. 54, p. 223-227.

Mikey, N. I., 1963, Fluorine content in atmospheric precipitation and in surface waters of various origin: Trudy Gos. Gidrolog. Inst., no. 102, p. 209-226.

Miller, A. R., Densmore, C. D., Degens, E. T., Hathaway, J. C., Mannheim, F. T., McFarlin, P. F., Pocklington, R., and Joelka, A., 1966, Hot brines and recent iron deposits in deeps of the Red Sea: Geochim. et Cosmochim. Acta, v. 30, p. 341-360.

Miller, J. P., 1961, Solutes in small streams draining single rock types, Sangre de Cristo Range, N. Mex.: U.S. Geol. Survey Water-Supply Paper 1535-F, 23 p.

Mirtov, B. A., 1961, Gaseous composition of the atmosphere and its analysis: Akad. Nauk SSSR, Inst. Prikl. Geofiz. Moskva. [Translated by Israel program for scientific translations, pub. in Washington, U.S. Dept. Commerce Office of Tech. Services, 209 p.].

Moore, G. K., 1965, Geology and hydrology of the Claiborne group in western Tennessee: U.S. Geol. Survey Water-Supply Paper 1809-F, 44 p.

Morey, G. W., Fournier, R. O., and Rowe, J. J., 1962, The solubility of quartz in water in the temperature interval from 25° to 300°C: Geochim. et Cosmochim. Acta, v. 26, p. 1029-1044.

———— 1964, The solubility of amorphous silica at 25°C: Jour. Geophys. Research, v. 69, p. 1995-2002.
Morgan, J. J., 1967, Chemical equilibria and kinetic properties of manganese in natural waters *in* Faust, S. J., and Hunter, J. V., eds., Principles and Applications of Water Chemistry: New York, John Wiley and Sons, p. 561-624.
Morris, J. C., and Stumm, Werner, 1967, Redox equilibria and measurements of potentials in the aquatic environment *in* Equilibrium concepts in natural water systems: American Chem. Soc. Advances in Chemistry Ser., v. 67, p. 270-285.
Muss, D. L., 1962, Relationship between water quality and deaths from cardiovascular disease: Am. Water Works Assoc. Jour., v. 54, p. 1371-1378.
Nair, V. S. K., and Nancollas, G. H., 1959, Thermodynamics of ion association, Part VI, Transition metal sulfates: Chem. Soc. [London] Jour., v. 1959, p. 3934-3939.
Nakai, N., 1960, Carbon isotope fractionation of natural gas in Japan: Jour. Earth Sci. [Nagoya Univ., Nagoya, Japan], v. 3, p. 65-75.
Nelson, J. L., Perkins, R. W., and Haushild, W. L., 1966, Determination of Columbia River flow times downstream from Pasco, Wash., using radioactive tracers introduced by the Hanford reactors: Water Resources Research, v. 2, p. 31-40.
Nichols, M. S., and McNall, D. R., 1957, Strontium content of Wisconsin municipal waters: Am. Water Works Assoc. Jour., v. 49, p. 1493-1498.
Noble, D. C, Smith, V. C. and Peck, C., 1967, Loss of halogens from crystallized and glassy silicic volcanic rocks: Geochim. et Cosmochim. Acta, v. 31, p. 215-223.
Oborn, E. T., 1964, Intracelluar and extracellular concentration of manganese and other elements by aquatic organisms. U.S. Geol. Survey Water-Supply Paper 1667-C, 18 p.
Oborn, E. T., and Hem, J. D., 1962, Some effects of the larger types of aquatic vegetation on iron content of water: U.S. Geol. Survey Water-Supply Paper 1459-I, p. 237-268.
Odum, H. T., 1951, Strontium in Florida waters *in* Black, A. P., and Brown, Eugene, Chemical character of Florida waters: Florida Board of Conserv., Div. Water Survey and Research, Water Supply and Research Paper 6, 119 p.
Oltman, R. E., 1968, Reconnaissance investigations of the discharge and water quality of the Amazon River: U.S. Geol. Survey Circ. 552, 16 p.
Orfanidi, K. E., 1957, Carbonic acid in underground waters: Akad. Nauk SSSR Doklady, v. 115, p. 999-1001.
Orlob, G. T., and Woods, P. C., 1967, Water-quality management in irrigation systems: Jour. Irrig. and Drainage Div., Am. Soc. Civil Engineers, v. 93, no. IR-2, Proc. Paper 5280, p. 49-66.
Osipow, L. I., 1967, Physical adsorption on solids *in* Faust, S. D., and Hunter, J. V., eds., Principles and applications of water chemistry: New York, John Wiley and Sons, 643 p.
Palmer, Chase, 1911, The geochemical interpretation of water analyses: U.S. Geol. Survey Bull. 479, 31 p.
Parker, R. L., 1967, Composition of the earth's crust *in* Data of geochemistry [6th ed.]: U.S. Geol. Survey Prof. Paper 440-D, 19 p.
Patten, E. P. and Bennett, G. D., 1963, Application of electrical and radioactive well logging to ground-water hydrology: U.S. Geol. Survey Water-Supply Paper 1544-D, 60 p.

Pearson, F. J., Jr., and White, D. E., 1967, Carbon 14 ages and flow rates of water in Carrizo Sand, Atascosa County, Texas: Water Resources Research, v. 3, p. 251–261.

Penchevia, E. N., 1960, Spectrographic study of trace elements in spring waters of the Chepinsk river valley: Izv. Geol. Inst., Bulgar. Akad. Nauk, v. 8, p. 193–222.

Peterson, D. W., 1961, Descriptive modal classification of igneous rocks (AGI Data Sheet 23a): Geotimes, v. 5, no. 6, p. 30–36.

Pettyjohn, W. A., 1967, Geohydrology of the Souris River valley in the vicinity of Minot, North Dakota: U.S. Geol. Survey Water-Supply Paper 1844, 53 p.

Piper, A. M., 1944, A graphic procedure in the geochemical interpretation of water analyses: Am. Geophys. Union Trans., v. 25, p. 914–923.

Piper, A. M., Garrett, A. A., and others, 1953, Native and contaminated waters in the Long Beach–Santa Ana area, California: U.S. Geol. Survey Water-Supply Paper 1136, 320 p.

Poland, J. F., Garrett, A. A., and Sinnott, Allen, 1959, Geology, hydrology, and chemical character of ground waters in the Torrance–Santa Monica Area, Calif.: U.S. Geol. Survey Water-Supply Paper 1461, 425 p.

Polzer, W. L., and Hem, J. D., 1965, The dissolution of kaolinite: Jour. Geophys. Research, v. 70, p. 6233–6240.

Polzer, W. L., Hem, J. D., and Gabe, H. J., 1967, Formation of crystalline hydrous aluminosilicates in aqueous solutions at room temperature *in* Geological Survey research: U.S. Geol. Survey Prof. Paper 575–B, p. B128–B132.

Pommer, A. M., and Breger, I. A., 1960, Equivalent weight of humic acid from peat: Geochim. et Cosmochim. Acta, v. 20, p. 45–50.

Price, Don, Hart, D. H., and Foxworthy, B. L., 1965, Artificial recharge in Oregon and Washington 1962: U.S. Geol. Survey Water-Supply Paper 1594–C, 65 p.

Pyatt, E. E., 1964, On determining pollutant distribution in tidal estuaries: U.S. Geol. Survey Water-Supply Paper 1586–F, 56 p.

Rainwater, F. H., 1962, Stream composition of the conterminous United States: U.S. Geol. Survey Hydrol. Inv. Atlas HA–61.

Rainwater, F. H., and Thatcher, L. L., 1960, Methods for collection and analysis of water samples: U.S. Geol. Survey Water-Supply Paper 1454, 301 p.

Rainwater, F. H., and White, W. F., 1958, The solusphere—its inferences and study: Geochim. et Cosmochim. Acta, v. 14, p. 244–249.

Rankama, Kalervo, and Sahama, T. G., 1950, Geochemistry: Chicago, Univ. Chicago Press, 912 p.

Reed, J. E., Deutsch, Morris, and Wiitala, S. W., 1966, Induced recharge of an artesian glacial-drift aquifer at Kalamazoo, Mich.: U.S. Geol. Survey Water-Supply Paper 1594–D, 62 p.

Reiche, Parry, 1950, A survey of weathering processes and products: Albuquerque, New Mexico Univ. Pubs. in Geology, no. 3, 95 p.

Reistle, C. E., Jr., 1927, Identification of oil-field waters by chemical analysis: U.S. Bur. Mines Tech. Paper 404, 25 p.

Renick, B. C , 1924, Base exchange in ground water by silicates as illustrated in Montana· U.S. Geol. Survey Water-Supply Paper 520–D, p. 53–72.

Roberson, C. E., Feth, J. H., Seaber, P. R., and Anderson, Peter, 1963, Differences between field and laboratory determinations of pH, alkalinity, and specific conductance of natural water *in* Short papers in geology and hydrology· U.S. Geol. Survey Prof. Paper 475–C, p. C212–C217.

Roberson, C. E., and Hem, J. D., 1969, Solubility of aluminum in the presence of hydroxide, fluoride, and sulfate: U.S. Geol. Survey Water-Supply Paper 1827-C, 37 p.

Roberson, C. E., and Whitehead, H. C., 1961, Ammoniated thermal waters of Lake and Colusa Counties, California: U.S. Geol Survey Water-Supply Paper 1535-A, 11 p.

Robie, R. A., and Waldbaum, D. R., 1968, Thermodynamic properties of minerals and related substances at 299.15°K (25.0°C) and one atmosphere (1.013 bars) pressure and at higher temperatures: U.S. Geol. Survey Bull. 1259, 256 p.

Robinove, C. J., Langford, R. H., and Brookhart, J. W., 1958, Saline-water resources of North Dakota: U.S. Geol. Survey Water-Supply Paper 1428, 72 p.

Robinson, B. P., 1962, Ion-exchange minerals and disposal of radioactive wastes— a survey of literature: U.S. Geol. Survey Water-Supply Paper 1616, 132 p.

Rogers, A. S., 1958, The physical behavior and geologic control of radon in mountain streams: U.S. Geol. Survey Bull. 1052-E p. 187-211.

Rossini, F. E., Wagman, D. D., Evans, W. H., Levine, Samuel, and Jaffe, Irving, 1952, Selected values of chemical thermodynamic properties: U.S. Natl. Bur. Standards Circ. 500, 1268 p.

Schnitzer, M., and Skinner, S. I. M., 1967, Organo-metallic interactions in soils; 7., Stability constants of Pb^{+2}, Ni^{+2}, Mn^{+2}, Co^{+2}, Ca^{+2}, and Mg^{+2} fulvic acid complexes: Soil Sci., v. 103, p. 247-257.

Schoeller, Henri, 1935, Utilitie de la notion des exchanges de bases pour la comparison des eaux souterraines: France, Soc. Geol. Comptes rendus Sommaire et Bull. Ser. 5, v. 5, p. 651-657.

—— 1955, Geochemie des eaux souterraines: Revue de L'Institute Francais due petrole, v. 10, p. 230-244.

—— 1962, Les eaux souterraines: Paris, France, Massio and Cie, 642 p.

Schofield, J. C., 1960, Boron in some New Zealand ground waters: New Zealand Jour. Geol. and Geophys., v. 3, p. 98-104.

Schweisfurth, Reinhart, 1963, Mikrobiologische Unterschungen auf manganoxydierenden mikroorganismen aus schwarzen Schlámmen: Deutsche Gewässerkendiche Mitt., v. 1963, p. 25-27.

Scofield, C. S., 1932, Measuring the salinity of irrigation waters and soil solutions with the Wheatstone bridge: U.S. Dept. Agriculture Circ. 232, 16 p.

—— 1940, Salt balance in irrigated areas: Jour. Agr. Research, v. 61, p. 17-39.

Scott, R. C., and Barker, F. B., 1961, Ground-water sources containing high concentrations of radium *in* Short papers in the geologic and hydrologic sciences: U.S. Geol. Survey Prof. Paper 424-D, p. D357-D359.

—— 1962, Data on uranium and radium in ground water in the United States, 1954-1957: U.S. Geol. Survey Prof. Paper 426, 115 p.

Setter, L. R., Regnier, J. E., and Diephaus, E. A., 1959, Radioactivity of surface waters in the United States: Am. Water Works Assoc. Jour., v. 51, p. 1377-1401.

Sever, C. W., 1965, Ground-water resources and geology of Seminole, Decatur, and Grady Counties, Georgia: U.S. Geol. Survey Water-Supply Paper 1809-Q, 30 p.

Shand, S. J., 1952, Rocks for Chemists: New York, Pitman Publishing Co., 146 p.

Shapiro, Joseph, 1957, Chemical and biological studies on the yellow organic acids of lake water: Limnology and Oceanography, v. 2, p. 161-179.

—— 1964, Effects of yellow organic acids on iron and other metals in water: Am. Water Works Assoc. Jour., v. 56, p. 1062-1081.

Seiver, Raymond, 1962, Silica solubility 0°-200°C and the diagenesis of siliceous sediments: Jour. Geology, v. 70, p. 127-150.

Sillén, L. G., 1967a, The oceans as a chemical system: Science, v. 156, p. 1189-1197.

——— 1967b, Gibbs phase rule and marine sediments, *in* Equilibrium concepts in natural water systems: Am. Chem. Soc. Advances in Chemistry Ser., v. 67, p. 57-69.

Sillén, L. G., and Martell, A. E., 1964, Stability constants of metal-ion complexes: Chemical Soc. [London] Spec. Pub. 17, 754 p.

Silvey, W. D., 1961, Concentration method for the spectrochemical determination of minor elements in water: U.S. Geol. Survey Water-Supply Paper 1540-B, p. 11-22.

——— 1967, Occurrence of selected minor elements in the waters of California: U.S. Geol. Survey Water-Supply Paper 1535-L, 25 p.

Simpson, H. E., 1929, Geology and ground-water resources of North Dakota: U.S. Geol. Survey Water-Supply Paper 598, 312 p.

Skougstad, M. W., and Horr, C. A., 1963, Occurrence and distribution of strontium in natural water: U.S. Geol. Survey Water-Supply Paper 1496-D, p. D55-D97.

Skougstad, M. W., and Scarboro, G. F., 1968, Water sample filtration unit: Environmental Sci. and Technology, v. 2, p. 298-301.

Slack, K. V., and Feltz, H. R., 1968, Tree leaf control on low flow water quality in a small Virginia stream: Environmental Sci. and Technology, v. 2, p. 126-131.

Slichter, C. S., 1902, The motions of underground waters: U.S. Geol. Survey Water-Supply Paper 67, 106 p.

Smith, R. L., O'Brien, W. J., Lefeuvre, A. R. and Pogge, E. C., 1967, Development and evaluation of a mathematical model of the lower reaches of the Kansas River drainage system: Lawrence, Kans., Univ. Kans. Center for Research, Inc., Engineering Sci. Div., 127 p.

Smyshlyaev, S. I., and Edeleva, N. P., 1962, Determination of the solubility of minerals; 1, Solubility product of fluorite: Izv. Vysshikh Uchebn. Zavedenii Khim. i Khim. Tekhnologiya, v. 5, p. 871-874.

Sniegocki, R. T., 1963, Geochemical aspects of artificial recharge in the Grand Prairie region, Arkansas: U.S. Geol. Survey Water-Supply Paper 1615-E, 41 p.

Stabler, Herman, 1911, Some stream waters of the western United States with chapters on sediment carried by the Rio Grande and the industrial application of water analyses: U.S. Geol Survey Water-Supply Paper 274, 188 p.

Steele, T. D., 1968a, Seasonal variations in chemical quality of surface water in the Pescadero Creek watershed, San Mateo County, Calif.: Ph.D. dissertation, Stanford Univ. (unpub.), 179 p.

——— 1968b, Digital-computer applications in chemical-quality studies of surface water in a small watershed: Internat. Assoc. Scientific Hydrology, UNESCO Symposium on use of analog and digital computers in hydrology, Tucson, Ariz., 12 p.

Stehney, A. F., 1955, Radium and thorium X in some potable waters: Acta Radiologica, v. 43, p. 43-51.

Stewart, B. A., Viets, F. G., Jr., Hutchinson, G. L., and Kemper, W. D., 1967, Nitrate and other water pollutants under fields and feed lots: Environmental Sci. and Technology, v. 1, p. 736-739.

Stewart, F. H., 1963, Marine evaporites *in* Data of geochemistry [6th ed.]: U.S. Geol. Survey Prof. Paper 440-Y, 52 p. Y1-Y52.

Stewart, G. L. and Farnsworth, R. K., 1968, United States tritium rainout and its hydrologic implications: Water Resources Research, v. 4, p. 273–289.

Stiff, H. A., Jr., 1951, The interpretation of chemical water analysis by means of patterns: Jour. Petroleum Technology, v. 3, no. 10, p. 15–17.

Stumm, Werner, 1961, Discussion of Hem, J. D., Stability field diagrams as aids in iron chemistry studies: Am. Water Works Assoc. Jour., v. 53, p. 228–232.

Stumm, Werner, and Lee, G. F., 1961, Oxygenation of ferrous iron: Indus. and Eng. Chemistry, v. 53, p. 143–146.

Swenson, F. A., 1968, New theory of recharge to the artesian basin of the Dakotas: Geol. Soc. America Bull., v. 79, p. 163–182.

Swenson, H. A., 1953, Geochemical relationships of water in the Powder River basin, Wyoming and Montana: Am. Geophys. Union Trans., v. 34, p. 443–448.

Swenson, H. A., and Colby, B. R., 1955, Chemical quality of surface waters in Devils Lake basin, North Dakota: U.S. Geol. Survey Water-Supply Paper 1295, 82 p.

Szalay, A., 1964, Cation exchange properties of humic acids and their importance in geochemical enrichment of UO_2^{++} and other cations: Geochim. et Cosmochim. Acta, v. 28, p. 1605–1614.

Tanji, K. K., and Doneen, L. D., 1966, Predictions on the solubility of gypsum in aqueous salt solutions: Water Resources Research, v. 2, p. 543–548.

Tanji, K. K., Doneen, L. D., and Paul, J. L., 1967, The quality of waters percolating through stratified substrata as predicted by computer analyses: Hilgardia, v. 38, p. 319–347.

Taylor, S. R., 1964, Abundance of chemical elements in the continental crust—a new table: Geochim. et Cosmochim. Acta, v. 28, p. 1273–1285.

Theis, C. V., and Sayre, A. N., 1942, Geology and ground water *in* The Pecos River Joint Investigation—reports of participating agencies: Natl. Resources Planning Board, Washington, 407 p.

Theobald, P. K., Lakin, H. W., and Hawkins, D. B., 1963, The precipitation of aluminum, iron, and manganese at the junction of Deer Creek with the Snake River in Summit County, Colo.: Geochim. et Cosmochim. Acta, v. 27, p. 121–132.

Todd, D. K., 1959, Annotated bibliography on artificial recharge of ground water through 1954: U.S. Geol. Survey Water-Supply Paper 1477, 115 p.

Toler, L. G., 1965, Relation between chemical quality and water discharge of Spring Creek southwestern Georgia *in* Geological Survey research 1965, U.S. Geol. Survey Prof. Paper 525-C, p. C209–C213.

Torrey, A. E., and Kohout, F. A., 1956, Geology and ground-water resources of the lower Yellowstone Valley between Glendive and Sidney, Mont.: U.S. Geol. Survey Water-Supply Paper 1355, 92 p.

Turekian, K. K., and Scott, M. R., 1967, Concentrations of Cr, Ag, Mo, Ni, Co, and Mn in suspended material in streams: Environmental Sci. and Technology, v. 1, p. 940–942.

Turekian, K. K., and Wedepohl, K. H., 1961, Distribution of the elements in some major units of the earth's crust· Geol. Soc. America Bull., v. 72, p. 175–192.

Tyler, P. A., and Marshall, K. C., 1967, Hyphomicrobia—a significant factor in manganese problems: Am. Water Works Assoc. Jour., v. 59, p. 1043–1048.

U.S. Department of Agriculture, 1955, Water—The Yearbook of Agriculture: 751 p.

U.S. Federal Water Pollution Control Administration, 1968 Report of the committee on water quality criteria· Washington, D.C., 234 p.

U.S. Geological Survey, Quality of surface waters of the United States.
Annual reports as follows: 1942, Water-Supply Paper 950; 1943, Water-Supply Paper 970; 1944, Water-Supply Paper 1022; 1946, Water-Supply Paper 1050; 1947, Water-Supply Paper 1102; 1949, Water Supply Papers 1162 and 1163; 1951, Water-Supply Paper 1198; 1952, Water-Supply Paper 1253; 1962, Water-Supply Paper 1945; 1963, Water-Supply Paper 1947, pts. 1 and 2; Water-Supply Paper 1948, pts. 3 and 4; and Water Supply Paper 1950, pts. 7 and 8.

U.S. Geological Survey, 1963, Summary of investigations: U.S. Geol. Survey Prof. Paper 475–A, 300 p.

U.S. Public Health Service, 1961, Proc. 1961 Symposium on Ground Water Contamination: R. A. Taft Sanitary Eng. Center Tech. Report W61-5, 218 p.

U.S. Public Health Service, 1962, Drinking water standards, 1962: U.S. Public Health Service Pub. 956, 61 p.

U.S. Salinity Laboratory staff, 1954, Diagnosis and improvement of saline and alkali soils: U.S. Dept. Agriculture Handb. 60, 160 p.

Valyashko, M. G., 1958, Some general rules with respect to the formation of the chemical composition of natural waters: Akad. Nauk SSSR Trudy Lab. Gidrogeol. Problem im F. P. Savarenskogo, v. 16, p. 127–140.

―――― 1967, Geochemistry of natural waters: Geokhimiya, v. 1967, p. 1395–1407.

Valyashko, M. G., and Vlasova, N. K., 1965, On the processes of formation of calcium chloride brines: Geokhimiya, v. 1965, p. 43–55.

Van Denburgh, A. S., and Feth, J. H., 1965, Solute erosion and chloride balance in selected river basins of the western conterminous United States: Water Resources Research, v. 1, p. 537–541.

Van Lier, J. A., de Bruyn, P. L., and Overbeek, T. G., 1960, The solubility of quartz: Jour. Phys. Chemistry, v. 64, p. 1675–1682.

Vinogradov, V. I., 1957, Migration of molybdenum in the zone of weathering: Geokhimiya, v. 1957, no. 2, p. 120–126.

Vogeli, P. T., Sr., and King, R. U., 1969, Occurrence and distribution of molybdenum in the surface water of Colorado: U.S. Geol. Survey Water-Supply Paper 1535–N, 32 p.

Von Buttlar, Haro, and Wendt, Immo, 1958, Ground-water studies in New Mexico using tritium as a tracer: Am. Geophys. Union Trans., v. 31, p. 660–668.

Wahlberg, J. S., Baker, J. H., Vernon, R. W., and Dewar, R. S., 1965, Exchange adsorption of strontium on clay minerals: U.S. Geol. Survey Bull. 1140–C, 26 p.

Wahlberg, J. S., and Fishman, M. J., 1962, Adsorption of cesium on clay minerals: U.S. Geol. Survey Bull. 1140–A, 30 p.

Warburton, J. A. and Young, L. G., 1968, Neutron activation measurements of silver in precipitation from locations in western North America: Jour. Appl. Meteorology, v. 7, p. 444–448.

Wayman, C. H., 1967, Adsorption on clay mineral surfaces in Faust, S. D., and Hunter, J. V., eds., Principles and applications of water chemistry: New York, John Wiley and Sons, 643 p.

Weart, J. G., and Margrave, G. E., 1957, Oxidation-reduction potential measurements applied to iron removal: Am. Water Works Assoc. Jour., v. 49 p. 1223–1233.

Weast, R. C., ed., 1968, Handbook of chemistry and physics [49th ed.]: Cleveland, Ohio, Chemical Rubber Co., p. A1–F286.

Weber, W. J., and Stumm, Werner, 1963, Buffer systems of natural fresh waters: Jour. Chem. and Eng. Data, v. 8, p. 464–468.

Wedepohl, K. H., 1953, Untersuchungen zur geochemie des Zinks: Geochim et Cosmochim. Acta, v. 3, p. 93–142.

Weyl, P. K., 1958, The solution kinetics of calcite: Jour. Geology, v. 66, p. 163–176.

White, D. E., 1957a, Thermal waters of volcanic origin: Geol. Soc. America Bull., v. 68, p. 1637–1658.

―――― 1957b, Magmatic, connate, and metamorphic waters: Geol. Soc. America Bull., v. 68, p. 1659–1682.

―――― 1960, Summary of chemical characteristics of some waters of deep origin *in* Short papers in the geological sciences: U.S. Geol. Survey Prof. Paper 400–B, p. B452–B454.

―――― 1965, Saline waters of sedimentary rocks *in* Fluids in subsurface invironments—a symposium: Am. Assoc. Petroleum Geologists Mem. 4, p. 342–366.

White, D. E., Brannock, W. W., and Murata, K. J., 1956, Silica in hot-spring waters: Geochim. et Cosmochim. Acta, v. 10, p. 27–59.

White, D. E., Hem, J. D., and Waring, G. A., 1963, Chemical composition of subsurface waters *in* Data of geochemistry [6th ed.]: U.S. Geol. Survey Prof. Paper 440–F, p. F1–F67.

White, D. E., and Waring, G. A., 1963, Volcanic emanations *in* Data of geochemistry [6th ed.]: U.S. Geol. Survey Prof. Paper 440–K, p. K1–K27.

Whitehead, H. C., and Feth, J. H., 1964, Chemical composition of rain, dry fallout, and bulk precipitation at Menlo Park, Calif., 1957–59: Jour. Geophys. Research, v. 69, p. 3319–3333.

Wiebe, A. H., 1930, The manganese content of the Mississippi River water at Fairfield, Iowa: Science, v. 71, p. 248.

Winchester, J. W., and Duce, R. A., 1966, Coherence of I and Br in the atmosphere of Hawaii, northern Alaska, and Massachusetts: Tellus, v. 18, p. 287–292.

Wollast, R., 1967, Kinetics of the alteration of K-feldspar in buffered solutions at low temperature: Geochim. et Cosmochim. Acta, v. 31, p. 635–648.

Woodward, R. L., 1963, Review of the bactericidal effectiveness of silver: Am. Water Works Assoc. Jour., v. 55, p. 881–886.

Yaalon, D. A., 1961, On the origin and accumulation of salts in groundwater and soils of Israel: Dept. Geology, Hebrew Univ. Jerusalem Pub. 255, 42 p.

Yamabe, Takeo, 1958, Anion exchange equilibria involving carbonate ions: Chem. Soc. [Japan] Jour. Indus. Chem. sec., v. 61, p. 1531–1533.

Yanat'eva, O. K., 1954, Solubility of dolomite in water in the presence of carbon dioxide: Izv. Akad. Nauk SSSR Otdel. Khim. Nauk, no. 6, p. 1119–1120.

Zelenov, K. K., 1958, Transport of dissolved iron into Okhotsk Sea from the hot springs of Ebeko volcano: Akad. Nauk SSSR Doklady, v. 120, p. 1089–1092.

ZoBell, C. E., 1946, Studies on the redox potential of marine sediments: Am. Assoc. Petroleum Geologists Bull., v. 30, p. 477–513.

INDEX

[Italic page numbers indicate major references]

A	Page
Accumulator plants, manganese	126
molybdenum	200
selenium	207
Accuracy, analyses	*233*
Acidity	*159*, 234
Activation energy	33
Activity	18
Activity coefficient	19
Adsorbed ions, load in streams	88, 191
Adsorption	*36*, 37, 145
calculations	*38*
Agricultural use, natural water	*324*
Albite, dissolution	17, 90
Algal growth	46, 185
Alkali	165, 329
Alkali metals	78, *193*
Alkaline earth metals	78, *194*
Alkalinity	78, 84, *152*, 159, 225, 234
ionic species	156
Alkyl benzene sulfonate	216, 322
Allegheny River, Pittsburgh, Pa	247
Aluminum	78, *109*
Amazon River, Brazil	10, 41
Ammonia	49, 183
Analysis of water	*74*
accuracy of data	*232*
application of quality measurements to quantitative hydrology	*315*
check of data	*233*
evaluation of areal water quality	*236*
evaluation of data	*231*
extrapolation of records	*270*
graphical methods for representation	*256*
minor constituents	*190*
organization of data	*230*
reported properties and constituents, significance	*86*
study of data	*230*
Anhydrous residue, in expressing analyses	*82*
Anion exchange	37, 178, 297
Anorthite, dissolution	131
Antarctica	48
Arrhenius equation	33
Arsenic	78, *206*, 322
Astragalus, relation to selenium	207
Atlantic Coastal Plain	124, 287
Atlantic Coastal Plain sediments, analysis of water sample	148
Atmosphere, composition	*10*
solutes, sources	*47*
Atomic alterations	*14*
Averages, for treating water-quality data	*239*
Averages, discharge-weighted	240

B	Page
Bacteria, chemical roles	124
sulfur oxidizing	161
Barite, solubility	197
Barium	78, *197*, 322
Basalt, analysis of water sample	106
Base flow of streams, quality	55
Basin and Range physiographic province	284
Bayerite, solubility	111
Beryllium	78, *194*
Bicarbonate	34, 37, 41, 59, 78, 99, 102, 152, 156, *157*, 159, 218, 331
Biochemical factors, relation to natural water	*42*
Biochemical oxygen demand	87, *223*
Biota, aquatic	185
aquatic, relation to natural water	*45*
Black alkali	329, 331
Blue Springs, Ariz	157
Boiler feed water, quality requirements	334
Boric acid, dissociation	188
volatility	187
Boron	48, 78, *187*
crop tolerance	329
Brandywine Creek, Del	*56*
Brazos River, Tex	286
Bromine	78, *206*
Buffer capacity	59
defined	92
Buffered solutions	*92*

C	
Cadmium	78, *204*, 322
Calcite, solubility	*22*, 133
Calcium	59, 78, *131*, 195, 224, 314, *325*, 329, *330*, 331
Calcium bicarbonate water	77, 237
Calcium carbonate	79, 224
concentration	*84*
Calcium chloride brine	175
Capitan Limestone, analysis of water sample	139
Carbon	6, 12, 14, 48, 214, 230, 315, 318
total organic	*223*
Carbon chloroform extract	322
Carbon cycle	158
Carbonate	78, 152, *157*, 159
Carlsbad Caverns, N. Mex	140, 144, 182
Carrizo sand aquifer, Tex	214
Castile Formation, analysis of water sample	138
Celestite, solubility	195
Cesium	78, 193, 210
Chattanooga Shale, analysis of water sample	123, 301
Chemical equilibrium	*18*
Chemical oxygen demand	*223*

359

INDEX

	Page
Chemical reactions	15
at interfaces	35
membrane effects	39
Chemical reaction rate, effect of temperature	35
Chemisorption	37
Chinle Formation, analysis of water sample	149, 301
Chloride	37, 48, 49, 78, 99, 170, 251, 314, 322, 324, 329
Chlorine	57, 323
Chlorophenols	323
Chromium	78, 199, 322
Claiborne Group, Tenn	284
Clays, semipermeable membranes	40, 147, 208
Clifton Hot Springs, Ariz	274
Climate, relation to natural water	41
Closed systems, water chemistry	51
Cobalt	78, 200, 210
Collection, minor constituents	190
water samples	60
Collins' diagram	257
Colloidal particles	86
Color, due to organic constituents	216
Colorado River, at Grand Canyon, Ariz	247
below Hoover Dam	270
Columbia River at Hanford, Wash	316
Composite samples	64
Composition of natural water, evaluation	60
principles controlling	12
Conductance, physical basis	96
range of values	102
Connate water	146
Constituents, natural water	78, 86, 188, 215
Copper	78, 202, 322, 324
Correlations, in treating water-quality data	250
Crenothrix, relation to oxidation of iron	125
Crust of the earth, composition	5
Cumberland River at Smithland, Ky	138
Curie, defined	211
Cyanide	180, 322

D

Dakota Sandstone, analysis of water sample	123, 139, 147, 298
Dead Sea	174
Debye-Hückel equation	19, 21, 23, 156, 167
Delaware River estuary, Delaware	312
Density, water	229
Detergents, content of organic constituents	216
Deuterium	229, 230
Diffusion	36
Dipolar nature of water	5
Discharge, relation to solute concentration	275, 277
Dispersion of wastes	310
Dissociation of water	89
Dissociation reactions	17
Dissolved-solids determination	218
Dissolved state of water constituents	86
Dolomite, solubility	142
Dolores River near Cisco, Utah	183
Domestic uses, natural water	321
Douglas basin, Arizona	284
Drinking-water standards, animals	324
humans	321
Drought, relation to water quality	246

E

	Page
Ecology, balance	41
relation to natural water	43, 45, 50
Eh	28, 30, 129, 226
Eh-pH diagram	117, 120
Electric logs, indicators of ground-water quality	69, 75
Electrodes, specific ion	19
English Lake District	48
Enthalpy	24
Entropy	24
Environment, influence of man	50, 309
relation to natural water	40, 242
Equilibrium, electrochemical	28
Equilibrium constant	21, 22, 23, 26, 27, 35, 90
Equivalent-weight units	81
Equivalents per million	82
Escherichia coli, in natural water	321
treatment with silver	203
Estuaries, water quality	312
Euphrates River, Iraq	326
Eutrophic water	46, 187
Evaporites	8, 164, 171, 187, 288, 306
Evaporites, relation to composition of water	306
Exchange capacity	37
Extrusive rocks	290

F

Factor analysis	255
Fallout	49
Feldspar, dissolution	146, 150, 151
Ferrous bicarbonate water	77
Field testing of water	74
Fisher Formation, analysis of water sample	196
Fluoride	78, 176, 189, 322, 323
Fluorite, solubility	177
Formality	82
Fort Union Formation, analysis of water sample	148, 301
Free energy	24
Frequency distributions, in treating water-quality data	247
Fresh-water-salt-water contact zone	312
Freundlich adsorption isotherm	38
Fulvic acid	217

G

Geochemical cycles	56
Geochemical prospecting	189
Geologic effects, relation to natural water	41
Germanium	78, 205
Gibbsite, solubility	111
Gila River, Ariz	71, 174, 178, 327
Bylas, Ariz	99, 252
Gillespie Dam, Ariz	271
Gold	78, 203
Goodwin Wash, Ariz	286
Grains per gallon	80
Grand Canyon, Ariz	140
Grand Canyon region, Arizona	305
Ground water, pollution	311
rate of movement	318
sampling processes	68

INDEX

361

	Page
Ground-water quality, variation with time	71
Ground-water systems, characteristics	54
Gypsum, solubility	136, 168

H

Half time	32
Halite, solubility	148
Hardness, water	79, 84, 99, 224
classification ranges	225
Heart River, N. Dak	269
Hill diagram	257
Hoover Dam	247
Humic acid	216
Hydrogen	6, 102
Hydrogen bonds of water	5
Hydrogen-ion activity. See pH.	
Hydrographs, water quality	270
Hydrologic cycle	47
Hydrolysis reactions	16, 90
Hydrolyzates	146, 289, 300
defined	8
Hydrosphere	10
Hyphomicrobium, relation to oxidation of manganese	130
Hypothetical combinations	85

I

Igneous rocks	8, 9, 42, 57, 58, 114, 141, 146, 150, 151, 162, 164, 171, 177, 187, 290
classification	291
relation to composition of water	294
Illinois River at Peoria, Ill	320
Imperial Valley, Calif	42
Industrial use, natural water	333
Intrusive rocks	290
Iodine	48, 78, 209
Ion-concentration diagrams	257
Ion-correlation diagrams	251
Ion exchange, calculations	38
reactions	16, 37, 54, 55, 143, 145, 147, 151, 175, 297
Ion inventory	255
Ion pairs	35, 167
Ion ratio	237
Ionic mobility	98
Ionic strength	20
Iowa River at Iowa City, Iowa	182
Irish Sea	48
Iron	5, 54, 78, 114, 322, 324
Iron bacteria	124
Irreversible reactions	289
Isogram maps	284

J

Jackson Group, analysis of water sample	115
Juniata River, Pa	63
Juvenile water	42

K

Kinetics, geochemical	30, 293
Knox Dolomite, analysis of water sample	144, 304

L

Lake Mead, Ariz.-Nev	139, 286, 337
Lake Whitney, Tex	286

	Page
Lakes, relation to natural water	56
sampling processes	68
Langelier index	24, 134
Langmuir adsorption isotherm	38
Lead	78, 205, 211, 322
Leptothrix, relation to oxidation of iron	125
Limestone, solution	133, 140
Lithium	78, 193
Little Colorado Canyon, Ariz	157
Little Colorado River, Ariz	94
Livestock, limits on quality of water for use	324

M

McMillan Reservoir, N. Mex	319
Magnesium	5, 48, 78, 140, 224, 314, 325, 328, 329, 330, 331
Management of water, concepts and problems	336
Manganese	54, 78, 126, 322, 324
Maps, water quality	280
Mass action law	18, 29, 117
Membrane effects	147
Mercury	48, 78, 205
Metamorphic rocks	57, 306
Metamorphic water	306
Meteoric water	42
Methyl-orange alkalinity	152
Methyl-orange end point	159, 234
Mica schist, analysis of water sample	106
Mineral surface effects on water quality	296, 297
Minor constituents	78, 188
Mississippi River near New Orleans, La	10, 271
Mixing of solutes, ground water	69
in stream water	61
Mobility, geochemical	57
Modeling of water quality	256, 273, 277
Mohorovicic discontinuity	6
Mole	81
Molecular structure of water	4
Molybdenum	78, 199
Monolayers	37
Monongahela River, Pittsburgh, Pa	247
Moreau River at Bixby, S. Dak	146, 166, 301
Movement, natural water	54

N

Neptunium	14
Nernst equation	28, 29, 30, 117, 226, 227
Niagara Group, analysis of water sample	144
Niagara Dolomite, analysis of water sample	145
Nickel	48, 78, 201
Niobrara River, Nebr	271
Nitrate	49, 78, 181, 322
Nitrite	180
Nitrogen	6, 78, 180
Nitrogen cycle	181
Nonmetal elements	78, 206
Nuclides, radioactive	9, 14, 32, 48, 209
stable	229
Nutrients, nitrogen and phosphorus	185

O

Ocala Limestone, Fla	133
Ohio River at Parkersburg, W. Va	131
Ohio River basin	130

	Page
Oligotrophic water	46
Olivine, dissolution	141
Olivine tuff-breccia, analysis of water sample	144
Open systems, water chemistry	51
Organic constituents	215
Osmotic pressure	39, 325
Oxidation reactions	17, 28, 32, 162, 172, 180, 298
Oxidation-reduction potential	226
Oxydates	9
Oxygen	6, 229, 230
dissolved	221
Oxygen demand	222

P

Partition coefficient	39
Parts per million	80
pE, defined	228
Pecos River, N Mex	319
Artesia, N Mex	139, 278
Pecos River basin, New Mexico	166, 174
Pescadero Creek, Calif	152, 277
pH	19, 23, 74, 88, 117, 129, 153, 159, 183
measurement and interpretation	95
range in natural water	93
Phase diagrams	53
Phase rule	52
Phenolphthalein alkalinity	152
Phenolphthalein end point	160
Phenols	322
Phosphate	78, 184
Phosphorus	185, 210
Photosynthesis	17, 36, 44, 45, 138, 158, 221
Physical properties of water	4
Pie diagram, plotting analyses	260
Piper diagram	268
Platte River, Nebr	320
Poising	226
Pollution	36, 46, 161, 164, 215, 309, 321, 336
thermal	335
Pollution load. organic, evaluations	222
Poncha Springs near Salida, Colo	271
Port Deposit granitic gneiss, analysis of water sample	115
Portland Arkose, analysis of water sample	139
Possum Kingdom Reservoir, Tex	286
Potassium	14, 38, 78, 150, 210
Powder River Basin of Montana	303
Precipitates	303
defined	8
Precipitation, atmospheric, composition	48
chemical	9
Procedures, water analyses	76
Profiles, water quality	286
Public supplies, natural water	321
Purpose of report	2
Pyrite, oxidation	124, 162

Q

Quality of water, management	336
relation to lithology	287
relation to stream discharge	271
relation to use	320

	Page
Quartz, rate of solution in water	108
solubility	107
Quartz monzonite, analysis of water sample	154
Quartzite, analysis of water sample	145

R

Radioactive elements	6, 12, 78, 209
drinking water	323
Radioactivity, induced	14
Radium	14, 78, 212, 213, 323
Radon	14, 79, 213
Rainwater, chloride composition	172, 173, 174
pH	91
sulfate content	164
Raoult's law	39
Rate constant	31
Rate law	33
Rates of ground-water movement	318
Reaction mechanisms	108, 112, 130
Reaction order	30
Reaction rate	30
Reactions in water chemistry	16
Recharge, artificial	337, 338
Recreational uses, natural water	336
Red Sea	42
Redox potential See Eh	
Reduction reactions	17, 28, 162, 172, 180, 298
Reduzates	9
Redwall Limestone, analysis of water sample	96
Reservoirs, effect of evaporation	337
relation to natural water	56
sampling processes	68
stratification of water	337
Residence time, elements in ocean	57
Residual sodium carbonate	331
Residue on evaporation	218
Resistates	146, 171, 289, 295
defined	8
relation to composition of water	298
Reversible reactions	16
Rhodamine B	316
Rhyolite, analysis of water sample	106
Rio Grande at San Acacia, N. Mex	66, 242, 244, 270, 278
Rio Grande River, Colo	327
Rio Grande valley at Albuquerque, N Mex	298
Rio Puerco River, N Mex	242, 301
Rio Saladdo River, N Mex	242
Rivers sampling process	61
River water, average composition	10, 12
River-water quality, onsite sensors	74
Rogue River basin, Oregon	174
Roswell basin, New Mexico	304
Rubidium	78, 193, 210

S

Safford Valley, Ariz	71, 178, 284, 318, 319
Saline River near Russell, Kans	277
Saline water, defined	219
Salinity, defined	79
Salt balance	327, 328, 330

INDEX

363

	Page
Salt disposal in irrigated land	325
Salt-dilution method for discharge measurement	316
Salt River, Ariz	137
Globe, Ariz	149
Shepherdsville, Ky	277, 278
Salt River valley, Ariz	175, 182
Salt-water intrusion	311
Salton Sea in Imperial Valley, Calif	140
Sampling, completeness	71
See also specific type of water body.	
San Andreas Limestone, analysis of water sample	145, 304
San Francisco River at Clifton, Ariz	274, 275, 277, 317
San Luis Valley, Colo	327
Santa Cruz basin, Arizona	147
Santa Fe Formation, analysis of water sample	106, 298
Scope of report	2
Sea water, composition	10, 11
intrusion in aquifers	313
minor elements	192
Sedimentary rocks	8, 9, 57, 58, 141, 150, 162, 171, 177, 294
Seeped land	326
Selenium	78, 207, 322, 325
Seneca anthracite mine, Duryea, Pa	95
Significant figures	256
Silica	78, 103
Silicate ions	107, 153
Silicate mineral structures	103
Silicates	5
Silver	78, 202, 322
Sodium	48, 57, 78, 145, 251, 324, 330, 331
computed	150
Sodium-adsorption-ratio	228
Sodium percentage	229
Softening of water, natural	137, 149
Soil, leaching	326
Soil chemistry, relation to water chemistry	44
Soil air	45
Soil moisture	45
Solubility concepts	34
Solubility controls, aluminum	111
iron	117
magnesium	141
manganese	128
quartz	107
sodium	147
Solubility product	34
Solute-balance equation	256, 272
Solute inventory	255
Solute transport	47
Sorption-desorption reactions	16
South Platte Valley, Colo	182
Specific electrical conductance	74, 96, 219
relation to dissolved solids	99
Spring Creek, Ga	278
Stability-field diagram	117
Standard free energy	25, 26, 27, 90
Sulphur Bank, Calif	183
Supai Formation, analysis of water sample	138, 305

	Page
Surface-water systems	55
Suspended solids, separation	87
Susquehanna River, Pa	127
Harrisburg, Pa	61
Sylvania Sandstone, analysis of water sample	139
Standard potential	29
Statistical treatment, water-quality data	239
Stiff method	259
Strontianite, solubility	195
Strontium	78, 195, 210, 323
Sulfate	37, 49, 78, 99, 161, 322, 324, 328, 329
reduction	169
Sulfides	168
Sulfur	48, 230, 330
stable forms	162

T

Technetium	14
Thermal stratification	56, 68, 127
Thermal water	42
Thermodynamic concentration	18
Thermodynamics, application to chemistry	24, 30
Thorium	6, 79, 210, 213
Tigris River, Iraq	326
Titanium	78, 198
Tons per acre-foot	80
Travertine, deposition	157
Trilinear plotting systems	264
Tritium	12, 48, 172, 210, 214, 230, 318, 319
half life	14
Tuscaloosa Formation, analysis of water sample	123
Tuscumbia Limestone, analysis of water sample	138

U

Units, conductance	96
radioactivity	211
water analyses	79
comparison	85
Uranium	6, 14, 78, 210, 211, 212

V

Vanadium	78, 198
van der Waals forces	37
Vector diagram	259

W

Waste disposal, deep-well injection	311
techniques	310
Waterlogged land	326
Weathering	9, 41, 57, 105, 288
Weight-per-volume units	79
Weight-per-weight units	79
Well exploration	75
Wellton-Mohawk area, Arizona	71, 271
Wheatstone bridge	74
White alkali	329
Wilcox Formation, analysis of water sample	123

Z

Zinc	78, 203, 322, 324

UNITED STATES DEPARTMENT OF THE INTERIOR
GEOLOGICAL SURVEY

WATER-SUPPLY PAPER 1473
PLATE 1

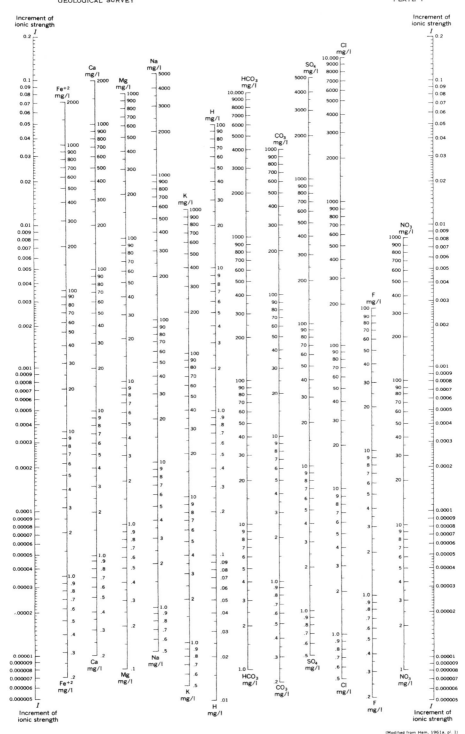

(Modified from Hem, 1961a, pl. 1)

NOMOGRAPH FOR COMPUTING IONIC STRENGTH OF NATURAL WATER

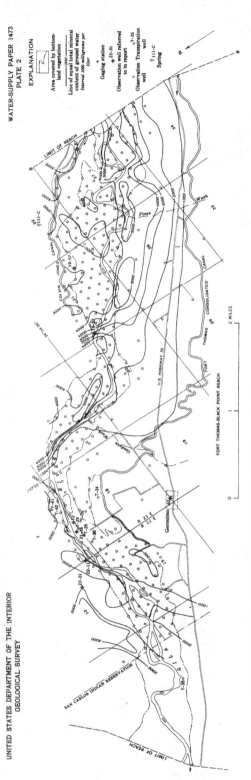

MAP OF PART OF LOWER SAFFORD VALLEY, ARIZONA, SHOWING DISSOLVED-MINERAL CONTENT OF GROUND WATER IN ALLUVIAL FILL OF THE INNER VALLEY, 1944

Printed in the USA
CPSIA information can be obtained
at www.ICGtesting.com
LVHW051924221123
764660LV00001B/56